P9-CFP-420

ALSO BY SCOTT WEIDENSAUL

The Ghost with Trembling Wings:
Science, Wishful Thinking, and the Search for Lost Species

Living on the Wind:
Across the Hemisphere with Migratory Birds

Mountains of the Heart:
A Natural History of the Appalachians

The Raptor Almanac:
A Comprehensive Guide to Eagles, Hawks,
Falcons, and Vultures

RETURN TO WILD AMERICA

RETURN TO WILD AMERICA

A Yearlong Search for the Continent's Natural Soul

SCOTT WEIDENSAUL

North Point Press

A division of Farrar, Straus and Giroux

New York

North Point Press
A division of Farrar, Straus and Giroux
19 Union Square West, New York 10003

Distributed in Canada by Douglas & McIntyre Ltd.
Printed in the United States of America
First edition, 2005

Portions of chapter 3 are adapted from an article that originally appeared in *Smithsonian.*

Grateful acknowledgment is made to the following for permission to reprint previously published material:
Excerpts from *Wild America* by Roger Tory Peterson and James Fisher, copyright © 1955 by Roger Tory Peterson and James McConnell Fisher, copyright renewed © 1983 by Roger Tory Peterson and Crispin Fisher. Reprinted by permission of Houghton Mifflin Company. All rights reserved.
"Daniel Boone" by Stephen Vincent Benét, from *A Book of Americans* by Rosemary and Stephen Vincent Benét, Holt, Rinehart and Winston, copyright © 1933 by Rosemary and Stephen Vincent Benét, copyright renewed © 1961 by Rosemary Carr Benét. Reprinted by permission of Brandt & Hochman Literary Agents, Inc.

Library of Congress Cataloging-in-Publication Data
Weidensaul, Scott.
Return to wild America : a yearlong search for the continent's natural soul / Scott Weidensaul.— 1st ed.
p. cm.
Includes bibliographical references and index.
ISBN-13: 978-0-86547-688-2 (alk. paper : hardcover)
ISBN-10: 0-86547-688-8 (alk. paper : hardcover)
1. Natural history—North America. 2. Weidensaul, Scott—Travel—North America. 3. Peterson, Roger Tory, 1908—Travel—North America. 4. Fisher, James, 1912—Travel—North America. I. Title.

QH102.W45 2005
578′.097—dc22

2005047720

Designed by Jonathan D. Lippincott

www.fsgbooks.com

1 2 3 4 5 6 7 8 9 10

In memory of Roger Tory Peterson
and James Maxwell McConnell Fisher

For Amy with love

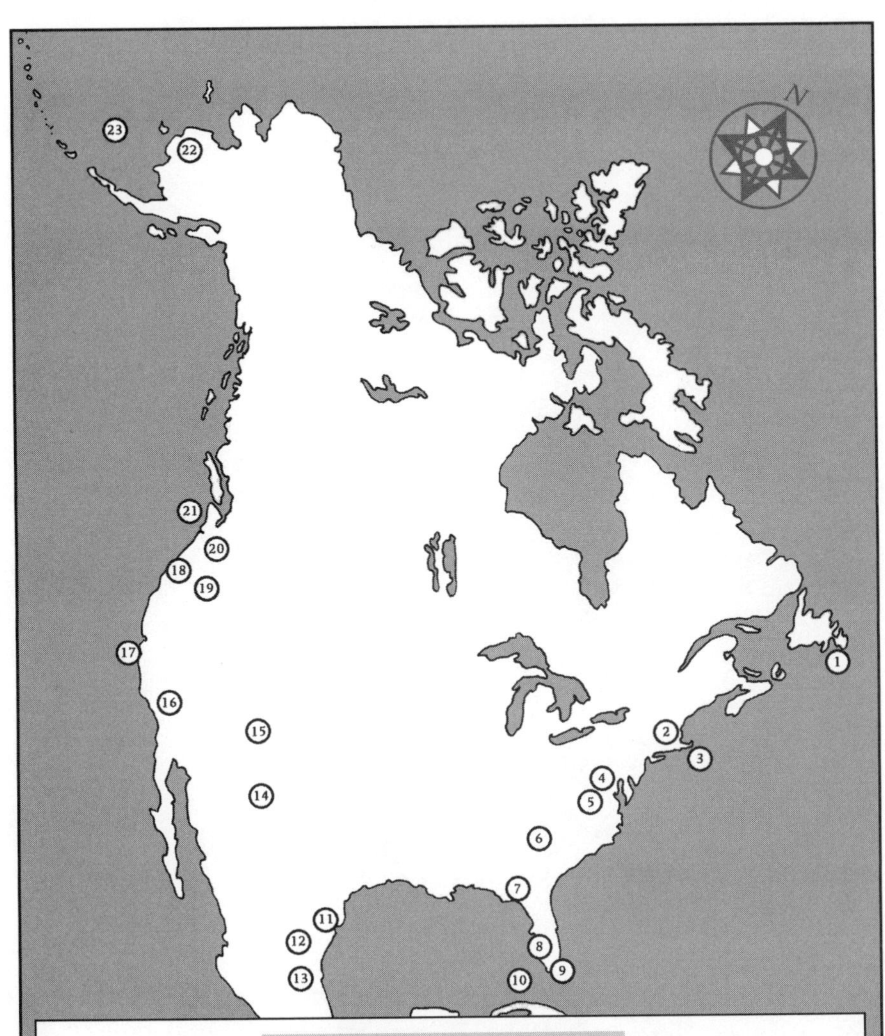

 **RETURN TO WILD AMERICA** 

1. Cape St. Mary's
2. Boston
3. Monomoy Island
4. Washington, D.C.
5. Shenandoah National Park
6. Great Smoky Mountains
7. Bradwell Bay Wilderness
8. Corkscrew Swamp Sanctuary
9. Everglades National Park
10. Dry Tortugas National Park
11. Lower Rio Grande Valley
12. La Soledad
13. Xilitla
14. Sky Islands Region
15. Four Corners Region
16. Hopper Mountain National Wildlife Refuge
17. Monterey Bay
18. Rogue River–Siskiyou National Forest
19. Klamath Basin
20. Opal Creek
21. Olympic Peninsula
22. Old Chevak
23. Pribilof Islands

Contents

Preface

The mind is so easily tricked. One moment I was balanced over an abyss, waves crashing almost three hundred feet below me, the rim of the cliff a hundred feet above my head; my fingers were locked and white on the safety rope as I tried not to shift even fractionally from the narrow ledge of loose rock and dirt on which I had paused. Vertiginous space yawned below. I haven't an especially good head for heights, and this seemed, very insistently, to have been a bad idea.

Then cold fog enveloped me, the kind of thick, cotton-batting mist that's common to summer in the North Atlantic and has a solidity all its own. No longer able to see the long, lurching drop below me, I unconsciously relaxed; I took slower, deeper breaths, and my fingers eased up a bit on the rope. I realized that Tony, the fellow who was leading me down this windswept cliff on Newfoundland's Cape St. Mary's, was already vanishing into the white, and I resumed my undignified, butt-scraping descent behind him.

Tony stopped some yards farther, down an even more precipitous slope, and with a final rain of small pebbles I eased in beside him on a ridiculously thin finger of dirt and tufted grass. My glasses were clouded with moisture, and with one hand I dug out a handkerchief and cleaned them. That's when I noticed six murres perching on the edge of the drop-off, barely ten feet away—slim seabirds, black and white in a tuxedoed, vaguely penguinish way, sitting stiffly erect like bowling pins. Their beaks were thin and pointed, with a bit of an upward tilt. They cocked their heads, shifting from foot to foot and looking, from each other to us and back again, with what seemed to be incredulity: Humans? *Here?*

So captivated was I by the sight that it took me a moment to realize that the fog had swept away. The chasm gaped below me again, but this time it scarcely registered.

Birds. There were birds *everywhere*—a vast city of them, close enough almost to touch.

As though the fog had deadened all my senses, I was suddenly overwhelmed by renewed sight, sound, and smell. On three sides, the cliff fell off vertically to the sea, where heavy green waves smashed to foam on the black rock. To our right, across a deep crevasse in which the breakers boomed and sucked, we faced a sheer wall just thirty feet away and stared right into the eyes of hundreds of nesting birds—murres jammed shoulder to shoulder, razorbills clinging to any purchase they could find, and the delicate little gulls called kittiwakes with their nests of piled seaweed. Gannets, huge and white, with rapier bills and tapered wings, wheeled above and below us, sailing through wraiths of mist snaking along the twisting breeze. The air reeked with the fish-guts smell of guano and reverberated with noise, a wall of cries and screams: the moaning *aaaarrr-aaaarrr* of the murres, the shrill *kitti-weeeeikk, kitti-weeeeikk* of the gulls, the rattling barks of the gannets, and the croak of an egg-stealing raven, whose sudden appearance panicked the colony into even greater crescendos.

In a lifetime of messing around with birds, I rank the next half hour, clinging to the side of Cape St. Mary's, among the finest moments. It wasn't just proximity; this was an astonishing degree of intimacy with species that have always been little more than blurs in a spotting scope. I could see the individual white feathers that ringed the eye of a bridled murre, the sheen of iridescent green and blue on the scales of a slender fish it held clamped in its beak, the flecking of tiny mist droplets collecting on the brown-black feathers of its throat. Looking closely, I realized I could see my own reflection in its dark eyes.

This was the first time I had been to Cape St. Mary's, but in a vivid sense it felt like a homecoming. In the spring of 1953, the field-guide author and birding guru Roger Tory Peterson set out with his friend, the noted British naturalist James Fisher, on what became a legendary journey—a thirty-thousand-mile, hundred-day marathon around the margins of North America, from the North Atlantic, down the Appalachians to the Florida Everglades, around the Gulf and coastal Texas and deep into Mexico, then up through the desert Southwest, along the Pacific coast into the Northwest, out into the tundra wilderness of western Alaska, and finally to the cold and remote Pribilof Islands in the Bering Sea. Two years later, in the fall of 1955, they published *Wild America*, a bestselling account of the travels that today remains a beloved classic.

Their journey began here, in Newfoundland, among the gannets and murres of Cape St. Mary's, one of the loneliest outposts of the continent's easternmost rim—a place of tundra barrens and twisted spruces where herds of caribou graze on headlands below which whales breach and roll and where seabirds by the tens of thousands crowd the rocky cliffs. For me, as for many naturalists who read that book over the past half century, Cape St. Mary's has thus always had an almost electric pull, and the patina of great adventure.

I first read *Wild America* as a kid, with a kid's understanding and with a kid's jealousy and longing. I saw it as an almost endless holiday adventure, two buddies chasing birds and exploring distant, empty places—the deep swamps of the Florida Panhandle where ivory-billed woodpeckers still lived; Native villages in Alaska where Eskimos paddled skin boats and lived in sod-covered shelters; cloud forests in Mexico full of colorful birds I'd never heard of, whose names dripped exoticism: Motmot. Potoo. Euphonia.

I came back to the book as a young adult who had begun his own travels, and with a fresh perspective. I was struck this time by the sheer magnitude of the task Peterson and Fisher set for themselves—the enormity of the distance they covered in just three and a half months, almost all of it in a big Ford station wagon that ran out of gas and blew tires in singularly inconvenient places. In spite of this, they still managed to average three hundred miles a day while setting a new record for the most species of birds seen in a single year, shooting miles of motion-picture footage and hundreds of still photographs, and maintaining voluminous field notes on the rich diversity of plant and animal life they encountered. And across all the miles, these two close but relatively new friends (who had met just three years earlier) admitted to but a single argument.

When I read *Wild America* now, it is with those past layers of excitement and admiration, but with something else keener and still more evocative—the realization that Fisher and Peterson saw a continent and a culture at a turning point. For a century before their trip, America's nascent conservation movement had been taking shape, sparked by disgust over the wholesale slaughter of wildlife, the pillaging of forests, and the waste of natural resources. It had, in its early days, been driven by moneyed and well-educated society, but in the years following World War II it began to emerge as a national force, drawing much of its strength from ordinary people who found enjoyment in amateur nature study.

"Wherever one goes, from one remote end to another of our two great Unions, this interest in natural history is growing, developing, hungrily seeking inspiration," Fisher wrote after attending a packed-house meeting of the Newfoundland Natural History Society. He was too polite, or too modest, to add that he and Peterson themselves deserved much of the credit. Peterson's bestselling field guide to birds was, in 1934, the first measurable manifestation of a growing thirst by the wider, increasingly urbanized public to reconnect with nature; in the years thereafter, his field guides sold more than 10 million copies. Fisher had done much the same for nature study in England, popularizing it through his writing and BBC broadcasts. In the 1950s, bird-watching and related pastimes were still considered slightly fussy, eccentric hobbies in America, but even those who didn't crouch in the bushes with binoculars had a growing interest in the natural world and were uneasy about humanity's impact on it. Conservationists, exerting the muscle that comes with broader support, began to pick—and win—high-profile battles, like their defeat of plans for the Echo Park Dam on the upper Colorado in Dinosaur National Monument, or the push for controls on water pollution, presaging the great strides to come in the 1960s and '70s, when the environmental movement became a national priority.

The wild America through which Peterson and Fisher traveled, and about which they wrote, was also changing. The first half of the twentieth century was the quiet nadir for conservation in North America. By the turn of the century, the continent's eastern forests had been laid waste, and the timbermen were turning to the old-growth stands of the West and the Northwest. The once-huge stocks of wildlife were on the verge of extermination—the bison almost gone, the elk and pronghorn rare, even white-tailed deer a threatened species. Passenger pigeons, Carolina parakeets, and heath hens all tumbled into extinction. Things began to improve, slowly, with the worst excesses of market hunting and the feather trade banned, with creation of the National Wildlife Refuge and the National Forest systems by Teddy Roosevelt, who also wielded the Antiquities Act to protect such treasures as the Grand Canyon and the Olympic

Mountains. Predators remained in the gun sights, though, with most western states (aided by the federal government) actively exterminating large predators like wolves and grizzlies in the 1920s and '30s, then turning their lethal attention to smaller species like coyotes and birds of prey. Even though Fisher and Peterson spent weeks traveling through Mexico, the Southwest, and the Sierras, they didn't see a single golden eagle—a frequent target of ranchers—until they reached Oregon.

It's easy to overlook how far we've come in America, where conservation is concerned, unless you take the long view that half a century affords. In the 1950s, there was no effective federal oversight of the air or water; although Congress tried to regulate urban water pollution in Chicago as early as 1910, even the most grotesque abuses were usually seen not as a stigma but as the inevitable price of economic growth. When Congress passed the Water Pollution Control Act of 1948, it was so toothless it provided not a single enforcement agent, nor any recourse to the courts. Endangered species, untouched wilderness, free-flowing rivers—they had no legal standing or substantive protection. When, in 1953, a Supreme Court justice championed the idea of preserving open space by preventing the construction of a highway along the historic Chesapeake & Ohio Canal near Washington, D.C., many people thought him a kook. Instead, he was in the vanguard of a change in public attitude that would revolutionize the way we think of the natural world.

It took a series of environmental crises, however, to galvanize Americans in the fight for environmental protection. An air inversion in the fall of 1948 over Donora, Pennsylvania, just south of Pittsburgh, trapped air pollution from the Monongahela valley's zinc smelter, steel plants, and coal-burning locomotives, killing nineteen people and hospitalizing many more and spurring the first attempt at federal clean-air legislation in 1955. Urban rivers around the country, loaded with debris and pollutants, were catching fire with depressing regularity—in Buffalo, Detroit, and most famously, Cleveland, whose Cuyahoga River repeatedly burst into flames. One such blaze, in 1952, caused $1.5 million in damage, though it was a smaller fire in 1969 that finally received national attention, along with a disastrous oil spill along the Southern California coast. The unfettered use of pesticides was also a growing worry. Roger Peterson had studied DDT's effects on birds for the Army during World War II and was an early critic of the burgeoning chemical arsenal, lecturing through the

1950s on the dangers of pesticides and pollution. It wasn't until the 1962 publication of *Silent Spring* by his friend the biologist Rachel Carson, however, that the country took alarmed notice, making the word "ecology" part of the national vocabulary and setting the modern environmental movement in motion.

All these threads came together in the late 1960s and early '70s, when the United States—in a remarkably short period of time—tried to reverse the course of two centuries of reckless exploitation. The Wilderness Act, proposed in 1955 in the aftermath of the Echo Park Dam fight, finally passed Congress in 1964 after being rewritten sixty-six times, one of the first links in a chain of federal and state environmental protections that owed much to the lobbying of increasingly potent conservation organizations like the Sierra Club, the Wilderness Society, and the National Audubon Society, and the fast-growing constituencies for which they spoke. Most of the seminal laws were forged in a span of just seven years. The National Trails and Wild and Scenic Rivers acts of 1968 were modest steps, but 1969's National Environmental Policy Act and the creation of the Environmental Protection Agency were major developments, followed the next year by the Clean Air Act (which greatly expanded protections first outlined in 1955). The first Earth Day, in 1970, showed the breadth of public support for such action, and politicians took note, passing the Clean Water Act in 1972, along with legislation giving protection to coastal areas and regulating pesticide use (that same year, the EPA banned the use of DDT). A robust Endangered Species Act followed in 1973, and the Safe Drinking Water Act in 1974. Never, before or since, had a nation so quickly and fundamentally rewritten its laws to give the natural world a measure of protection.

As they set out on their journey, Peterson and Fisher may have sensed that the tables were starting to turn in conservation's favor. And though they may not have known it, the two men also saw America at a cultural crossroads. The postwar economic prosperity was kicking in, and the baby boom was under way. Not only was 1953 the fiftieth anniversary of the Wright brothers' first flight; it was the anniversary as well of the first cross-country automobile trip. The modern car culture was taking off, and serious work on an interstate highway system was in the offing; color television was about to hit the stores, and suburbia was rapidly expanding. Yet there is a nostalgic quaintness to *Wild America* today. As

Fisher and Peterson flew into Boston from their first stop in Newfoundland, "various never-before-seen objects were identified from the air by James, including a clover-leaf crossing, a baseball field, and Boston's only skyscraper. He failed at a drive-in motion picture theater, which had to be carefully explained to him." And to their readers, the authors felt they had to explain what motels and air-conditioning were.

Wild America touched a nerve with Americans, as had several other recent books. Aldo Leopold's epochal *A Sand County Almanac*, with its revolutionary ideas on ecology and humanity's relationship with the wild world, had appeared in 1949, giving the environmental movement its philosophical framework through Leopold's notion of a "land ethic." Then, in 1951, Edwin Way Teale had published *North with the Spring*, the first of an eventual quartet of seasonal nature travelogues, which at least partially inspired *Wild America*. What set the latter book apart, though, was Peterson's extraordinary knowledge of American flora and fauna coupled with Fisher's unusual perspective of being an experienced naturalist in a completely new land. Fisher had never set foot in North America until he joined Roger Peterson in Newfoundland, and his journals, which formed the meaty core of the book, showed a engaging combination of wide-eyed boy and seasoned scientist.

Coming back to *Wild America* in my middle age, I realized that this book, whose meaning for me has evolved with the decades, might serve yet another purpose—that half a century was a pretty good benchmark from which to gauge the land Peterson and Fisher saw. I could retrace their journey and see how the continent's natural soul—and ours—were faring at the dawn of the twenty-first century.

It was something Peterson and his third wife, Virginia, were planning to do themselves (Fisher having died in a car accident in 1970, leaving a void in his American friend's world). "There is another book to be written about Wild America," Peterson wrote in 1980, shortly after the publication of the fourth edition of his legendary field guide. "Ginny and I will repeat this odyssey at a more leisurely pace and try to describe the changes that have taken place in 30 years." The Petersons never made the trip, though a few years later the Canadian naturalist Lyn Hancock did, for a book she called *Looking for the Wild*. Roger Peterson himself died in 1996, at the age of eighty-seven, his plans to redo the 1953 journey unfulfilled.

So, well aware of the hubris of following in such big footsteps, I pulled out stacks of maps and my old, thumb-worn copy of their book and started running my finger along the route Roger Peterson had laid out for them. Fisher was a seabird biologist, so their journey hugged the coast for the most part. This gave me pause; it would seem a fool's errand to search for America's wilderness heart not in the high Rockies or the North Woods but largely along the densely populated, rapidly sprawling rim of the continent where they traveled—from the northeastern megalopolis to the Sunbelt and Southern California. And it's true that many places that were at least nominally wild in the 1950s have not been paved, parceled, and Wal-Marted out of existence.

Yet I found that wild America is still out there, its tenacity and strength surprising. It's on the long, lonely sandspits of South Monomoy Island off Massachusetts, where gray seals weave intricate knots in the clear water beneath one's boat as flights of shorebirds stream overhead. It's in the deep mountain forests of the Smokies, where elk roam again after 150 years' absence; in the orchid-draped, panther-haunted Fakahatchee Strand in Florida, in the sky islands of the southwestern deserts, and in the alpine meadows of the Olympic Mountains.

Much of this is due to nature's inherent resiliency, but a lot of the credit goes to the more enlightened policies that have reshaped our relationship with the natural world in the past half century—to aggressive attempts to restore the land and bring back long-vanished species. The bedrock environmental laws passed in the 1960s and '70s did much to stanch the worst wounds, but in recent decades we have begun to try our hand at restoration, at actually rebuilding what had been almost wholly destroyed—an imperfect process of trial and (thus far) more error than success, but the mere fact that we're attempting it is cause for hope.

It is both an irony and an obscenity, therefore, that against this historic arc of progress we're now confronting the most environmentally hostile administration and Congress in generations, which have made dismantling those fundamental environmental safeguards a national priority—though they window-dress their actions with claims of streamlining bureaucracy or "balancing" the needs of nature and commerce. Because most Americans, regardless of their political views, consider the environment an important issue, few of these battles have been waged openly. Far more often they have been the result of administrative rule changes that receive scant attention and are relatively immune to public debate, like the rule announced a few days before Christmas, 2004, that eviscerated the 1976 National Forest Management Act, making it far easier for national forest managers to open 191 million acres to logging or drilling, and severely restricting public participation in the forest planning process. By reinterpreting how laws should be applied and enforced, they can appear to leave intact the environmental regulations that the vast majority of Americans support while reducing them to a hollow and ineffectual shell.

It has been a debilitating period for conservationists. But as grim as the current political scene may be, once I took to the road, following the faint trail Peterson and Fisher laid down five decades ago, I came to see that there is still much to celebrate—more, perhaps, than a conservationist may realize, unless he or she takes the long view that such an anniversary provides. The National Wildlife Refuge System (which marked its hundredth year in 2003) has nearly quintupled in size since the early 1950s, to more than 93 million acres, including some places—like the Yukon River delta in Alaska, which was added to the system in 1960—that Peterson and Fisher urged be protected in just such a fashion. The federal wilderness system, which didn't even exist until 1964, now encompasses more than 106 million acres.

Along the lower Rio Grande, which Fisher and Peterson mourned as a lost wilderness, native forest is being resurrected from old farmland to create a 275-mile corridor for rare birds and wildcats; the widespread slaughter of birds of prey has been replaced by universal protection, and large carnivores like wolves and even jaguars are returning to areas where they've long been absent. America's supple, resurgent wilderness has created wholly unexpected dramas that would have awed Peterson and Fisher. They had to travel to the Coronados Islands of Mexican Baja to see a handful of surviving elephant seals; today the huge pinnipeds are found by the tens of thousands along the California shore—and with them have come leviathan hunters. Today scientists off California's Farallon Islands watch, awestruck, as eighteen-foot great white sharks grapple with the three-ton seals, the most extraordinary contest between predator and prey in all of North America.

One of the things that strikes me most forcefully today as I reread the original *Wild America* is how remarkably prescient Roger Peterson and James Fisher were about many of the emerging threats to North America's natural systems. They already recognized the damage of what we now call sprawl development (though I doubt even they could foresee the colossal growth in suburban and coastal zones). And long before others were concerned, they voiced warnings about the impact of invasive alien species and foresaw the increasingly bitter fight over scarce water resources in the West and over questions of forest management like protection for old-growth stands on federal land and the proper role of fire.

Yet the modern world is also bringing pressures to bear that they scarcely imagined—environmental contaminants that are far more subtle (and potentially far more damaging) than the pesticides like DDT they worried about, for example. Nor could they foresee the overarching specter of global climate change. The high Cascade glaciers they admired are retreating toward swift extinction, and Alaska, which served as a climax for their continental odyssey, is showing the effects of global warming in startlingly unambiguous ways.

I went to the great gannetry at Cape St. Mary's to pick up the trail of two of the twentieth century's most accomplished naturalists. Because my goal wasn't merely to run a valedictory lap for their book, I would be using their original itinerary as a broad framework, but I wouldn't follow it with religious precision; I'd deviate a bit from their route to visit places that best illustrate the changing landscape or to see sights that weren't even possible for them to enjoy in 1953.

The pace would also be much different. Where they sprinted, I would amble. Peterson and Fisher pushed themselves relentlessly for three and a half brutal months, taking just two days off in all that time, often driving through the night to keep to Roger's meticulously plotted schedule. The breathless quality that comes from this perpetual motion is one of *Wild America*'s charms, but I needed time to dig deeper, to look more contemplatively at a continent that has changed dramatically in fifty years. And so I would make most of the journey over the space of about nine months, gaining in insight, I hoped, what I lost in perfect replication.

I found a continent changed—for the better in some places, for the worse in others. Yet the land, the rugged heart of natural America, retains an essential timelessness, which on my own journey I discovered again and again. Ours is still, at its core, a wild country. "Never have I seen such wonders," James Fisher wrote at the end of their trip, "or met landlords so worthy of their land. They have had, and still have, the power to ravage it; and instead they have made it a garden."

RETURN TO WILD AMERICA

ONE

Atlantic Gateway

In the early 1950s, just getting to Cape St. Mary's was an adventure. The Avalon Peninsula is the easternmost prow of North America—a vaguely H-shaped chunk of land that is very nearly an island itself, attached to the rest of Newfoundland by the slenderest of threads. It is rimmed by sheer cliffs, by beaches of dark quartz-shot cobblestones and wave-smashed capes. Where there is forest, it is somber and mossy, spruce and balsam fir hung with long pale sheets of lichen dangling from the branches like rotting curtains. But much of the Avalon is tundra, known locally as barrens—an open, windswept land home to flocks of ptarmigan and the southernmost wild caribou herd in the world, where the trees, if they grow at all, cower in dense, waist-high thickets known as tuckamore.

When Fisher and Peterson met here to begin their journey, Newfoundland was very much a world apart, sparsely populated and isolated from the rest of the country not only geographically but also politically; it had confederated with Canada only four years earlier, ending its long history as a separate dominion of Great Britain. Most of the people lived in remote fishing villages called outports accessible only by sea, and beyond the handful of large towns like St. John's the few roads were largely dirt and gravel, and at times all but disappeared into the spruce bogs.

Accompanied by the local ornithologist Leslie Tuck, the two travelers spent a long day bouncing south from St. John's on awful roads. Fisher, keyed up to see new birds, found himself "seeking the differences and finding the similarities"; the first North American species he saw was a gannet, which was also the last British species he'd seen as his plane crossed the Scottish coast. This isn't surprising; few places in North America have as strong an Old World flavor, at least in terms of natural diversity, as Newfoundland. The landscape, Fisher thought, was strikingly similar to the spruce forests he'd known in Sweden, and of the forty-six species of birds they saw, almost two-thirds were ones he knew from Europe. He was stunned to find that the most common birdsong in the dark conifer woods was "a voice as familiar to me in my own English garden as on the cliffs of St. Kilda and the remote Shetland Islands"—that of the tiny winter wren. Though a common backyard bird in Great Britain, in North America it inhabits only the boreal forests of the North or high elevations.

They spent the night in the fishing hamlet of St. Bride's, where the navigable road ended, and the next day—with a local guide and a pony to carry Peterson's heavy camera gear—they set off for the great seabird colony near the Cape St. Mary's lighthouse, ten miles down the coast. The special attraction would be thousands of northern gannets, majestic seabirds with six-foot wingspans and an extraordinary means of fishing in which they plunge, like white lances, straight down from more than a hundred feet in the air, hitting the water like Olympic high divers to intercept fish far beneath the surface. Fisher was perhaps the world's leading authority on gannets, and he was anxious to see them on this side of the Atlantic.

"The most spectacular New World [gannet] colony, the one at Bonaventure Island off the Gaspé, is visited by scores of bird students every year," Peterson noted. "It has become a profitable thing for the innkeeper to cater to—and even advertise to—an unending succession of summer gannet watchers . . . On the other hand, the colony at Cape St. Mary['s] sometimes goes for several years unvisited by any of the field-glass fraternity." It was easy to see why. The exhausting hike, first along a muddy, rutted cart track, and then overland around treacherous bogs, took them until early afternoon, and they didn't make it back to St. Bride's until well after midnight, Peterson limping from a badly strained leg muscle.

It's a safe bet that Peterson and Fisher would find Cape St. Mary's today a strange, perhaps disquieting blend of the familiar and the unfamiliar. I flew into St. John's with my fiancée, Amy Bourque, who was taking some time off from the Audubon sanctuary she ran in Maryland to get me started on my way and would join me at a couple of other points in the year ahead. Our drive from St. John's to St. Bride's took a couple of easy hours in a rental car, with just one stretch of gravel track that skirted bogs bounded by stands of pink rhodora azaleas. It was a cold day in the middle of June; the sedges and irises made an emerald splash along the coffee-colored rivers, where a few blackish spruce grew, but higher up, the still-brown tundra rolled inland beyond the limit of vision, empty of any sign of humanity.

The reason for this little slice of the Arctic, at the same latitude as Montreal and Seattle, is the ocean. Pack ice surrounds much of the Avalon Peninsula until April or May, and throughout the summer icebergs that originated in Greenland are a common sight, floating south on the Labrador Current. Especially along the immediate coastline, summer fogs are an almost daily event, bathing the land in damp cold that counteracts the wan heat of the sun. This creates an ecosystem known as hyperoceanic barrens—a nearly horizontal plant community dominated by ground-hugging Arctic species like crowberry, several species of cranberries, hardy grasses, and pale reindeer lichen in spongy green-white mats, into which a hiker sinks ankle-deep. Dwarf Arctic willows, with trunks barely as thick as a finger, spread over rocks like a green cascade, raising maroon flowers five or six inches above the ground, their diminutive stature masking the fact that these trees are often three or four hundred years old, as deserving of awe as any craggy old-growth pine.

From the air, the center of the Avalon Peninsula, and its southern capes in particular, appear pale brown, edged with black; the brown is the tundra barrens, while the darker rim is the spruce woods, which cling to the lower elevations and deeper river valleys. We spent part of one day hiking along some of the rivers that cut across the barrens, providing a convenient pathway in the otherwise featureless landscape. The only green was down along the water, where thick grass and irises were sprouting, but when we looked closely, the tundra was coming into great flower as well.

To my eye it all looks like classic Arctic tundra, but there are enough floral differences that a botanist would immediately peg this as Newfoundland barrens: a spongy, wet mat of sphagnum moss and crowberry studded with pitcher plant, the diminutive white flowers of goldthread, and the pink of pale laurel, barely an inch high. Cloudberry, Newfoundland's famous "bake-apple," was just opening its single white flowers that would, by August, produce the delicious clear orange raspberries that are a provincial obsession. Much of the ground was covered in creeping juniper or, where there was a bit more shelter, patches of spruce tuckamore that were barely waist-high but might be four or five hundred years old.

We were watching for caribou, and there was plenty of sign—a lot of droppings, some quite fresh, and tracks in the mud, each print slightly wasp-waisted in the middle. We also found the remains of a calf that had died over the winter, down along the stream—a corona of white hair and the clean, bleached bones, along with piles and piles of mammal scat, much of it fox, all of it containing great quantities of caribou fur. When we later saw a live caribou, a large bull, it was albescent, as white as the weathered bones; even its nose was pale gray, so that the only color was the dark eyes and the black hair that grew inside its ears.

The wind howled, and for five minutes a cold, heavy downpour soaked us before the sun peeked back out apologetically. We hiked an hour or so down the main river, then looped back and struck a smaller tributary, which had cut a confused series of low gorges in the rock—a creek maybe thirty feet wide, the water so stained with tannin it was nearly black. It was fed by a series of small pools and ponds on the higher ground to either side; some of these were jewel-like, surrounded by sedges and irises, set in little steep-sided dells where the wind couldn't reach and the dark water reflected the sky like glass. In one, a yellow-rumped warbler alit on a rocky shelf above the water, blue-gray against the yellow lichen, flew down to splash with its reflection, then whirled off with a single chip.

More than a thousand square kilometers of the central Avalon are protected as a wilderness reserve, but even the margins are largely wild, largely empty land. The only paved road is a loop that skirts the edge of the sea, linking the small towns like St. Shott's and Trepassey, with a handful of dirt tracks like the one that runs down to the old lighthouse at Cape Race, where we were paced by swift-flying horned larks and found the remains of a freshly killed murre on a low bluff beside the sea, its feathers gently waving in a long plume to the leeward where a peregrine falcon had sat and plucked it. Its skeleton was still wet and bloody, the two black wings untouched, but the falcon was gone.

On the forty-mile-long peninsula that ends at Cape St. Mary's, the road dipped and soared, moving down into sheltered hollows and then up across great headlands where the wind lashed the long grass furiously. The cove villages were small—Great Barasway, Patrick's Cove, Gooseberry Cove—a few dozen houses, maybe a church, old fishing boats hauled up. The beaches were all of rounded stone, not sand, formed in a series of wide steps rising fifteen or twenty feet, testament to the heights of the tides and the strength of the surf; the highest levels had the biggest cobbles, gray and purplish rocks the size of moderate pumpkins, dark and tiger-striped with thin layers of quartz.

At one point we got out of the car for a stretch, heading for a tidal pond a quarter mile down the coast, where gulls and terns darted into the choppy outflow. The tide was well out, and while there were no tidal pools per se, we could walk carefully over the flat cobble beach, trying to stay on the larger, drier rocks, lifting up mats of rockweed to peek underneath. The water was polar, and my fingers began to ache, but there was too much stuff under there to stop: limpets cruising slowly over the pink-crusted stones; anemones the size of fifty-cent pieces wedged in the crevices, their striped tentacles spread wide; hordes of mating amphipods like tiny, half-inch shrimp, the males bigger and bright red, cupping the gray-brown females with their hooked front legs. Along the high-tide line, among the brittle wine-brown pieces of old kelp, there were uncountable millions of the exoskeletons of these creatures, each dried a delicate purple-pink, and so fragile that no matter how carefully I tried, they fell to pieces when I picked them up.

We found the bones of eiders and seabirds, as well as dozens of dead skates, reduced to spiny tails a foot long, curling and dry, and a few bony pieces set in the framework of their cartilaginous skeletons. The jaws looked alarmingly like large, fleshy lips from some nightmare, not soft but harsh and covered with rows of tiny, sharp teeth. The tide line was also festooned with junk, most of it the detritus of commercial fishing: mesh bait bags; long skeins of tangled half-inch nylon rope in blue, yellow, and green; scraps of netting; and fragments of wooden lobster traps. There were also lots of shotgun shells, the brass bases long gone but the plastic casings still looking quite fresh; twelve-gauge, mostly, and in numbers I found hard to credit—hundreds and hundreds of them strewn along the shore. Even granting that plastic lasts basically forever, I couldn't imagine what anyone would be shooting in such quantity. It would be some days before that particular mystery was solved for me.

St. Bride's sits at the end of the cape road, bigger than its neighbors to the north—big enough, at least, for a gas station/general store/tackle shop/motel known as the Bird Island Resort, which, as its name suggests, owes much of its custom to tourists anxious to visit Cape St. Mary's, an official wildlife reserve since 1964. A cluster of efficiency units perched on the edge of the high cliff, neat as pins and looking out over the water by a wee miniature golf course on an equally wee patch of lawn. (Our room, I found, came thoughtfully equipped with a golf club.)

It was not a day for mini-golf, though. The sea was leaden, and it was ear-pinching cold; even the locals were huddled inside their parkas with the hoods up, and snow flurries were in the forecast. The wind was really cooking, shaking the motel, ripping at the still mostly brown grass, plunging over the cliff and fanning out to sea in furious gusts. I could only imagine how much colder and gloomier it must have been for Peterson and Fisher, arriving as they had in early April.

St. Bride's spreads out over a mile or so of flat land above and to either side of a small creek, anchored by two large graveyards near the highest points on either side of the valley. The houses were generally small and boxy, devoid of yard trees, painted mostly white but with the flashes of gaudy color I find delightful in Newfoundland towns; one square building was brilliant purple, others had their shutters, trim, and a wide band below their rooflines painted kelly green or cobalt. It's a scene that has changed little since the 1950s, except that the road is better and now the village is wired for electricity. But one change my predecessors would have noticed: here and elsewhere along the cape there are far fewer boats hauled up these days in the village coves, and the racks of drying codfish, with their all-pervasive smell, are long gone.

Timeless as Newfoundland may appear, in recent decades it has been swept with changes that are little short of tectonic. Fifty years ago, cod fishing was the engine that drove this region, as it had for five hundred years before that. Most Newfoundlanders made their living from the sea, and for much of the year that meant fishing for cod, which were once so plentiful they could be caught simply by lowering a weighted basket over the side. As late as the 1960s, cod populations were robust and harvests were high, but mechanized factory trawlers were siphoning off the schools in unsustainable numbers, and government subsidies meant more and more fishermen chasing what was left; by one estimate, more cod were taken between 1960 and 1975 than in the 250 years prior.

By the 1980s, inshore fishermen were seeing cod dwindle in numbers and size, and some of the government's own top scientists warned that the official stock estimates were hugely inflated. In 1986, one scientist, speaking at a conference, politely suggested (in Latin) that his department's perennially rosy assessments were *non gratum anus rodentum*, "not worth a rat's ass."

By 1992, the situation had become so dire that a moratorium was imposed on cod fishing off parts of Newfoundland; the following year, it was extended to all groundfish. An estimated thirty thousand people were thrown out of work and into government retraining programs, and a culture wedded to the sea was gutted. Rural Newfoundland, already struggling economically, began to hemorrhage its population as people moved away to find work. When cod showed some small signs of recovery in the late 1990s, the government prematurely reopened some fishing, driving stocks to even lower levels than before while putting most of the blame on seals, colder-than-normal water, and other natural pressures. In April 2003, the Canadian fisheries minister finally declared a complete closure of cod fishing in the country's Atlantic waters.

Even while traditional rural livelihoods like fishing were reeling, Newfoundland's inherent beauty had been stoking a different industry—tourism, which has been increasing steadily in recent decades. The great bulk of that is directly attributable to outdoor and nature tourism, to folks coming to hike and camp, watch whales, fish, sea-kayak among icebergs, or go birding. Cape St. Mary's, which, as Peterson noted, once went years without a visitor, now gets more than eighteen thousand tourists each summer, drawn by one of the world's most accessible large seabird colonies.

There is no need to traverse bogs with a pack animal these days, or to struggle with the sea, as the builders of the original lighthouse did; it took them three months to get the first lamp out here in 1860, failing over and over again to safely land the massive glass orb. Today, not far from the lighthouse, there is a large new visitors center, a red-roofed white building named in honor of Les Tuck, with big windows that look east half a mile to the fingered headlands where most of the birds nest and above which white gannets perpetually swirl. Such is the view, that is to say, when you can see it; fog may blanket the cape three days out of four in the summer.

We were lucky; it was a bright if windy day, and I was anxious to see the birds up close. Inside, we were greeted by Tony Power, the reserve's manager, who was expecting us and who introduced us to Chris Mooney, one of his interpreters—both of them quietly friendly in the way I've come to expect from Newfoundlanders. Chris was a big, bluff man in his thirties, with premature gray in his dark hair and goatee, who grew up in the nearby village of Branch; like many on the so-called Irish coast of Newfoundland, he had a pronounced brogue. "Tony and me, we're both talking slow so you can understand," he said with a smile, after we'd been chatting a while. "It hurts me jaw to speak so slow."

The trail leading out to the colony, a shaley path through sheep-cropped grass, wandered close to the edge of the cliff and was marked along the way with small rock cairns every few yards; this is not a place where you'd want to lose your way in the fog, and even the sure-footed sheep sometimes make a mistake and fall to their deaths. Chris led us off the trail and out onto the cape, the wind snatching words from our mouths and the hat from my head, which I saved with a frantic lunge. The ground was hummocked and broken, purplish red rock split into sharp-edged blocks over which grew an almost solid cover of blazing orange lichen known as *Xanthoria*, which thrives on the nutrient-laden air near seabird colonies. We hopped a little trickle of a stream, where the fuzzy buckskin-colored fiddleheads of cinnamon fern were just uncurling, then sat down near the edge of the cliff, out of the wind a bit, as Chris set up a spotting scope.

During most of the year, seabirds roam the oceans hundreds, even thousands of miles from land and are all but invincible on the wing. But they're vulnerable when they come to land to nest, so they tend to pick the most inaccessible places to breed—lonely islands, remote cliffs, unclimbable rock stacks. At Cape St. Mary's, the murres, razorbills, and kittiwakes nest all along the mainland cliffs, forming a layer-cake effect typical of seabird colonies: The white icing of the kittiwakes was highest, each one sitting on a neat, mounded little nest of grass and wrack, most with their heads turned and bills tucked into their back feathers, dozing. Below them were the dark murres, standing in little mobs wherever the ledges were wide enough to support them, facing the wall as they balanced their single eggs beneath their bellies on the bare rock—the view through the four-week-long incubation period must leave a great deal to be desired.

Most of the ten thousand pairs at Cape St. Mary's are common murres, but a few—maybe one pair in a thousand—are thick-billed murres, a slightly stockier, beefier species that is among the most abundant seabirds in the Arctic, but rarer at these more southerly latitudes. With Chris's scope, we could see the diagnostic white streak along the upper mandible of the thick-billeds, but he, with long familiarity, pointed out an identification shortcut that you won't find in any field guide: their guano. The ledges below a common murre nest were stained with white droppings, while those of the thick-billeds were pinkish, perhaps from a dietary difference.

While gulls and murres can make do with narrow cliff ledges, the gannets need more room. At Cape St. Mary's, their activity focuses on Bird Rock, a four-hundred-foot-tall square-sided block with a wide, flat top, separated from the mainland by a gap of fifty feet and protected by sheer sides that killed several of the early climbers who tried to scale it. It is here that most of the gannets stake out their tiny nesting territories, defending just as much ground as they can reach with their long, sharp bills; from a distance, the white birds appear to be precisely spaced like geometrically arranged snowballs.

And a lot of snowballs—eleven thousand pairs of gannets crowd the rock, up from fewer than forty-five hundred pairs in the 1950s, a time when fishermen still killed them, thinking them to be competitors. (In a neat bit of turnabout, today the gannets have incorporated so many bits of fishing tackle and nylon rope into their messy grass-and-seaweed nests that the colony has a pervasive plastic green tinge.) Wheeling above those incubating their eggs, always moving in a clockwise flow, flew hundreds more, a constant coming and going from the fishing grounds, birds jockeying the air currents to execute very precise landings on their spot, and their spot alone—for even inadvertent trespassers were treated to bites and beak stabs.

Any birder has his or her own favorites, and gannets have always been among mine. I see them most often in winter, when they hunt the sea south as far as the Gulf of Mexico, usually far from the sight of land unless the wind pushes them toward shore; I remember many frigid days, staring into an easterly wind through teary eyes, watching legions of white gannets spear down into the water for fish. The dry ornithological term "plunge-diving" doesn't do this justice; the gannet's hunting technique has a lean ferocity that is utterly breathtaking.

For one thing, these are big birds, with the wingspan of a small eagle, stretched to a needle's point in all directions—long, tapered wings tipped in black and a three-foot body that ends in a wedge-shaped tail at one end and a swordlike beak at the other. The plumage is gleaming white, except for a wash of gold on the head. The eyes and beak are rimmed in heavy outlines of black, which I've always thought give the bird a piercing, implacable look.

When hunting, a gannet flies slowly into the ever-present sea wind, hanging as much as a hundred feet in the air, pivoting on its angular wings, head down, watching. A gannet's eyesight must be extraordinary to spot a fish among the shattered reflections and windblown spume of the sea, but when it sees the shadow of its prey, it tips forward and begins to drop headfirst, often flapping to accelerate, quickly reaching speeds of sixty miles per hour. A second before impact, it throws its impossibly long wings straight back, becoming a shaft. It makes almost no splash as it goes in, plunging up to thirty feet deep; little wonder the old Cornish name for the gannet was *saithor*, "arrow."

Now, take that aggressive, flung-javelin effect and multiply it by hundreds of birds—for gannets tend to hunt in great flocks, seeking deep shoals of fish that they alone, among the seabirds, are heavy and strong enough to reach. To look out through such a frenzy of diving, flashing bodies, their trajectories crisscrossing, hurtling past each other, then rising from the waves to plunge again, is to see something truly elemental.

Gannets on their nests, however, reveal a much more endearing, almost goofy quality. Like most seabirds, they sacrifice a large measure of their grace when they land, plodding on big webbed feet, black but with an improbable streak of bright green running down each toe like nail polish. And while the gap between Bird Rock and the mainland is wide enough to keep predators away, it is narrow enough that a visitor, watching with binoculars, feels almost as though he's sitting among the gannets—a rare chance to watch their endlessly fascinating behavior.

Gannets are highly demonstrative birds, with an array of stylized displays linked to social behavior and breeding. A gannet may threaten another with a scooping movement of its open bill, and the intruder will react by showing the nape of its neck in appeasement. When one gannet returns to the nest from hunting, the reunited pair stands a-tiptoe, breast to breast, their bills held high and slightly crossed. At other times, they bow low to each other, wings partially raised and necks demurely curved; biologists, showing an unexpected poetic side, call this the "curtsy" display. A gannet about to depart the rock stands, wings open and ready, bill aimed skyward, poised like a flagpole eagle for long moments as the wind ruffles its feathers. Nor are they all show; a gannet that misses its footing and tumbles into another's territory may be killed. We watched in painful silence as one bird, tripping over the edge of a rock, tumbled down through a gauntlet of biting, stabbing gannets that ripped at its head and wings, minute after violent minute, leaving it to finally stumble, bloodied and dazed, onto neutral ground.

In the days before radar or even lighthouses, the fishermen who jigged for cod off Cape St. Mary's listened through the fog for the sound of the gannets and kittiwakes on Bird Rock, to warn them if they were running too close to the shore. Even on a sunny day like this one, the noise was remarkable, loud enough that my eardrums began to rattle, as though they were short-circuiting on an overload of sound. The wind howled past us, combing the grass, which was festooned with white feathers, as though a pillow factory had exploded.

In many respects, we were seeing a far more dramatic show than the one Fisher and Peterson enjoyed in 1953. The number of birds has risen substantially since then; they counted barely four thousand gannets, whereas today the population has almost tripled, and keeps growing. Fisher, who returned in later years to help Les Tuck monitor the murres, put their population at about twenty-five hundred pairs, a quarter of what it is today. Kittiwakes, cormorants—all are increasing with each passing year.

It was appropriate that *Wild America* began on the gannetry of Cape St. Mary's, because seabirds first brought Roger Tory Peterson and James Fisher together. They met in 1950 during an ornithological conference on the Swedish island of Gotland, in the Baltic Sea, on a field trip to a murre colony. Peterson was nervous about the meeting, which was no accident;

Fisher was the nature editor at the British publishing house William Collins Sons, and Peterson needed to convince him that Collins should publish a field guide to European birds that he and several colleagues were working on. Peterson needn't have worried; Fisher was anxious for Collins to publish the book, but, more important, the two men immediately struck up a firm friendship.

At the time of the Gotland trip, Roger Peterson was forty-two years old and something of an international celebrity within the world of natural science. Sixteen years earlier, in 1934, he'd published *A Field Guide to the Birds; Giving Field Marks of All Species Found in Eastern North America*, the first modern identification guide, which brought bird study out of its largely academic, specimen-oriented past and made it easily accessible to anyone with a pair of binoculars.

It's hard to convey these days just how important an advance Peterson's compact book was. Through the end of the nineteenth century, ornithology was literally a shotgun science; if you wanted to study or even simply identify a bird, the first thing you usually did was shoot it. By the time Peterson was born in 1908, some ornithologists had moved away from this approach, and a few books had popularized bird study for the general public, like Chester Reed's small *Bird Guide*, with its limited selection of idealized portraits, which Peterson devoured as a child. He also read *Two Little Savages*, a 1903 book by the naturalist Ernest Thompson Seton, in which one of the young protagonists draws simple pictures of the unique "uniforms" of mounted ducks so he can later identify them in the field. But most bird books were either densely written scientific texts given to tedious verbal description or lightweight efforts like Reed's that covered only a few common species.

By the time he finished high school, Peterson had become a rabid birder, but he'd decided to pursue a career in commercial art; in 1927 he traveled to New York City to attend the Art Students League, and then the National Academy of Design. He also fell in with a group of young, equally hard-core birders under the tutelage of Ludlow Griscom, an ornithologist at the American Museum of Natural History who took to new heights the concept of "field marks"—easily seen patterns, plumages, or behaviors that could distinguish birds at a distance. Everyone who looks at birds uses field marks, consciously or not—a robin's orange breast, a male mallard's green head. But Griscom was a genius at it, applying the technique to diverse, confusing groups of birds that heretofore required in-the-hand identification, and Peterson soaked it up like a sponge.

In *A Field Guide to the Birds*, Peterson took what was then the cutting edge of identification and translated it into small, schematic drawings of birds as they appear through binoculars, with lines pointing at the most salient features. Similar-looking birds were grouped together for ease of identification, while the text was brief and snappy, but it covered all the species of eastern North America, not just the common ones. It was not a fancy book; the illustrations were in black and white to save money, and the publisher, Houghton Mifflin, was chary enough of its chances to print just two thousand copies. (Three other publishing houses had already rejected it, and Houghton Mifflin only signed the book after Peterson agreed to forgo royalties on the first half of the print run, even though the editor, Francis Allen, was another of Griscom's protégés.)

The book sold out in less than a week, and over the past seventy years the guide (soon expanded to full color, and now in a fifth edition that bears scant resemblance to the original) has, with an equally revised western counterpart first published in 1941, sold something in excess of 7 million copies. The Peterson Identification System was soon being applied to the gamut of natural diversity, and the bird book anchored a wildly popular field-guide series that eventually covered ferns, fungi, reptiles, and the stars, among much else. The success of the Peterson guides, in turn, fostered the appearance of similar, competing books, all of which made it much easier for beginners to find their way into the fascinating world of nature study. But in the popular mind (so far as the popular mind worried about such things in the 1950s) Peterson had *become* birds and birding, the personification of the class Aves and the slightly odd people who chased them.

I only met Roger Peterson a couple of times, all within the last decade of his life, so my mental picture of him is of a lean, somewhat frail man with snowy hair and a craggy face—quiet, intense, unfailingly polite to the crowds of birders that besieged him, always willing to autograph copies of his field guides with a red felt-tip marker. So the jacket photograph on the first edition of *Wild America* is always a bit of a shock to me. Peterson then was exactly my own age, a wiry man of early, vital middle age, with a long nose set in a triangular face, narrow at the neck and high in the forehead, his hair graying. He's wearing a shirt buttoned to the neck and a string tie. His eyes are level and very serious, with only a hint of a smile on his face.

James Fisher, on the other hand, I knew only from his words and this one photograph, which shows a slightly portly man—fewer edges than Peterson, a more open face that is rounded by a loose grin, a thatch of unruly dark hair that flops in loose curls over his forehead. In some ways, the picture is a weird reversal of national stereotypes; Peterson looks more like the starchy, buttoned-down image of a Brit, while Fisher could pass for an easygoing Yank, ready with a joke and a backslap.

When they met in 1950, James Maxwell McConnell Fisher was thirty-eight, a few years younger than Peterson, but at least as well known in Britain as his friend was in North America. Like Peterson, Fisher had been mad about birds from almost his earliest memory (his father and his uncle were both keen naturalists, and he later said he began birding at age two), but unlike Peterson, Fisher followed a prestigious academic track, attending Eton and then Magdalen College at Oxford with an eye toward medicine. But in 1933 he joined an Oxford expedition to the Norwegian Arctic island of Spitsbergen and lost his heart to the cliff-rimmed landscape and the teeming seabird colonies of the Far North. When he returned to school, he switched to zoology, and seabirds became his professional focus.

But Fisher was also a graceful communicator, and he rapidly made a name for himself among the wider public—first with his popular book *Watching Birds* in 1940, which did much to fan Great Britain's latent interest in birding, then through a series of BBC Radio programs beginning with *Nature Parliament* in 1946, which attracted huge audiences. By 1950, Fisher had also moved into television, making him easily the most widely recognized naturalist in the United Kingdom, a place where nature study has always been a more general passion than in the United States.

Both men were, in Peterson's words, "obsessives" about natural history, especially birds, and they hit it off immediately. Peterson was spending a great deal of time in England in those days, working on his European guide, and he and Fisher were often in the field together, roaming from Lapland to the Alps and the Mediterranean. The Fisher home, a chilly old rectory in the small Northamptonshire village of Ashton, became a base for Peterson and his second wife, Barbara, though Fisher also saw to it that Peterson became a member of his London club, the Savile, where Fisher spent much of his working week.

The *Wild America* tour was Peterson's idea, the culmination of a long-held dream that crystallized after he met Fisher. "So much had I seen of wild Europe, and especially wild Britain in the company of my colleague, and so much had I learned under his tutelage . . . that I had a growing desire to reciprocate, to show him my own continent," Peterson wrote. The tour promised to be "a more complete cross section of wild America than any other Englishman, and all but a few North Americans, have ever seen." Through late 1952 and early 1953, Peterson was busy cobbling together what he often referred to as the Grand Tour of Wild America—an immense logistical jigsaw puzzle that had to balance work deadlines, travel schedules, seasonal timing, and the availability of local guides, biologists, and wildlife managers who would help along the way. "Roger recently spent a long Saturday afternoon at my house, with a map spread out on the floor, planning your journey day by day," the Houghton Mifflin editor Paul Brooks wrote to Fisher in December 1952. "It is extremely exciting and makes me absolutely pea-green with envy. I don't see how the combination of subject and authorship can fail to produce a unique and magnificent book."

Houghton Mifflin offered the pair an advance of a thousand dollars each—a respectable sum in those days—and promised to pay the bulk of Fisher's share at the beginning of the tour to offset his expenses. But finances remained a sticking point; Fisher and his wife had six young children, including an infant girl born shortly before the trip. Just a few months before their scheduled departure, in fact, Fisher suggested holding off a year. Peterson, who had spent months arranging planes, boats, lodging, guides, even ordering a new Ford station wagon, was adamant: it was now or never. "The trip is only three months away," he wrote in late January 1953. "You have just got to come as we shall not be able to do this another year. Our well-timed chance will be gone. Therefore, if the squeeze at home is too tight and a small loan would help out, I stand ready." Whether the loan was necessary isn't clear from their subsequent correspondence, but Fisher decided to take the plunge.

Peterson leaned on a continent's worth of friends, colleagues, and contacts to smooth the way and provide local guidance. Looking back from this day of e-mail and wireless communication, one might see the task of arranging three and a half months of daily travel, lodging, and rendezvous with experts as almost insurmountable, but Roger pulled it off with aplomb. Fisher commented on this a number of times during the trip, as when they made a quick stop in the piñon forests of northern Arizona and within moments found two birds restricted to that habitat. "I was gradually beginning to realize the planning and care and experience behind Roger's conducted tour of North America," Fisher marveled. "Our apparently casual stop in these pinelands . . . had, I suspected, been contrived, quite deliberately, at Roger's desk in Maryland the previous winter."

Because Fisher was a seabird biologist, the route Peterson devised rarely strayed more than a few hundred miles from the coasts, skirting the perimeter of North America and taking them to the breeding grounds of the majority of North America's seabirds—Newfoundland, the Dry Tortugas in Florida, the islands of Baja, the Pacific Northwest, and the Bering Sea. This, they thought, would also pay professional dividends, because the men were hoping to collaborate on a guide to the seabirds of the world. The route had the further advantage of hitting many of the most biologically diverse ecosystems on the continent, like peninsular Florida, the thornscrub forest of the Mexican borderlands, and Southern California. This gave Fisher and Peterson the chance to see the greatest possible variety of birds, but whatever the rationale, it raises an obvious criticism: How could they have called it "Wild America" when they missed most of the truly wild, truly empty parts of the land—places like the northern Rockies and the upper Great Lakes, to say nothing of the vast bulk of subarctic and Arctic Canada?

After the trip was finished and Fisher had flown back to the U.K., Peterson stopped in the northern Rockies on his way home. He admitted in a letter to his friend, "If I had realized how smashing Yellowstone is I would have made every effort to have you see it." Yet accidentally or by design, the route they followed managed to include many of the remaining pockets of truly wild land along the continental margin—the ones that survived almost under the noses of most of America's populace. For most of the trip they did not, it must be said, ever stray far from their car; their schedule was too demanding, and Peterson was shooting movie footage of the trip for the National Audubon Society's lecture circuit, which required a lot of heavy, bulky gear. But they boated deep into the

Apalachicola River basin in northern Florida, looking for the possibly extinct ivory-billed woodpecker; they climbed into untouched cloud forest in Mexico and sailed to the Coronados Islands off the coast of Baja, home of the then-rare northern elephant seal. They wandered through the vast empty quarters of the Southwest, to ghost towns along the upper Rio Grande and ancient Anasazi ruins in New Mexico. And the final weeks of their journey, through the tundra of western Alaska and the far Pribilofs, took them to some of the most remote wilderness left on the planet.

But much of their time was spent in the preserved yet accessible pieces of North America's wild heritage—in national wildlife refuges and national parks and monuments. These made a deep impression on Fisher, used to the thoroughly domesticated English landscape, who longed for a similar preservation ethic back home. "Sometimes in Britain, there seem parks in name only," he wrote in *Wild America*:

> their boundaries unknown to the public, their dedication still flaunted and their use abused for schemes for hydroelectric power and exploitation which they were supposed to have been created to prevent. Perhaps we in Britain think it's vulgar, or something, to make a park act as a park, but until we do, there won't *be* parks. A park is not nature wrapped in a plastic bag and filed away in the freezer, but nature served up, to nature's best customer. A customer does not defile or waste the goods he saves and works to buy.

The book came together in 1954 and early 1955, with drafts moving back and forth across the Atlantic, or with the Petersons bunking at the Fisher house in Ashton, where Barbara Peterson typed up the longhand manuscripts in an upstairs room so cold and damp she had a hard time getting her fingers to work. The format was for Roger to introduce each chapter, then hand it over to James, whose journal entries form the core of the narrative. There was in Fisher's writing a dry humor that counterbalanced Peterson's fairly sober approach, and a fresh-eyed enthusiasm that was hard to resist. Brooks, their editor, had initially worried that the book would be little more than a catalog of birds hardly worthy of their lofty title, but once the chapters began coming in, he quickly changed his mind. "James' account of the Grand Canyon is really magnificent," he wrote to Peterson. "It also points up one of the unique features of this book; i.e., here we have a completely fresh impression—the shock of discovery—of the most written about and photographed phenomenon in America . . . the title becomes more and more justified as you go along."

Wild America was published in October 1955 to great acclaim and very strong sales. Within a month it was on the bestseller lists, and fifty years later Houghton Mifflin still offers a paperback edition, illustrated with more than a hundred of Peterson's ink-and-scratchboard drawings.

There are enough facets to the original book that many people have found inspiration in it over the years—some harder to explain to the non-birder than others, like the concept of a Big Year. As Peterson was laying the groundwork for the trip, he wrote, "It occurred to me that we might as well do things up brown and try for a record." In 1939, a New York banker named Guy Emerson, who always timed his business trips for the best birding, set a benchmark that stood for thirteen years—497 species of birds seen in one year, all found north of the Mexican border, a region with about 750 to 800 possible species. Peterson figured he and Fisher could better that mark, and to his surprise James—who was normally disdainful of what he called "tally hunting," simply treating birds like inventory—went at it with great enthusiasm. Every time he saw a new species, which was an almost daily occurrence through the entire one hundred days, he shouted, "Tally ho!"*

In fact, by the end of the trip it was the Englishman, Fisher, who held the title, with 536 species, having picked up a few extra species on a layover in Anchorage while Peterson was tied up in a hotel room finishing a drawing. (Peterson later passed him and ended the year with 572 species, not including 65 more in Mexico.) The duo dedicated their book to Emerson, who, when he learned of the honor, told Peterson it felt something like being knighted.

Since then, the Big Year idea has snowballed among birders into the most zealous, highly competitive permutation of the hobby. Men (this seems to be largely a testosterone-driven obsession) have spent enormous

*Fisher's youngest daughter, Dr. Clemency Thorne Fisher, an ornithologist in her own right who has become a good friend through my research for this book, tells me that yelling "Tally ho!" was something of a family tradition—and one that she continues. Clem, a huge fan of American baseball, tends to scream it instead of the traditional "Charge!" during mid-game pep rallies, earning bewildered or hostile stares from the other fans.

sums of money and endured months of grueling travel zigzagging back and forth across the continent, racking up lists of more than seven hundred species. I, however, did not plan to do my own Big Year while following the old *Wild America* trail. I know when I'm out of my league, not only in comparison with today's super-birders, but especially in trying to match two superb naturalists like Peterson and Fisher.

The next morning we were back at the cape, and we checked with Tony Power, who wanted to show us a few of his favorite places on the reserve. But he had some things to attend to first, so he suggested we head off on our own for a couple of hours. Should we be back at a particular time, or meet him somewhere? "No, no, I hate schedules," Tony said. "I'll find you when I need to."

We were lounging around at False Cape about lunchtime, stretched out on the dry, springy tundra, watching a razorbill preen on the orange-splashed rocks, when Tony appeared over the hill, a billed cap set a bit cockeyed on his head, dark hair escaping in every direction, a walkie-talkie clipped to his belt, and a small digital camera and notebook in his jacket pocket. Forty-five, with wire-rimmed glasses and a bit of a brogue, Tony Power is clearly a man born to his job, which includes oversight not only of Cape St. Mary's but also of the ecological reserves at Witless Bay, Baccalieu Island, and Funk Island, the last famous as the home of the now-extinct great auk. Tony's grandfather came to Newfoundland from the east of Ireland and settled just up the coast from Cape St. Mary's at Golden Bay, an isolated beach hemmed in by five-hundred-foot cliffs on either side and tangled spruce woods just beyond the few small meadows; the family were the only inhabitants of the bay, living by fishing the cod-rich waters and raising a few cattle, horses, and sheep.

Tony sank down beside us and pointed out Golden Bay, an hour's hard hike eastward up the coast, the cliffs on either hand of it red-black in the sun. When his mother was little, she often walked barefoot the ten or fifteen miles from the bay to St. Bride's or the village of Branch, where Tony himself grew up; later, after both her parents died, she and her siblings were packed off to an orphanage, and Tony says she still recalls those earliest days in that wild, remote place as the best of her life. Tony grew up working the family's small bit of pasture and farm, scything the hay by hand, then ricking it up—enough for half a dozen cows and five or six horses, just what they needed for personal use. He and his siblings collected seaweed to fertilize the potatoes and cabbages, and his parents' gentle attitude toward wildlife rubbed off on him; once he started fishing, he always enjoyed the birds around the boat, often feeding them. His later job at the reserve seemed a natural transition, and he eventually rose through the ranks to manager. The old family homestead at Golden Bay is now part of his ecological reserve, and he hikes back there often, though nothing is left of the old place but corner posts and cellar holes in the ground.

Tony, Chris, and the other reserve staff had spent the morning counting nests along the coast, checking survey plots that the province has been monitoring since the early 1970s, and we tagged along as Tony finished up the sites around False Cape. When Chris had brought us out here the previous day, he'd been more circumspect about where he led us; Tony, on the other hand, was edging us right out to the crumbly lip of the cliff, leaning out over four-hundred-foot drops, and I could only trust to his knowledge of the place to pick the most secure spots.

Moving carefully, Amy and I edged out behind him, all of us scraping along on our backsides until we could see over the edge of the cliff, from where Tony would glass the rocks below, make a mental tally of the birds, and jot the total down in his notebook. We looked at False Cape from four different vantage points, each one opening up unexpected new neighborhoods of birds. At one spot, Tony pointed out a harbor seal hauled out—a thick, short sausage of an animal, so plump it was hard to believe it was even a mammal—and while we watched, a much larger gray seal popped up nearby and hauled itself out, dwarfing the harbor seal. Tony was thrilled; he's rarely seen gray seals, which were persecuted by fishermen for years, and never before in the reserve.

The excitement grew as we watched at least five fulmars coming and going on stiff wings from a hidden portion of the cliff, gray gull-like birds that are actually members of a group known as tubenoses, more closely related to albatrosses. James Fisher was in his day the world's authority on fulmars, and he noted somewhat wistfully that as lovely as Cape St. Mary's was, it lacked the hoarse cackling of his favorite seabird. Fulmars, which have been expanding their range in the North Atlantic for decades, began to nest at the cape a few years ago, then disappeared. Now they were clearly back, though to be certain they were nesting, Tony said, he'd have to bring a boat around to look from the water.

Tony was also looking for three pairs of gannets, which the previous year had established a new nesting site half a mile from the main gannetry at Bird Rock—further evidence of the colony's robust growth. But he looked in vain; though we found many gannets loafing on the flat, guano-splashed rocks, none had brought in kelp and grass for nesting material. "I wonder if they died. They found a lot of dead gannets in Tampa Bay this winter—more than five hundred of them," he said. "I've been in touch with some people down there, rehabilitators, who did necropsies on them, but they found nothing. Three new breeding pairs like this, I could see that they would travel together in the winter."

Things have certainly improved for the seabirds of Newfoundland since the 1950s. Confederation with Canada meant that national bird-protection regulations came into effect in Newfoundland, sharply curtailing the killing of seabirds like gannets and shearwaters for food or bait—but not murres, or "turres," as they are known here.* Wildlife managers, faced with harsh opposition from rural Newfoundlanders who hunted murres for food, quickly rescinded protection on the two murre species, and through the 1980s murre hunting was something of a free-for-all, with no limits, seasons, or restrictions. The annual kill, which had been about 80,000 in the 1960s, eventually climbed to perhaps 900,000, with hunters using powerboats and semiautomatic shotguns to go after the rafts of mostly thick-billed murres that wintered offshore; a license was not even required, and a third of the kill was being illegally sold. Tighter controls were finally instituted in 1993, and the harvest has fallen to about 200,000 birds, but that still explained the windrows of old shotgun shells I'd found along the cape shoreline.

And there are more significant dangers than the shotgun. "The biggest threat is oil," Tony told us as we walked along the cliffs. "As there's an increase in traffic, there's an increase in accidents, and a lot of bilge pumping—not only tankers, but also container ships." Inside the federal twelve-mile limit, ships aren't allowed to pump bilgewater containing

*Newfoundlanders have a rich store of colloquial names for their birds, some still in common use: kittiwakes were "tickle-ace" or "tickle-ass"; Atlantic puffins were "sea-parrots"; razorbills were "tinkers"; and red-throated loons were "whobbies" or "whabbies," probably because they walk so poorly on land.

more than one part per million of oil. The trouble is, he said, it's very hard—not impossible, but hard—to monitor bilge pumping and prosecute violators, and so a lot of oily bilge is pumped close to land. "Rough seas and stormy nights—that's when it happens. Fines are increasing, but getting convictions is hard—you need better evidence for that. More money needs to be poured into surveillance."

Tony admits that the number of oiled birds is down from some of the worst years, like 1989, although even as recently as 1997 he found many oiled birds on his beach surveys—three hundred dovekies, murres, and puffins, which probably represented thirty thousand or more dead birds just in the Cape St. Mary's area, since he estimates only about one in a hundred washes up on the beach to be found. "Lots of years you don't see spills, but you'll still find oiled birds, so bad it looks like they were dipped in tar." And he isn't kidding, as I saw from the photographs he showed me later, including dovekies that were unrecognizable even as birds—just globules of glistening, shining oil.

Even though seabird numbers are rising, Tony believes the potential for a catastrophic oil spill is getting worse, not better. "We're one of the foggiest areas on this island, the ship traffic is increasing, and the tanker, container ship, and fishing traffic is so concentrated—it's all right here," he said, gesturing out at the gray ocean. "And there's the increasing oil exploration on the Grand Banks, where they're doing sonic testing on a great grid—we can hear the booms here on shore. What are the effects on the cod stocks, on the whales?"

He's also uneasy about the impact fishing may have. With the cod gone, fishermen (including two of Tony's brothers) have shifted to other species like snow crabs or to capelin, eight-inch smelt that swarm off the North Atlantic coast, coming onto beaches by the millions in early summer to spawn. Capelin is a bedrock of the marine food chain, feeding everything from puffins and gannets to humpback whales, but the past several years the run has been a shadow of its former self, Tony said. "In my father's time, my grandfather's, the capelin run on the beach always lasted a month. The last two years it was only a few hours, and the fish were small, no bigger than a sardine." Perhaps not coincidentally, there are now 124 boats seining for capelin off the cape that never fished for them before. If after seining the fishermen discover the market price is bad, or if they discover the capelin have been feeding on the wrong food and the taste is off, they'll dump the whole catch overboard.

Newfoundland is changing, and even the new, nature-oriented tourism poses risks, if it's not well planned. "What happens if someone says, I can employ fifty or sixty people if I can open a restaurant next to this seabird colony? We have to realize what we have here—we're a step ahead, because we're so pristine." There's also still an undercurrent of tension with some in the local community who resented the creation of the preserve, which cut into the traditional community pasture when it was created in 1983. The boundary was later changed, and the grazing sheep are a compromise, even though the animals damage the rare dwarf Arctic willows. It's an incremental process, Tony said, but he's made progress. "I'd put up signs, they'd tear them down. I'd put them up, they'd tear them down. It didn't happen overnight, but now it's pretty good. I'm in touch with the schools, and the kids are in touch with their parents. It's slow, but there's a sense of ownership—a sense of pride in what they have here."

We finished the surveys at False Cape and began the easy walk back toward Bird Rock and the headquarters. With the speed for which it is notorious in these waters, the fog rolled in thick; we could hear the gabble of the gannets, but the only ones we saw materialized out of the heavy white for a second, then wheeled and vanished again. The observation point had a few silhouetted tourists, one of whom had crept out too close to the edge for Tony's liking, and he left us in the swirling mist for a minute to walk over and caution the man back. This was a bit of unintended hypocrisy, given what came next.

"I have just one more stop to make," Tony said when he came back. "CM6, common murre plot six." He looked at us sideways, a hint of mischief in his eyes. "And how would you be with heights?"

We followed Tony through the fog, out along the edge of Bird Rock Cove several hundred yards, then down a steep turf slope and over loose rock slides that clattered underfoot. Here and there were murre eggs, some deep blue, others buff, all heavily speckled and all cracked open by hungry ravens, the insides of the empty shells clean and white. The hill got even steeper, and we came to a bed-sized boulder around which was tied a long, inch-thick nylon rope, knotted every few feet and neatly coiled.

As Tony paid out the rope, we followed him carefully down what was now a roughly sixty-degree slope of rock, sparse grass, and loose dirt, beneath which we could hear the booming waves. If I am uncomfortable with heights, Amy is much worse, and she said little as we descended the cliff. Hanging on to the lifeline—the word was rarely more appropriate—we skirted the edge of a deep crevasse, then inched out a narrow finger of rock to find ourselves smack in the midst of the colony as the fog parted and we saw thousands of birds on every hand.

It was, as I have said, one of the finest birding experiences of my life, but the best of it was being in the company of someone like Tony Power, who knew the inhabitants of CM6 with an intimacy born of a lifetime's acquaintance. "Last year, when I did this survey, there were sixty-one pairs of murres along that ledge," he said, his finger pointing, "and this year there are sixty. But they've moved into new ledges. That thick-billed up there"—with his lilt, the word came out "tick-billed"—"moved in two years ago, and those on the little ledges higher kicked out kittiwakes that had been there."

"See how the common murres like the long, deep ledges but the thick-billeds like the little ledges, like steps?" Tony continued. To my eye, the "long, deep" ledges couldn't have been more than six or seven inches wide, and made for a pretty dicey place to raise a baby, but the spots the thick-billeds picked weren't ledges, for crying out loud, they were *bumps*, mere pimples on the cliff, and I couldn't see how even the famously pointed eggs of a murre wouldn't roll straight off into empty space.

Yet when one of the black-and-white birds straightened up a bit as though to stretch its back, we could see the single huge blue-green egg peeking out from beneath its breast. High above the cold Atlantic, as they have since the glaciers departed, the birds of Cape St. Mary's were poised to bring another generation into this wild and lovely place.

TWO

Woods in the City

I grew up in the hills of the Northeast, and from an early age I found myself drawn to the region's colonial past, when the ridges and long valleys that I knew were not tame farmland and quiet towns but the frontier—home to panthers and wolves, to flocks of passenger pigeons that blackened the sky in fact and not merely metaphor, to lynx and elk. Like many eastern naturalists, I was seduced by the idea of this gentle, settled land as a vibrant wilderness—a connection that gets more tenuous with each passing year and every new housing development.

But scratch beneath the surface and you'll find that the story of the Northeast is more complicated than this linear march from untouched hinterland to sprawling megalopolis would suggest. Waves of settlement, clearing, and regeneration have rolled across this region several times since Europeans arrived, each time creating a new union of wild, pastoral, and urban in the restricted landscape between the northern Appalachians and the Atlantic.

Nor has this process ended. In the past half century, twin forces have been at work in the Northeast, at once softening and rending the region. On one hand, the forest that had been stripped over the two preceding centuries reasserted itself as old hill farms were abandoned and mountains laid bare by clear-cutting in the nineteenth and early twentieth centuries regrew. With the slowly maturing woodland came many of the once-lost species of wildlife.

But even as the forest ripened, the landscape as a whole steadily grew more ragged: more and more roads, tract housing, utility corridors, and the like; more fragmentation, less cohesiveness and unity to what remained. Peterson and Fisher never used the word "sprawl," but even in 1953 they could see the megalopolis linking its tendrils from Boston south to Washington, D.C. They titled their chapter about the nation's capital "City in the Woods," but today that would be wishful thinking; the city and suburbs have become so ubiquitous that today the woods are clearly the islands, set within a matrix of cul-de-sacs, strip malls, and overlapping highways.

"The Northeast can hardly be called *wild* America, but tucked away here and there are small remnants of wilderness that have withstood three centuries of ever-expanding civilization," Peterson wrote. That's obviously even truer today, but it makes the lingering pieces all the more intriguing, especially since, paradoxically, some of them are demonstrably wilder and more diverse than they were fifty years ago. These relicts of an older time range from the Blue Hills of Boston, with their urban rattlesnakes, to the Potomac Gorge just a few miles from the Capitol dome, which boasts one of the largest assemblages of rare species in the East. Some, like Rock Creek in Washington, D.C., where Fisher and Peterson birded, are now bisected by beltways and seem hopelessly hemmed in, but even in these more urban settings wildlife shows a surprising elasticity—sometimes with human help, like the screech-owls that have been returned to Central Park, and sometimes on their own, like the beavers that now threaten Washington's famous cherry trees or the wild turkeys that have colonized Manhattan.

After their sojourn on Newfoundland, Fisher and Peterson flew into Boston, where Roger's movie camera gear attracted the scrutiny of U.S. Customs and the city's only skyscraper caught James's attention. Over the following week the pair moved frenetically from Boston to Rhode Island to New York City, down the New Jersey shore, and west to Washington, D.C. Much of their time was spent in the cities proper—visiting the National Audubon Society office and the American Museum of Natural History in Manhattan, for instance—but mostly they were in the field. By design, they started in the Concord valley in Massachusetts on April 19, the anniversary of the Battle of Concord Bridge that started the Revolutionary War. The next couple of days they birded north of Boston, near Newburyport, with a cadre that included Peterson's old mentor Ludlow Griscom, and then down into Rhode Island, before heading to New York.

"The little pockets of wildness in New York City are going, one by one, as they are in London and in every metropolis. In this mobile age everything within fifty or a hundred miles of most large cities is undergoing change," Peterson observed. You can see that clearly around Boston, where since the

1950s the population of many suburban communities has quadrupled, even as the number of urban residents dropped from 800,000 in 1950 to fewer than 600,000 today. While some parts of the state, like Cape Cod, enjoy relatively strict land-use protections, other areas are disappearing fast; by one estimate, Massachusetts loses forty acres of forest, farmland, or open space every day, mostly to new housing. And the homes themselves are half again as large as they were in the 1970s, as are the lots on which they're built—even as the number of people living in them, on average, has dropped.

One conservation group calls it the "mansionization" of Massachusetts, but the trend is the same throughout the Northeast and the mid-Atlantic region. Pennsylvania may be one of the worst offenders. In the fifteen years ending in 1997, its population grew just 2.5 percent, while its urbanized footprint jumped by 47 percent, in what the Brookings Institution called "one of the nation's most radical patterns of sprawl and [urban] abandonment." The price paid by wildlife and wildlands has been especially steep. Between 1990 and 2000, the state lost a million acres of land to development—four hundred acres per day, mostly farm and forest habitat—while its road system, already more than 120,000 miles long, grew by another 3,000 miles of paved highway.

It's happening everywhere; commercial and housing development, along with the road network needed to support it, is the single greatest pressure on natural landscapes in the United States, and by its very pervasiveness the hardest to control. Between 1982 and 1997, developed land in the forty-eight contiguous states increased by 25 million acres—meaning a quarter of all the open land lost *since European settlement* disappeared in just those fifteen years. This isn't a trend, it's a juggernaut, and the worst may be yet to come. At this pace, by 2025 there will be 68 million more rural acres in development, an area about the size of Wyoming, and the total developed land in the United States will stand at a Texas-sized 174 million acres. Already, just the impervious covering we put on the land, the things like roads, sidewalks, and buildings we pave with asphalt or concrete, adds up to an area the size of Ohio.*

*The rampant pace of development has led Dolores Hayden, a professor of architecture and American studies at Yale, to compile a new lexicon by which to describe it, including "boomburb" (for a fast-growing suburb), "zoomburb" (for one that is growing even faster), LULU (Locally Unwanted Land Use), and "litter on a stick" (billboards). *The New York Times*, perhaps inevitably, called Hayden "the Roger Tory Peterson of sprawl."

Nor is the change all inland. "When James and I drove south along the New Jersey coast, I was shocked to see how things had altered since the war's end. The sandy coast, the dunes, and the estuaries are now one continuous line of flimsy beach dwellings, bathhouses, billboards, and hot dog stands," Peterson said. But the difference between what he bemoaned and the current state of the northeastern coast is almost exponential in terms of the density and permanence of that development: the Jersey shore, as elsewhere, is lined no longer with flimsy shanties but with tightly packed condos, luxury vacation homes, and all the attendant visual detritus of rampant coastal development.

Never mind that a single big Atlantic hurricane would render it all expensive flotsam, and that a true monster storm, a Category 4 or 5 whopper, is all but certain as the Atlantic moves out of a decades-long period of relative quiet and into one with more and stronger storms. Americans apparently believe they can safely ignore that reality, in part because federally subsidized flood insurance often underwrites this mindless excess and tax dollars will be used to compensate anyone whose home or business is harmed, even if storms damage or destroy it repeatedly. This makes precious little sense, and the result has been to render much of the Atlantic coast one linear avenue of vulgarity where the region's natural beauty is almost gone.*

But just as Peterson said, there are still pockets of wildness in the Boston-to-Washington corridor, if you know where to look—some coastlines that speak of an earlier age, some forests that persist in the heart of the city. And so despite "mansionization," I would start, as they did, in Massachusetts.

Though one of the largest population centers in the East, Boston and its outlying communities have always blended urban congestion with a

*Lest any indignant reader think I am attacking that most sacred of American cows, private property rights, I am not. If you are fool enough to build a million-dollar home on a sand dune beside a large, restless, and predictably dangerous ocean, that's your business—but it ought to be your nickel, too; don't expect the rest of us to foot the bill when nature inevitably hammers it into kindling. In 1982, Congress passed the Coastal Barriers Resources Act, which limits federal subsidies, including transportation money and flood insurance, on about 1.3 million acres of coastline while still permitting owners to develop their land if they bear the risks. The program, despite many shortcomings, has succeeded in restricting growth on sensitive sites—except that Congress also routinely exempts prime coastal real estate from the CBRA umbrella, eroding its purpose.

surprising degree of wildness not far away, from the snowy owls that congregate each winter on the runways of Logan International Airport, the fallouts of migrant songbirds in the city's parks and wooded cemeteries, the islands in Massachusetts Bay, the cranberry bogs and pine barrens of Plymouth, and the long beaches and windswept bluffs of Cape Cod and Monomoy Island.

Just south of Boston, and within its beltway, rise the Blue Hills—buckled ridges, forests, and lakes that constitute a seven-thousand-acre reservation. The Blue Hills aren't high—barely six hundred feet—but sitting so close to the city and the ocean, they seem more impressive than that; on a clear day on Chickatawbut Hill, you can count the windows in the downtown skyscrapers with binoculars. The Blue Hills were purchased by farsighted Boston officials in 1893 as a back-door park for the growing city. But besides forming an oasis for city dwellers, the reservation is also home to a number of species that have vanished elsewhere in eastern Massachusetts—most famously, one of the most northerly surviving colonies of timber rattlesnakes.

There may be more serious birders per capita in Boston and eastern Massachusetts than anywhere else in the country, and while Philadelphia may rightly claim to be the cradle of American ornithology, the Boston area has been its nursemaid. It is home to the oldest bird group in the United States, the Nuttall Ornithological Club, which was founded at Harvard in 1873 (and which Fisher addressed during the pair's brief stop). The Nuttall club gave rise to the influential American Ornithologists' Union a decade later, and in 1896 a group of Boston Brahmin ladies formed the Massachusetts Audubon Society to fight the slaughter of birds for the hat trade. From the welter of state Audubon societies that arose around that time, the National Audubon Society coalesced, but Mass Audubon remains an independent organization, with tens of thousands of members and responsibility for dozens of sanctuaries, including the Blue Hills Trailside Museum, the interpretive center for the state reservation.

My friend Norman Smith manages the Blue Hills museum for Mass Audubon, but he is probably best known for his twenty-five-year study of the snowy owls at Logan airport, another juxtaposition of modernity and wilderness that non-birders find surprising. Every winter, in larger or smaller numbers depending on the supply of small rodents in the Arctic (with heavier flights south in the lean years), the huge ghostly owls come down the coast, where the open dunes and tidal marshes are a good substitute for the treeless tundra. Airports fit the same bill, and many of the owls wind up at Logan, with its flat, grassy expanses on a peninsula jutting into Boston Harbor. They perch next to the runways, unmoved and apparently unbothered by the appalling noise and exhaust blasts of the passing airliners, hunting rats, gulls, and even great blue herons and small songbirds like snow buntings.

Norman catches the owls, bands them, and releases them with a color mark on the back of their heads, so observers can track them as they move around the northeastern coast; lately he's even been attaching sophisticated satellite transmitters to a few to follow their movements to and from the Arctic. He's found that instead of being dead-end birds doomed to wander until they meet a premature death in hostile southern areas, as had long been assumed, most of the owls survive to return to the Arctic—and perhaps to come back to the airport another year.

A few years ago, I took some of the volunteers from my own banding project in Pennsylvania, where we work with tiny saw-whet owls, to spend a weekend with Norman and the snowies. In the uneasy world of post-9/11 security, he was no longer allowed to take us out onto Logan's tarmac to trap owls with him, so we contented ourselves with setting up spotting scopes across the quarter-mile-wide channel from the airport, on the upper deck of a low parking garage beside a shipping terminal.

It was a January day, bitterly cold, but we had plenty to watch: there were four or five owls perched on taxiway lights or the rocky riprap above the channel, as well as a young peregrine falcon on the superstructure of a dockside crane, mergansers and a few seals in the water, and a variety of winter gulls. We looked in vain for a dark gyrfalcon that had been hanging around (and that, a few days later, killed and ate the peregrine as observers looked on). After a while, we heard the *whomp-whomp-whomp* of distant helicopters and saw a line of four choppers off to our left, coming down over the heart of the city. We paid them little attention at first, but soon someone with a scope noticed that these weren't just any helicopters; there were two massive, double-rotored Chinooks bracketing two smaller choppers that were, from the markings, clearly Marine One, the presidential helicopter, and a backup.

At this news, the owls were forgotten, and all binoculars and scopes swung around to watch as the line of helicopters turned to the east and came right toward us, straight down the channel to make a landing at the airport.

"Um, take a close look at that first helicopter," someone eventually said. "At the *door* of the helicopter."

Despite the frigid weather, the big side door of the lead Chinook was open, and we could see a figure hanging half out of the doorway, a safety harness around his body and a very large scoped rifle in his hands, shouldered and aimed. Aimed at us—at the group that was obviously pointing shiny, undoubtedly suspicious objects at the presidential party.

All of us immediately leaped back from the scopes and tripods, trying to look as harmless as possible. The choppers whirred in for a landing directly opposite us, and we saw that the second Chinook was likewise bristling with armament. They were on the ground for only a short time; we later learned it was a test run for a visit the next day by George W. Bush. The snowy owls, however, were nonplussed; one was perched only thirty or forty yards from the four screaming machines and merely hunched low in the violent rotor wash.

Like much of the northeastern coast, eastern Massachusetts has experienced tremendous growth in recent decades, and even once-remote areas are being engulfed by a tide of new development that washes right up to the boundaries of the state's many protected tracts. But in a few places, by a happy accident of geography, the current has actually run the other way, and humanity's footprint is much lighter than it was a generation ago. Monomoy is such a one.

Sometimes, a place you've heard about but never visited takes on an almost mythic quality, and so Monomoy was for me. For years I'd heard about this pair of islands just off Cape Cod—indeed, they had once been part of the mainland—but Monomoy was often referred to in ways that made it seem as remote and wild as Sable Island far to the north. The waters surrounding it were said to be dangerous, full of treacherous shoals, powerful currents, and ever-shifting sandbars, the graveyard of literally thousands of ships and boats over the centuries; almost all traces of Monomoy's more settled past had vanished beneath the dunes, where today only birds call. The empty beaches were now the haunt of great northern seals, themselves once a rarity in New England waters. So out of space and time had it become, here in the bustling Northeast, that it was long since designated a federal wilderness.

Such preconceptions, shielded by distance and a lack of firsthand experience, cannot withstand close scrutiny; when one at last goes there, the daydreams are replaced by a more pedestrian reality. And so when I finally visited Monomoy, I knew there was no way it could measure up to my expectations.

Except that it did.

It was a calm day in the middle of September, with heavy fog over Nantucket Sound, so that the run down from Chatham was cold and dreary, the swells looking oily in the gray light. No one on the boat—a small group from Massachusetts Audubon and the famed Manomet Center for Conservation Sciences—said much on the way out. I mostly remember that those of us wearing glasses wiped the mist off of them a lot.

But as the captain throttled back the boat, signaling that despite the wall of featureless white we must be getting close to South Monomoy, we emerged from the mist into soft sunshine. Before us lay a long, low ripple of sand, golden white in the hazy sun, with dune grass like brushstrokes in a Japanese painting. Flocks of shorebirds rose on fast wings, calling, but my eyes were on the water. All around us were seals—not the pudgy little harbor seals I am used to seeing, but huge gray seals with heads like horses, Roman-nosed, massive, each seven or eight feet long. They were, as I said, everywhere—dozens of them on the beach, caterpillaring into the surf with blubber rippling, and dozens more swimming around the aluminum boat that would land us. The water was just a few yards deep, and against the pale bottom the seals moved with startling clarity, their dark coats flecked with pale gray, as though marked by sunlight through heavy leaves.

Two of the big seals crossed just beneath us in the shallow water, then rose in unison to either side, less than a dozen feet away, staring at us with breathtaking calm and gravity. There was none of the beguiling, puppy-dog look that a harbor seal has, but this went beyond the accident of nature that makes one animal cute to our eyes and another not. The gaze was neither hostile nor disinterested, but it was cool, assessing, almost commanding.

We waded ashore, slinging daypacks on our backs and balancing spotting scopes and tripods over our shoulders. The kid running the skiff gunned the outboard and planed back through the curious seals to the big boat, barely visible in the mist; they would meet us later in the day, at the north end of the island. With the boats gone, the outside world vanished. We sat in a long lozenge of sunlight, an island-sized hole in the dense fog that covered the sound—and so it remained all day, sealing off even the glimpse of boats or a distant shore. With isolation, the enchantment was complete, and I felt like a child playing hooky—not from school, but from my own century.

It was past the peak of the fall shorebird migration, but there were still hundreds of sandpipers and plovers resting and feeding on Monomoy, and as we ambled up the beach, they rose and fell before us, an undulation of wings and movement that lasted for hours. Roseate terns, starkly white with long forked tails that twisted gently in the breeze like streamers, flew past us, as did the more numerous common terns, all screeching their high, discordant calls. Once, a merlin—walnut-colored, head like a bullet, wingtips like knife blades—slashed out of the haze and down the beach, scattering shorebirds like leaves behind a fast car, but it took no notice of them, intent on some invisible southern horizon. Yet even after the falcon had passed, the birds were restless, leaping into the air with no reason or warning, fretting shrilly to themselves in agitation.

There were semipalmated sandpipers by the score, that day all drab and gray, but with one or two laggards still in the rusty plumage of summer, which stood out among the monochrome flocks. We ticked off species after species—dowitchers, knots, black-bellied plovers, white-rumped sandpipers, a few phalaropes. Higher, against the pale green-gray dune grass, four buff-breasted sandpipers stood motionless, the color of fresh biscuits with buttery legs, their eyes wide and dark, giving them an air of faint bemusement. When I think of that day, this is the moment I recall most clearly: those four birds standing in the gauzy sunlight, poised, all facing in the same direction with a contained energy in their tapered bodies, then springing into the pellucid air on long wings, making a quiet little trill as they flew south, out into the ocean and the waiting fog, the wet pampas of Argentina pulling them across the hemispheres.

In ways both concrete and spiritual, Monomoy is a vastly wilder place today than it was when Peterson and Fisher came through Massachusetts in 1953. Peterson's old mentor Ludlow Griscom often birded on Monomoy, but the pair did not stop here on their *Wild America* tour, focusing instead on Boston, Concord, and Plum Island, up along the state's northern coast. And I'm not surprised; in those days, Monomoy wasn't even an island at all, just an eight- or nine-mile-long peninsula jutting south from the "elbow" of Cape Cod.

But then a fierce spring storm in 1958 bulldozed a channel through the neck of the peninsula, isolating Monomoy from the mainland. That isolation changed everything. Buffered from mainland predators, terns and gulls could nest without raccoons and skunks raiding their eggs. No longer were the dunes and beaches accessible to fishermen and picnicking families in off-road vehicles (though, it must be said, Mass Audubon ran popular beach buggy tours of the island through the 1960s and early '70s). The devilish shoals kept casual boaters away; harbor seals, still much persecuted in those days, learned it was a safe place to haul out to bask; and in winter even a few rare gray seals came down from the far North Atlantic to hunt flatfish and skates in the waters of Nantucket Sound.

An infamous blizzard in 1978, which hit Cape Cod with nearly the ferocity of a hurricane, reshaped the island once more, carving it into the two-and-a-half-mile-long North Monomoy and its much larger sister, South Monomoy, five and a half skinny miles long with a bulbous, mile-wide dollop of sand at the south end. Such transience defines this coast; the old peninsula was itself of recent vintage, having built up steadily from a series of barrier islands in the nineteenth century. It even supported a small fishing hamlet, Whitewash Village, around its harbor; the settlement's schoolmaster, writing in *Harper's* magazine in 1863, described its main building as "a weather-beaten, barracky, amphibious structure, [with] fishermen and coasters' fitting-store on the first floor, lodging-house and excursionists' inn on the second . . . and a motley array of storage and packing sheds perfumed with fish-oil."

Monomoy's only crop, he wrote, was beach grass, "and doubtless even the less favored portions of its soil might be made to yield the same if they could but be tied down in one position for a day or two at a time—a rest which wind and wave have for centuries unknown refused." That restless sand was the village's demise; by the Civil War, as the little harbor was filling in, many of Whitewash Village's inhabitants had begun to leave, even going so far as to dismantle their homes and fireplaces, hauling the pieces off to the mainland until nothing was left but the lighthouse.

In 1944, the federal government took over Monomoy as a national wildlife refuge. Twelve years after the first storm severed Monomoy from the mainland, Congress designated it one of the few federal wilderness areas in the East. It became a place apart, a fact recognized by animals as well as people.

Gray seals, which had been hunted for bounty in New England waters as late as the 1960s and were all but extirpated there, finally received federal guardianship with the Marine Mammal Protection Act of 1972. A decade later, they began to appear in increasing numbers off Maine, New Hampshire, and Massachusetts, probably moving south from Canadian populations. At first they were merely seasonal visitors to Nantucket Sound, but in the winter of 1988—for this is a species that gives birth in the depths of winter—five pups were born on Muskeget Island off Nantucket. Breeding seals later appeared on Monomoy itself, and the Nantucket Sound population had risen to more than five thousand by the end of the 1990s, increasing at the astounding rate of more than 20 percent a year.

While Monomoy's isolation has made it a wildlife paradise, there are some thoroughly modern problems and controversies. The waters just to its west are the focus of a rancorous debate over wind energy, with developers proposing to put the world's largest offshore wind energy development, with 130 turbines rising 430 feet high, on a twenty-four-square-mile area known as Horseshoe Shoals in the middle of Nantucket Sound. Most of the opponents are fighting the wind farm on aesthetic grounds: the structures would, in the words of one group, destroy "the ungoverned natural beauty, solitude and wildness of its coasts." Proponents, including some prominent environmentalists, note that global warming and rising sea levels will sink those wild coasts completely unless we stop pumping carbon dioxide from fossil fuels into the atmosphere.

Others worry about the more immediate impact of so many turbines on birds. These include the half-million sea ducks, including scoters, long-tailed ducks, and eiders, that winter in Nantucket Sound and move back and forth through the wind-farm zone each day, as well as seabirds like gannets, which migrate through the region at an average pace of ten thousand to fifteen thousand birds per day each fall. But the greatest hazard, some ornithologists believe, would be to migrating land birds like sparrows and thrushes, which fly at night from Nova Scotia south to the New England coast and would presumably come right through the turbines. Even worse, the turbines, which would be several hundred feet tall, would require aircraft warning lights—the same strobes that are known to disorient nocturnal migrants around communication towers, often with fatal results.

One species of particular concern is the roseate tern, a federally endangered species, half of whose North American population nests near Nantucket Sound. But roseates have other problems besides the threat of windmills. Monomoy hosts one of the largest roseate colonies in the United States, but the islands are also home to the biggest nesting aggregation of herring and black-backed gulls in North America—species that prey on tern eggs and chicks and that, thanks to an astounding amount of human provender, have become far more common than they were in the past.

The story of gulls and terns along the northeastern coast echoes the give-and-take of Monomoy's sands. In precolonial times, it's likely that terns were the more numerous group—not just roseates but common, Arctic, and least terns, each of which prefers a slightly different nesting habitat along the coast, on offshore islands or (in the case of least terns) on sandy beaches. Around the time of the Civil War, common terns were nesting by the hundreds of thousands on islands in Nantucket Sound, and even roseate terns were considered abundant. But starting around 1870, fashionable women began wearing hats festooned with feathers or whole birds. The most famous victims of the millinery trade were the egrets and herons of the southern swamps, but seabirds like terns and gulls were also devastated. By 1890, perhaps as few as five thousand pairs of common terns and two thousand pairs of roseates still nested on a couple of the less-accessible islands near Nantucket, and herring gulls had been extirpated from the entire coast below eastern Maine.

Legal protection, thanks to pressure from the various state Audubon societies, professional ornithologists, and local bird-watching clubs, brought the terns back from the brink; in Massachusetts, there were an estimated thirty thousand common tern pairs in 1930, and about seventy-five hundred roseates just before World War II. But gulls, no longer hunted for their feathers or their nests raided for eggs, also resurged. New colonies of nesting herring gulls leapfrogged down from Maine, eventually reaching North Carolina, with populations that by 1960 had probably exceeded their original levels. In their wake came the bigger, more predatory great black-backed gulls, which sometimes displaced the herring gulls from nesting sites.

The reasons for the gull explosion are complex, ecologists say: not merely protection from harassment, but also a growing abundance of human garbage; increasing fishing activity, with its discarded bycatch, waste, and bait; and the great reduction in North Atlantic seal and whale populations that eased competition for smaller fish and invertebrates on which the gulls fed. Every lobster boat in New England waters uses vast quantities of bait each year, every fishing boat unloading a net discards unwanted fish and invertebrates, much of which goes into the waiting maws of gulls that circle each boat by the dozens. Think of the universe of gulls that swirls above every town landfill, the flocks that scour the sloppy mall Dumpsters of coastal communities, the hordes that snag unguarded sandwiches from beachgoers, and you begin to appreciate that we could scarcely create a more ideal world for gulls if we tried.

Or one worse for terns. In 1961 there were no nesting gulls on Monomoy, but thirty years later there were more than five thousand pairs of herring gulls and more than seven thousand pairs of great blackbacks; statewide, there were an estimated thirty-two thousand pairs of both species. As gull numbers climbed, the terns began a second slide toward oblivion on Monomoy, and wildlife managers felt they had to step in to give smaller birds a chance.

Fearing for Monomoy's tiny population of federally endangered piping plovers, a shorebird that nests on windswept beaches and can fall prey to gulls, the U.S. Fish and Wildlife Service decided to reduce the gull population on the refuge by 15 percent and create a 125-acre gull-free zone. They chose to use an avicide called DRC-1339, a fast-acting toxin. The first attempt, in 1996, did not go as planned, though; federal employees distributed poison-laced croutons among selected nests, but the dosage wasn't quite high enough. "The gulls took to the air and went into their death throes, circling over the town of Chatham, falling into the streets, creating an enormous, predictable uproar," recalled the former interior secretary Bruce Babbitt, in whose lap the mess landed. Animal-rights groups were howling, and even more sober conservation groups that supported the underlying goal of protecting the plovers were sickened by the spectacle. Under intense public pressure, Babbitt called a halt to lethal control and replaced it with a much more laborious program in which staffers patrol portions of the gull colonies, harassing them and puncturing eggs.

Although the target of the gull reduction was plover protection, it had the unintended effect of creating a safe haven for terns as well, so that Monomoy today has the state's largest tern colony, including ten thousand pairs of common terns, and its only nesting black skimmers. The gull hazing is an ongoing process, as is the removal of occasional coyotes that make it over from the mainland, and even a black-crowned night-heron or two. Such harsh intervention will probably always be necessary, so badly have we tipped the balance in favor of gulls.

At Monomoy, the wind and the waves are always modeling the shoreline, pinching away and piling up. At the north end, friends tell me, the channel between the mainland and North Monomoy is narrowing and someday soon may close off, reuniting the island with the rest of Cape Cod. If that happens—if Monomoy's storied isolation ends—how will it change the character of the place? All the legal protections with which it is girded will remain; no development will come to the sandspit, no beach houses to replace long-lost Whitewash Village. There will be more visitors, I'm sure, and the seals may find their patience tested. But in the end, the only casualty may be the perceptions of those who would hear of Monomoy—splendid, empty, and remote—yet never see it, except in daydreams.

From New England, Fisher and Peterson headed to New York City, where they made the best of a bad situation. Roger had planned a number of urban wild activities, including a visit to the Bronx Zoo, but Fisher's American publisher insisted he spend two days in their office working on outstanding projects, and there simply wasn't time for everything. What's more, they were disappointed by the one field session they did have, a jaunt out to Long Island that was ultimately dropped from the book "because it was dull," as Peterson later confessed. Instead, the authors decided to focus the short New York chapter on nature "in strong, distilled form," squeezing in Fisher's brief visit to the American Museum of Natural History and even a trip up the Empire State Building and a screening of the newfangled Cinerama in Manhattan, which gave them a 3-D preview of some of the scenic wonders, like the Grand Canyon and Crater Lake, that they'd be seeing later in the trip. It was a bit of a stretch; Peterson even mentioned that John James Audubon is buried in the city and that his original watercolors are in the historical society.

But lately the Big Apple has had a wilder side that has nothing to do with the club scene. Wildlife—not just pigeons and house sparrows, or even the hordes of migrant songbirds that stop off in Central Park each spring and fall, but animals once considered wilderness species—is moving into the suburbs, and increasingly into the urban core itself, a turn of events that would surely have pleased the two naturalists. In some cases, this has been a return to an older condition, as with peregrine falcons, perhaps the fastest and certainly the most storied of all the raptors. Until the middle of the twentieth century, peregrines nested just to the north of New York City, on the rugged cliffs of the lower Hudson River. But after World War II, when the pesticide DDT came into wide use, the falcons began to disappear there and elsewhere. While a few naturalists noticed the failed nests and vanished birds, no one put the pieces together, including Peterson, who had, ironically, studied DDT's effects on birds for the Army in the waning days of the war and had begun speaking out pointedly by the early 1950s on the dangers of pesticide use. When Rachel Carson published *Silent Spring* in 1962, she didn't realize one of the most potent examples of DDT's danger was already nearly gone; two years later, a frantic search by ornithologists at hundreds of traditional nest sites across the East failed to turn up a single peregrine. The eastern race was essentially extinct.

In the 1970s, peregrines were the focus of a highly publicized restoration effort spearheaded by the private Peregrine Fund, which used an ancient falconry technique known as hacking to release captive-bred chicks into the wild. The program was not without controversy; the breeding stock was an amalgam of peregrine subspecies not just from the East but from the Arctic, the Pacific Northwest, Scotland, Spain, and Chile, and the resulting hybrids neither look nor act exactly like their native predecessors. Also, many of the places where the raptors were released, like metropolitan skyscrapers and special towers in salt marshes, never had breeding peregrines to start with. (This was in part because the traditional cliff nests often proved too hazardous for hacking, as great horned owls took a toll on the unprotected falcon chicks.)

Today there are more than two hundred pairs of peregrines nesting east of the Mississippi, with the total rising every year. Although they have been reintroduced to wilderness cliffs in New England, the Adirondacks, and the southern Appalachians, most of the nesting pairs use urban bridges and buildings, perhaps because young birds seek out nest sites similar to those on which they were raised. Since the first two pairs appeared in 1983 on the Verrazano-Narrows and the Throgs Neck bridges, New York City's nesting peregrine population has climbed to more than a dozen pairs, all on man-made structures. While relatively safe from predators, urban nests offer their own suite of hazards to the birds, from collisions with glass windows to electrocution to secondary toxins from eating poisoned pigeons.

Inspired by the peregrine success, the city's Urban Park Rangers in 1998 initiated a project to see if eastern screech-owls could be reintroduced to Central Park, where the diminutive raptors had disappeared in the 1960s. Owlets brought into wildlife rehab centers elsewhere in the state were raised to adulthood and released; this being the big city, the owls shared their first press conference with the actress Isabella Rossellini. By 2005, several nests had been found, suggesting the owls were adapting. But not everyone was pleased. One group of critics contended the use of captive-reared owls somehow made the park less wild—a logical lapse that overlooks the fact that Central Park is itself a wholly artificial construct, with every hillock and pond built to the specifications of the landscape architect Frederick Law Olmsted.

Even some raptor biologists, however, have questioned the city's latest scheme, an attempt to bring nesting bald eagles back to Manhattan. Despite criticism that the modern city is simply too crowded and polluted for the huge fish-eating birds, eaglets from northern Wisconsin were placed in a hack tower in Inwood Hill Park, near the Hudson and Harlem rivers, beginning in 2002; one of the eagles was subsequently injured in a train accident and died. Wildlife management aside, even city officials admitted one goal of the release was restoring local spirit in the wake of the 2001 terrorist attacks.

If Manhattan has an avian celebrity, though, it is a red-tailed hawk called Pale Male. In 1991, a very light-colored young redtail showed up in Central Park, but unlike the migrants that sometimes pass through, he stayed; that spring he attracted an adult female, and they tried several times to nest in the park itself, with no success—too many people, too many aggressive crows. Then, with a new mate, Pale Male moved to the ledge of an apartment building on Fifth Avenue, facing the park, and he has remained there for years, with a succession of mates—celebrities first in Manhattan, and then, thanks to Marie Winn's bestselling book *Red-Tails in Love* and a subsequent documentary film, nationally. When the building's managers removed the nest and its structural supports in December 2004, claiming it posed a health hazard, they triggered candlelight protest vigils and condemnatory press coverage until, several weeks later, they relented and allowed the redtails to return.

While it may seem remarkable to think of large hawks nesting in the same swank neighborhood as Mary Tyler Moore and Woody Allen, several species of raptors have moved into the suburbs and cities of the Northeast. Red-tailed hawks are now common in only slightly less congested areas than Manhattan, and some American kestrels, a much smaller relative of the peregrine, have also adapted to nesting on enclosed ledges and cubbyholes on some urban buildings. Cooper's hawks—large, agile hunters of birds and small mammals—have taken the suburbs by storm, building their stick nests in backyard spruces and snatching house finches and mourning doves from feeders, thus providing homeowners with an often more realistic view of nature's drama than they might prefer with their breakfast bagel.

In fact, the Northeast has witnessed an odd collision of trends where wildlife is concerned. As development has moved out from the urban centers over the past fifty years, it has displaced many sensitive species of animals and plants that cannot handle the loss and fragmentation of habitat that accompany modern sprawl, even as a few hardier, more adaptable animals have moved right into the subdivisions. That relative handful of adaptable species mask how serious the net loss to wildlife has been.

Woodland songbirds that require large, intact blocks of forest in which to nest have suffered profound and region-wide declines; few such forests are left, and the tracts that remain are isolated and tattered, overbrowsed by white-tailed deer (one of the most successful of suburban creatures), and overrun with predators like raccoons and opossums that thrive close to humans, not to mention free-running pet cats and dogs and a parasitic bird, the brown-headed cowbird, which avoids deep forest but is abundant around homes. When a cowbird finds a songbird nest, it tosses out the owner's eggs and lays its own in their place, leaving the unsuspecting bird to raise the cowbird chicks. With nowhere else to go, songbirds such as wood thrushes congregate in the fragments, drawn like moths to a flame toward habitat that is often neither safe enough from predators nor productive enough in terms of insect food to support a growing nest of chicks. Such islands of forest thus become black holes for the species, reproductive sinks dragging the population down.

Less-mobile creatures are hit particularly hard by the tightening web of roads and mounting traffic in developed areas. This is especially true of amphibians and reptiles, which often travel long distances each year in order to breed. The large, colorful group known as mole salamanders are a good example. The most common, the eight-inch-long spotted salamander, is glossy black with half a dozen or so bright yellow polka dots on its back. Living underground most of the year, the salamanders emerge on the first rainy nights of spring, often when the temperature is barely above freezing, and make their slow-motion way to seasonal ponds where generations before them have bred. Here they quickly pair up and, amid a great deal of tail waving and gentle nuzzling, lay their eggs in the frigid water, leaving the pond just a few nights later.

When roads are built near the traditional vernal ponds where they breed, the toll on various mole salamanders, frogs, and toads can be horrific, often leading to the extirpation of the entire local population. But it can also be prevented, as a few farsighted municipalities have proven. The western Massachusetts college town of Amherst made quite a record for itself in this regard, first by closing a road near a spotted salamander breeding pond on mild, rainy spring nights and organizing "bucket brigades" to help the amphibians across, and in 1988 by building tunnels beneath the road so the animals could cross on their own. The salamanders have become such a symbol of civic pride that a large sculpture of one has been erected in a park. There was a local rock group known as Salamander Crossing, and the Amherst Brewing Company produces a popular lager of the same name.

Such concern, much less expensive action, remains rare, although it is spreading. On spring nights when conditions for amphibians seem right, the National Park Service closes a five-mile stretch of busy road through the Delaware Water Gap National Recreation Area in the Poconos; one evening, along just a one-hundred-yard stretch of the silent road, naturalists found one hundred spring peepers, twenty to thirty wood frogs, twenty spotted salamanders, and four or five much rarer Jefferson salamanders, most of which would have been pureed by traffic were it not for the barricades.

Nor are the problems only among salamanders and frogs. Box turtles are disappearing from many parts of the East where they were once common—and little wonder. As slow to reproduce as it is to cross a road, a box turtle does not mate until it reaches five or ten years of age, and a female will thereafter lay just six or eight eggs a year, most of which are dug up and eaten by the army of raccoons, opossums, and skunks that thrive near humans. Given its full life span of as much as fifty years, a box turtle would normally be able to overcome this reproductive hurdle, but automobiles pose a deadly danger, as do all manner of man-made activities and inventions, from backyard pesticides to the inground pools into which the turtles tumble and cannot climb out. One scientist, radio-tracking box turtles in his own subdivision, realized that concrete road curbs posed an insurmountable obstacle for the reptiles.

Not long ago, I came across a battered old box turtle, its shell a history book of nicks and scars, an aged female with brown eyes and pale yellow markings on her carapace. It was the first one I'd seen all summer, in a place where once they had been common. I wondered, as I inevitably do, just how old she might be. Centenarians are routine, and one remarkable box turtle is known to have reached age 138, so there's a very good chance that this turtle has been clumping slowly through these woodlots and fields since the land was plowed by horses. I put her back in the patch of frost-nipped blackberries and wished her luck. I figured she'd need it. As a culture, we're not very good at respecting our elders.

If the modern landscape in the Northeast is inhospitable for some species, it has been a boon to others, many of which were once restricted to wilderness. Black bears, which had probably been driven from the New York City area centuries ago, are now regular visitors to suburban New Jersey and Connecticut and have been tranquilized and removed from neighborhoods within sight of the Manhattan skyline. Here, as elsewhere in the country, bears are becoming bolder about foraging in backyards and Dumpsters; scientists have even tracked a "fast-food effect" as the urban bears become heavier and more sedentary. Black bears are native animals, if long absent; coyotes were never easterners at all, so far as can be determined, but rather colonized the Northeast some time after World War II, probably by way of southern Canada, where they picked up a little genetic flavoring from the small Algonquin timber wolf. Consequently, eastern coyotes are markedly larger and more robust than their western counterparts, weighing up to seventy pounds—so different, in fact, that some mammalogists use the term "eastern canid" instead of coyote.

By whatever name, this is one adaptable dog. Studies have shown that they eat whatever's available, from rotting pumpkins and windfall apples to road-killed deer and unguarded cat food. (They also have a fondness for the cats themselves.) The bulk of their diet, though, appears to be small and midsized mammals like woodchucks, opossums, and rabbits, so food is clearly not a limiting factor. Nor is living space; they are as comfortable in suburban neighborhoods as in rural countryside, and today coyotes are found in every county east of the Mississippi. And that includes the County of New York, better known as Manhattan; in 1999, a coyote showed up in Central Park, having presumably found its own way there. It was soon cornered, tranquilized, and moved to a zoo, but other coyotes were reported in the five boroughs—cavorting on a Bronx golf course, for example.

Wild turkeys, which were all but exterminated in the Northeast during the first half of the twentieth century, have made a dramatic comeback. Game managers tried and failed to reestablish turkeys by stocking captive-raised birds, which proved too naive and disease-ridden to survive. But then they hit on a technique known as trap-and-transfer, in which a wild flock is captured with rocket nets and then trucked to a new locale, where with their native wariness they settle in—and spread well beyond the release point. It was a huge success, so that today turkeys, like the coyotes that prey upon them, are commonplace in the bedroom communities ringing New York City. A few have even made it into the city itself, to places like Staten Island and the eleven-hundred-acre

Van Cortlandt Park in the Bronx, and to Inwood Hill Park at the northern tip of Manhattan.

In the spring of 2003, though, a hen turkey showed up in the least likely place of all—the balcony of an apartment high above West Seventieth Street. For several months, the huge bird had been seen on the Upper West Side, then down in Chelsea, on the heavily planted grounds of the Episcopal seminary between Ninth and Tenth avenues, and eventually in Greenwich Village. But even ornithologists, who assumed the bird had crossed the Harlem River from the Bronx and followed the Hudson shoreline south, were at a loss to explain how she had reached the twenty-eighth floor of the Lincoln Towers apartments. Perhaps, one guessed, by flapping from balcony to balcony? However the turkey arrived, the couple living in the apartment told *The New York Times* they had decided to rename their place "The Butterball Roost," and a year later birders knew of at least three turkeys living in Manhattan.

On April 25, about two weeks after they'd met in Newfoundland, Peterson and Fisher pulled in to Roger's driveway in Glen Echo, Maryland, just outside Washington, D.C., for a two-day respite before the serious traveling began—three uninterrupted months on the road. But amid the unpacking, laundry, shopping, and repacking, Peterson was able to show Fisher a little of Washington. This included a picnic shad roast on Plummer's Island in the Potomac with the Washington Biologists' Field Club, a who's who of the nation's top wildlife experts, which for James was a happy mix of social pleasantries and new bird species. (By the next evening, however, when Peterson and his wife, Barbara, threw a dinner party for still more local luminaries, Fisher's enthusiasm had begun to flag a bit; he'd been feted in Boston and New York, and this was enough. "Another party," he grumped in his private notes, though he recorded that among the guests were the renowned mammalogist Ernest Walker and a then-little-known biologist with the U.S. Fish and Wildlife Service named Rachel Carson.)

Peterson called Washington "a very satisfying city for a naturalist—if he does not mind the humid, almost tropical summer climate," with "wild country almost at the city's door." He enthused to his *Wild America* readers about the wealth of night sounds in the D.C. area—whip-poor-wills endlessly calling their names, night-herons squawking along Rock Creek, barn owls with their weird screams in the "castle" of the Smithsonian, various frogs and toads, nighthawks, barred owls, and even woodcock.

The night-herons still nest at the National Zoo, and it's possible to find barred owls in the wooded suburbs, but the night is far less melodic here than it used to be. Whip-poor-wills are all but gone (though the larger chuck-will's-widow has moved into the region), and nighthawks, which used to nest on flat gravel rooftops in the heart of the city, have likewise declined drastically. The barn owls, which occupied the Smithsonian attic for nearly a century—and made quite a mess in the process—were finally evicted in the 1950s by groundskeepers, who nailed up their entrance holes. (An attempt to reintroduce the owls by releasing a pair from the National Zoo in 1974 failed.)

It's no surprise the avifauna has changed. Few areas along the Eastern Seaboard have seen such extraordinary growth in population and development as Washington and its suburbs. Montgomery County, Maryland, where the Petersons once lived, experienced a more than fivefold increase in population between 1950 and 2000, from barely 165,000 people to almost 900,000, while Fairfax County, Virginia, which borders D.C. to the west, went from fewer than 100,000 residents in 1950 to 1.5 million a century later.

But if Washington has become more of an urban jungle and less of a literal one, it retains an uncommon number of parks and green spaces, whose survival was often the result of men or women of vision. For instance, Peterson's home in the 1950s was the onetime slave quarters of an old farm overlooking the Chesapeake & Ohio Canal, which was first proposed by George Washington as a way to link the coast with the Ohio River frontier and was built, at staggering cost, between 1828 and 1850. But as Roger and James pulled into town, the canal was under the gun; plans were being drawn up to run a scenic highway along the length of the historic waterway. Where some saw a new transportation corridor, the Supreme Court justice William O. Douglas saw a national treasure, "a refuge, a place of retreat, a long stretch of quiet and peace at the Capital's back door . . . not yet marred by the roar of wheels and the sound of horns." In 1954, Douglas, a strong advocate of wilderness protection, launched a quixotic fight to save the C&O. Responding to a series of newspaper editorials in favor of the road, Douglas challenged the editors who wrote them to walk the 189-mile length of the canal with him and see for themselves what was at stake.

Probably much to Douglas's surprise, they accepted. On March 20, 1954, a group of fifty-eight people, led by the justice and including print and broadcast reporters, conservation leaders like Sigurd Olson, and (according to Peterson) two stray dogs, set out from Cumberland, Maryland. When they reached Washington eight days later, a cheering crowd of five thousand greeted them, and Douglas seemingly had won the debate. Congress, however, was unimpressed. It dickered and delayed for years, until at last in 1961 President Dwight D. Eisenhower used the broad powers of the Antiquities Act to declare the canal a national monument. The chairman of the House Interior Committee, in a snit because of Eisenhower's end run, managed to block any further action on the C&O for another decade, but in 1971 Congress finally passed legislation designating the canal a national historic park.

Though he might not have realized it at the time, Douglas wasn't crusading only for a place of serene beauty and historical significance; the C&O Canal and the adjacent floodplain of the Potomac are also of tremendous ecological value. This is, surprisingly, one of the most biodiverse units of the entire national park system, with nearly two hundred species listed as rare, threatened, or endangered; its habitats range from moist hardwood forests painted with spring wildflowers to grassy savannas akin to western prairies. And the fifteen-mile stretch known as the Potomac Gorge is among the wildest rivers flowing through a major urban center; here the river tumbles through a series of roaring cataracts and waterfalls, braids around islands, and smooths into long, slick runs, and yet its lower end is only four miles from the Washington Monument. The gorge contains thirty distinct natural communities, holding twelve species of plants and four animals that are globally rare.

One of the places Peterson was anxious to share with his friend during their brief stop in Washington was Rock Creek Park, which runs due south into the heart of the capital. About a thousand acres of the Rock Creek valley, once dotted with gristmills and defended by part of a ring of forts during the Civil War, were acquired by the federal government in 1890 for a park, while more of the creek, outside the District of Columbia, was protected by the state of Maryland. Peterson, who often birded there, quoted John Quincy Adams and Theodore Roosevelt about Rock Creek's charms, but then complained that "this park which has served as almost a place of worship for statesmen is threatened with a new express highway."

While opponents, including the National Park Service, managed to prevent that highway, which would have run up the length of the valley and essentially gutted it, they could not stop the Capital Beltway, which crosses Rock Creek just north of the park's border. The beltway opened in 1964, just two lanes in each direction; today, some portions carry almost a quarter-million cars a day, and there are plans to upgrade (if that is the word) to twelve lanes.

During the morning rush, the beltway is more of an obstacle than a conveyance, and as I sat in jammed traffic on an unseasonably hot April day, trying to get to Rock Creek, I did my best not to dwell on the fact that the last few miles took nearly an hour. When I finally squeezed off the highway, however, the contrast couldn't have been more dramatic. Nice homes crowd the boundary of Rock Creek along Sixteenth Street on the east side, near Walter Reed Army Medical Center, and Oregon Avenue on the west, but in between were deep hardwood forests dominated by tulip-trees, oaks, basswoods, and sycamores, many of them big enough that you'd need several people, arms outstretched, to encircle their trunks. Early wildflowers were abundant; the bloodroot was finished, but white violets, mayapples, spring beauties, and many more were going strong. Some of the morning traffic spilled over onto winding Military Road (its name another Civil War holdover), which cuts across the park, and a few joggers and walkers went by on the paved footpaths, but on this weekday the park had a pleasantly empty feel.

At one of the picnic areas, I met up with two dozen students from a D.C. charter school, who crowded around National Park Service education specialist Maggie Zadorozny as she gave them an introduction to the park. Using dog-eared topo maps, she showed the kids—most of whom had never been in the woods before, even though many lived within fifteen minutes of the park—how the city had completely enveloped the Rock Creek watershed, a sea of pink on the maps with a wide green dagger down the middle. "This miniature oasis is quite an extraordinary home for lots of birds and other wildlife. There are deer, possums, raccoons, even fox here"—the kids murmured in surprise—"and you'll see a lot of birds today, because this is the spring migration and they're coming back from the tropics." As she talked, I could hear wood thrushes and a red-eyed vireo singing over the soft backdrop of the stream.

I slipped away, spending several hours on the trails that meander through the park. Although it has a golf course, ball fields, and a tennis stadium, among other amenities, some of the park is free from development, except for hiking trails. I walked up Pinehurst Branch, a small tributary, through palisades of old trees and singing ovenbirds, and it was hard to remember that this had all been open farmland a century and a half ago. There were a few gnawed saplings, the work of beavers, which recolonized the park in 1981 after an absence of probably two centuries or more. (Not everyone was pleased by this bit of resurgent wild America, especially when beavers started chewing down Washington's celebrated cherry trees along the Potomac Tidal Basin.) In this small stream valley, I was finally able to leave behind all human noise—except, that is, when a flight coming into Reagan National Airport made a low-altitude pass overhead.

Because it's a big, empty spot in an urban landscape, Rock Creek sometimes attracts darker activities; most Americans first heard of the park when the body of Washington intern Chandra Levy was discovered there in 2002, more than a year after her disappearance. For the same reason, Rock Creek has also been the focus of scientific study for much of its history. Annual breeding-bird censuses dating back to the 1940s show what happens when even a large piece of woodland becomes an island—a steady erosion in the diversity of nesting songbirds, especially those that require unbroken tracts of forest. At Rock Creek, six species have disappeared entirely, while others, including the wood thrushes and vireos I'd heard, have dropped to a fraction of their former abundance. And botanists have tracked the incursion of non-native plants, many of them escapees from surrounding gardens. Of the 656 plant species recorded in the park, 238 are exotics, of which 42 are considered aggressive. These include garlic mustard, which reminded Fisher of his home in England, and lesser celandine, which blankets much of the park to the exclusion of native wildflowers. Porcelain-berry vine, Japanese honeysuckle, Japanese stiltgrass, English ivy, Asiatic bittersweet—all have taken a toll on the park's native flora. Staff and volunteers pick, cut, slash, and spray the invaders in an ongoing war with at least a few victories; kudzu, the rampaging Asian vine that has buried much of the Southeast, has been fought to an uneasy draw in Rock Creek after decades of effort.

In early afternoon, I looped back to my starting place, where the charter-school students—having spent the morning surveying bird habitat and then learning how to test water quality from a team with Audubon's Maryland-D.C. office—were trying their hand at fishing, a new experience for virtually all of them. It had taken some convincing before they'd warmed to the idea of pulling on chest waders and climbing into the murky, fast-moving water, but now most of them were standing thigh-deep in the stream, or were perched out on a fallen tree that spanned the creek, their lines in the current.

I asked how it had been going, and one of the instructors grinned. "No one's caught anything, but three of them fell in and just laughed about it. And no one's hooked anyone else. So I'd say it's going fine." Suddenly there was a shriek from one of the girls, whose rod tip was bouncing. "I got a fish, I got a fish!" she hollered, cranking on the reel. When she lifted her catch clear, it was a fat silver creek chub about nine or ten inches long, flashing drops of spray in the sun—no trophy, unless of course you're a city kid who's never held a fishing rod before, feeling the press of a living stream around your legs for the first time, on a sweet spring day in a forest that must have seemed to her as exotic as the Amazon. The fish looked like a dandy to me.

THREE

Forests of Loss and Resiliency

When Daniel Boone goes by, at night,
The phantom deer arise
And all lost, wild America
Is burning in their eyes.
—Stephen Vincent Benét (the epigraph that introduced *Wild America*)

Drive west from Rock Creek Park and it'll take you a good twenty-five or thirty miles to finally clear Washington and its suburbs, but eventually you get beyond the subdivisions and hit the open farmland of the Piedmont, rich in Civil War history. Not long after that, the hills become high enough to warrant their own names—the Bull Run Mountains, Pignut Mountain, Little Cobbler Mountain—until about sixty miles from the city the land rears up in the crumpled, spellbinding rampart of the Blue Ridge, the front range of the Appalachian Mountain system.

Even after April has washed across the coastal plain and Piedmont, the mountains are still cold and brown; what is a chilly April rain in the lowlands, soaking the redbuds and the dogwoods, can be sleet and ice pellets on the ridgetops, rattling through the bare branches and last year's dead oak leaves. But like a rising tide, spring leaves its mark on the mountains—the "green line," the highest point at which the trees have begun to leaf out, which creeps steadily up the slopes, day by day, in a visible march of the advancing season.

Peterson took a particular delight in showing his English friend the many faces of Appalachian spring, from first blush to full roar—and all within a day's travel along the Skyline Drive, the scenic road that threads its circuitous way along the summits and valleys of Shenandoah National Park.

> Our journey along its serpentine length on this late April morning was a demonstration in vertical magic. Repeatedly we went in and out of spring. When our road dipped into valleys between the mountains we were plunged into a lush green fairyland of spring at floodtide; climbing out again we found ourselves in a more austere landscape, the landscape of late March or early April. In one or two spots where the road attained an elevation of nearly 4000 feet the aspect was quite Canadian, somber with the dark spires of red spruce. Here, the deciduous trees . . . were naked, still devoid of leaves.

The Appalachians have always been home for me, but growing up among the low, sway-backed ridges of Pennsylvania, I find that these higher, far more rugged peaks of the Blue Ridge exert an especially strong attraction. Even more than my home mountains, they are a source of mystery and an avenue of escape when the modern world gets to be too much: a promise, on the horizon, of hidden hollows and deep-shaded trout streams. The Blue Ridge province runs from southern Pennsylvania to northeastern Georgia, starting low and narrow in the north (at Harpers Ferry, where the Potomac squeezes through, the mountains are barely four miles wide and not much more than a thousand feet high) but reaching a magnificent crescendo on the North Carolina–Tennessee border, where enormous ranges crowd shoulder to shoulder against one another, forming a serrated barricade that includes the Great Smokies, the Unicois, the Unakas and Snowbirds, the Great Balsams, and the Black Mountains. Here, many of the peaks are five or six thousand feet in elevation, with Mount Mitchell, at 6,684 feet, the highest mountain east of the South Dakota Black Hills.

Yet however high and remote they may seem, these mountains have a battered past and a murky future. So completely were they stripped of their timber in the nineteenth and early twentieth centuries that little of the old-growth stands remain. Hill farmers worked slopes no sane man would try to till, if he had a choice, but that was the only land available to the poor; erosion followed the plow, and both the highlanders and the mountains suffered badly for it. Yet these were resourceful, self-reliant people, not the insular, backward clods of hillbilly stereotype, and when they were forced off their land to make way for Shenandoah and Great Smoky Mountains parks, the government money they received was cold consolation for the loss of a unique way of life.

It's hard to reconcile photographs of Shenandoah in, say, the 1930s with what one sees today. When the park was established in 1935, a third of it was completely bare, and virtually all the rest was brush, newly sprouted from the stumps and slash piles of recent logging; Peterson was well aware of how young and weedy was the landscape he was showing Fisher. The same was true of the Smokies, though here larger areas of virgin woods survived the saw; still, even Clingmans Dome, now a scenic mainstay, was in the 1920s corrugated with railroad spurs and skid roads, its forests clear-cut. That was the story up and down the southern highlands. "If the East is to have wilderness it must restore it," Peterson said. "The second growth, now thirty, forty, or fifty years old . . . will, while our sons are alive, become trees eighty, ninety, or one hundred years old. Our grandsons may see a forest approaching its climax."

Such is the resiliency of the Appalachian forests that today his prediction has largely come true, and one has to have a knowing eye to see the old wounds—the hiking trails that started as logging roads, for instance, or the preponderance of multiple-trunked hardwoods that are the sure sign of trees that sprouted from stumps instead of seeds. Now approaching their second century, the Blue Ridge forests are starting to attain not only the heft but also the more complex ecological workings of a mature woodland.

If given time and peace, they will regain some, if not all, of the features that made this one of the mightiest and most diverse temperate forests on the planet—but that's a big if. Two-thirds of the nation's timber is now cut in the Southeast, and chip mills, which chew entire logs into fragments for particleboard or paper, account for 4.6 million acres of trees per year, much taken in vast, indiscriminate clear-cuts. Until recently, most of that cutting had been on lowland pine forests south of the mountains, or in mixed forests on the Cumberland Plateau, but the mills are pushing into the higher, biologically richer, far more sensitive Blue Ridge forests, too, clearing timber at unsustainable rates.

Paradoxically, even as timbering ramps up again on second-growth land, there is far more of the original, virgin forest left out than we once realized. Most of us hear the words "old growth" and imagine stands of huge, mossy old trees whose towering canopy shuts out the light, and for some virgin forests that was the case. But not every ancient eastern forest was visually remarkable or dominated by massive trees, and as ecologists have learned how to recognize the signs of a forest free from human interference, they've come to realize that old growth is a lot more widespread than anyone suspected. Some of these newly appreciated stands include dwarf oak and pitch pine forests on Massachusetts mountaintops, red pines in West Virginia, and small surviving parcels in such unlikely places as New York City parks and the metro Washington, D.C., area.

Specialists have also found virgin stands in remote parts of the southern Blue Ridge that were overlooked, simply because they are in such inaccessible places—forests that would fulfill anyone's mental image of old growth, with white pines more than two hundred feet tall and sassafras trees two centuries old and more than a hundred feet high. In all, a recent tally put the amount of old growth in the East at 2.2 million acres, almost a fourfold increase in what was once so classified.

But even in the most distant hollows, the isolation of the central and southern Appalachian highlands is more illusion than reality, and the mountains suffer from their proximity to the most densely populated part of the United States. Half of all Americans live within a day's drive of Shenandoah National Park, besides the 1.5 million who visit it each year. Shenandoah is also a victim of its topography and history. Although Congress authorized the park to exceed half a million acres, less than 40 percent of that area was eventually acquired, so that the park, though it is more than a hundred miles long, is in places barely a mile wide, hemmed in by lower, less-protected private land and with development creeping right to the boundary in many spots. Nor are those porous borders a barrier to many of the worst hazards the park now faces, from exotic species to air pollution.

In the 1950s, visitors to the highest points in Shenandoah routinely could, with a good pair of binoculars, make out the tiny white spire of the Washington Monument, sixty-five miles away. Now visibility averages three to twelve miles in the park, and sometimes you're lucky to see the valley, much less Washington, D.C. Air pollution is one of the gravest threats to the Blue Ridge, especially the outfall from distant coal-fired power plants that have been built over the past half century and automobile exhaust from the megalopolis just to the east. While much of the fine particulate pollution comes from older power plants in the Midwest, Virginia has approved sixteen new power stations since deregulation in 1998, with one proposal for a plant just five miles from the park's north entrance.

Acid deposition, which occurs here at some of the highest rates in the country, has damaged mountain streams, many of which lack much natural buffering, so that some waterways have seen declines in everything from aquatic insects to fish. The acidity is so bad that even native brook trout, which have a natural tolerance for water with a low pH, have been disappearing. The wind-borne acidity also frees up naturally occurring aluminum in the soil, which poisons the roots of trees, but air pollution is damaging plants in other ways. More than forty species of plants in Shenandoah have shown signs of illness from ground-level ozone, and the park is one of the worst places in the country for this pollutant.

The story is even grimmer to the south. Great Smoky Mountains National Park has the dubious distinction of having the worst overall air quality of any national park, and it records more days in which the air violates federal standards (forty-two in 2002 alone) than many eastern cities, including Atlanta. In 2004, the federal Environmental Protection Agency went so far as to declare both Shenandoah and the Great Smokies as officially unhealthy, based on ozone levels, and the nonpartisan National Parks Conservation Association has listed both parks as among the ten most endangered in the country, in part because of pollution issues.

One of the most worrisome trends in the Blue Ridge is the toll that invasive alien species are taking on native plants and animals. Humans have been bringing exotic species with them ever since the first colonists arrived on this continent, and that includes domestic dogs that followed the Paleo-Indians here from Asia. (In fact, one can argue that we ourselves are an invasive alien.) But as the global economy grows, and international trade increases in both speed and volume, the unintended hopscotching of noxious organisms around the planet has worsened—with dire consequences to many ecosystems. This is a theme to which I will return a number of times in the course of this book, for invasives are, in the opinion of many experts, the greatest emerging threat to North America's biodiversity, but there is a singular poignancy about the subject here, in the Appalachian hills.

As they drove through the Blue Ridge, Fisher and Peterson saw stands of large, dead trees, their upright trunks bleaching in the sun—American chestnuts, doomed by the great blight that swept the Appalachians starting in 1904 and that killed virtually every mature chestnut in the East. This was the first time an introduced organism had all but wiped out a widespread, commercially valuable native, and it was a disquieting omen of what was to come. "No greater catastrophe has ever befallen a tree in our times," Peterson wrote, mourning that his own sons would never have the experience of gathering chestnuts in the fall, as he'd done with his father.

In the century since the chestnut blight hit, hope continues to flicker that the species may be brought back. The almost indestructible rootstock continues to sprout, decade after decade, but the trees rarely get bigger than saplings before the blight kills them again. Scientists and volunteers have searched for naturally blight-tolerant chestnuts and have laboriously attempted, over decades, to hybridize resistant Asian chestnuts with the native Americans, then backcross the offspring to weed out all but the gene for disease protection. But while they have worked to salvage the chestnut, a wave of new dangers has swept across the Appalachian forests, so that today it seems that almost every member of this incredibly diverse assemblage of trees is under threat.

The standing dead chestnuts that Fisher and Peterson saw have long since fallen, but visitors to Shenandoah today see great swaths of mountaintop forest that were defoliated by invading gypsy moths in the late 1980s and early '90s; in some parts of the park, more than 80 percent of the oaks died. With time, the dead trees have shed their outer branches and now their bark, so that they stand, white and truncated, against the jungle of smaller species crowding up in the unexpected light. Many of the ghost trees are draped in mile-a-minute vine, an exotic that usually thrives on sunny roadsides and suddenly finds itself freed from the deep shade, with entire hillsides to conquer. Brambles and blackberry canes grow in great disheveled thickets. In places, the view from the Skyline Drive can seem more like a vacant lot than a national park.

It had been several years since I'd been in Shenandoah, and while I've grown used to seeing the dead oak forests, I had a nasty shock when I hiked away from the Skyline Drive, down into some of the deep gorges that vein the mountains. It was a bright, sunny day after a frigid night, when I'd burrowed deep into my sleeping bag; spring was just reaching the higher elevations, with coltsfoot and hepatica, the wildflower vanguard, blooming beneath the still-bare hardwoods. But strangely, as I descended into the canyons, it was still bright and sunny. I remembered these as dim green places, heavy with the shade of old-growth hemlocks, trees that were four or five hundred years old. But now those same trees stood stark and dead, their branches bare to the sun, which poured down into streams normally well protected from its heat.

They had fallen victim to the hemlock woolly adelgid, a minute, aphid-like insect accidentally brought in from the Orient and first seen in the East in the 1950s. Since reaching Shenandoah in the 1980s, the adelgids have infested almost all of the park's eastern hemlocks: the needles turn yellow or grayish, their undersides flecked with the white fuzzy egg masses of the adelgids; the foliage thins, dropping off from the bottom up; and the trees—even majestic ancients like those along the park's popular Whiteoak Canyon Trail—are often dead within three or four years.

As bad as this is for the hemlocks, and for those of us who consider them among the most graceful of all eastern trees, the adelgid onslaught has tripped an ecological domino effect, much as the loss of the chestnut did, and no one knows how far the ripples will reach. Hemlocks create a shady, damp, acidic environment for many wildflowers; for aquatic organisms, their loss means that summertime stream temperatures may reach lethal levels in heretofore chilly mountain brooks. Birds are losing their habitat; in the park, hemlocks had been the almost exclusive choice for nesting blackburnian and black-throated green warblers, among others. A few years ago, in the Peaks of Otter region south of Shenandoah, I saw a blue-headed vireo sitting in her small, neatly woven nest, which was built, as is often the habit of this species, between the forked twigs of a hemlock branch. Unfortunately, the tree was nearly dead, and the small cluster of wan gray needles that still clung to the branch gave the poor bird no cover at all.

The hemlock adelgids are so tiny they can hitchhike on larger bugs, on the feet of birds, or on the wind, but usually they reach new areas on infested nursery material, which has spread them across the East. In 2002, the hemlock adelgid was found in Great Smoky Mountains National Park, which had already lost virtually all of its mountaintop Fraser firs to a related European pest, the balsam woolly adelgid, which moved into the southern Blue Ridge in the 1950s. The Smokies hold the largest stands of old-growth hemlocks in the East, so the park staff, hoping to get ahead of the curve this time, attacked the two new adelgid infestations, one in North Carolina near Fontana Lake and one in Cades Cove, the scenic valley on the Tennessee side of the park. They used injectible pesticides and even insecticidal soap sprays, but these are useless in remote locations, so managers also have used a small black predator beetle that researchers discovered in Japan and that feeds (so far as anyone can tell) exclusively on hemlock adelgids. But the beetles are difficult to rear in captivity and expensive to buy, and while the park estimates it needs several million to effectively fight the pest, there's only been money to buy a fraction of that. So even as the adelgids have spread to more than sixty new sites in the Smokies, making eventual control that much less likely, the park—which, like virtually every national park, is hopelessly underfunded—has been forced to rely on donations in order to buy beetles, like the eleven hundred dollars it received from a Michigan Brownie troop's can-and-bottle drive.

Sadly, the adelgids are the tip of the iceberg when it comes to invasive species in the East. Dutch elm disease, a fungal infection, famously wiped out the country's favorite street tree but also eliminated most elms in the wild. Another fungus, an anthracnose, has killed millions of flowering dogwoods in Appalachian forests, robbing migrant songbirds of the fat-rich berries the trees used to produce (fortunately for tourists, sunlight and air circulation curtail the fungus, so the dogwoods out along the roads have fared better), while still another anthracnose is now attacking sycamores. As is the case with the two anthracnoses, no one knows if a canker that has all but exterminated butternut trees is native or foreign, but beech bark disease, which is assaulting American beeches, definitely came from Europe.*

*As if all this weren't bad enough, the real disaster may have just arrived in the Appalachians—sudden oak death syndrome, an alien disease caused by the fungus *Phytophthora ramorum*, which since the mid-1990s has killed enormous numbers of oaks and other species in coastal California and Oregon. Despite attempts at quarantine, in March 2004 infected nursery stock from California was shipped to seventeen states, including many in the Appalachians, and additional infected nurseries were detected in the following months. Lab tests suggest that the red oak family—a critical component of Appalachian hardwood forests, including red, scarlet, black, and pin oaks—may be even more susceptible to the fungus than California oaks. "This is the scariest thing I've seen in my lifetime," one U.S. Forest Service researcher was quoted as saying.

Then there are the insects. Sugar maples are besieged by pear thrips, gypsy moths continue to attack oaks, and the Asian long-horned beetle—whose rapacious, wood-boring grub is as big as a child's finger and eats a remarkably catholic list of more than two hundred hardwoods—came in via untreated wood-packing material from China. This last one is giving forest managers nightmares; the only way to control the beetle is to cut down and burn infested trees, which has been done in parks and urban neighborhoods in Chicago and New York City, where there are infestations. Whether the bug is loose in the East's wild forests is anyone's guess, but it has been detected in dozens of import locations.

Insects and pathogens aren't the only threats to forests. Wild European boars, which were released decades ago in the Southeast as game animals and have few natural enemies here, churn up wildflowers, seedlings, and everything else in their path as effectively as rototillers, eat all manner of small animal life, and compete with black bears and other species for food. But ironically, one of the biggest threats to the continuing health of the Blue Ridge forests is a native, the white-tailed deer. Even in 1953, Peterson warned that deer were overpopulating parts of the East, and in recent decades burgeoning whitetail numbers have become a crisis. Across many areas of the Appalachians, deer overbrowsing is so bad that plant diversity has plummeted, and tree reproduction is so low that the only way to get any regeneration after a timber cut is to wire off the plot with electric fencing.

Faced with so many challenges, public land managers in the southern Blue Ridge are being pressed to come up with aggressive plans of action to confront the region's many environmental problems. The days of "letting nature take its course" are long gone, and a strictly hands-off approach would only exacerbate most of the difficulties. The balancing act, of course, involves making certain that a management plan to address one issue doesn't inadvertently create a new set of problems—a real risk when you're fooling around with one of the most complex temperate ecosystems in the world.

The first precaution in intelligent tinkering, Aldo Leopold famously observed of ecological management, is to keep all the cogs and wheels. But the fact is, we don't know what all the pieces are. No one knows, within several orders of magnitude, how many species there are on earth; scientists have cataloged about 1.5 million, but estimates for the total range from 10 million to 30 million, to as high as 100 million, with new species uncovered constantly.

Most of those new discoveries, not surprisingly, are made in the tropics. You'd expect that a popular, long-studied park like Great Smoky Mountains—close to civilization, generally accessible, and the focus of scientific research and amateur naturalizing for more than a century—would have been thoroughly inventoried by now. And you'd be wrong, as I found out for myself, not in some remote backcountry hollow, but in a mud puddle next to a parking lot in Cades Cove.

Charlie Staines's business card, which I had tucked in my shirt pocket, reads "Entomologist at Large." After twenty-five years of doing plant quarantine work for the U.S. and Maryland departments of agriculture, he decided to strike out on his own in 1997, doing research on beetles in the jungles of Costa Rica, the Dominican Republic, and New Caledonia; collecting across the desert Southwest; and scouring collections in the museums of North America and Europe for specimens of his specialty, the huge family known as leaf beetles. How huge? "If you take all the species of birds and all the species of mammals in the *world*, and multiply by two, you'd have as many species, about fifty thousand, as are in just one family of leaf beetles," Charlie said. That number keeps growing, thanks in part to him; he's just described two new species, bringing to about seventy-five the previously unknown beetles he's introduced to science.

Charlie Staines is in his early fifties, a cheerful man with a salt-and-pepper beard and a wide-rimmed hat. He pushed his feet into a battered pair of black knee-high boots and unhitched a dip net from the roof rack of his car while his wife, Sue, grabbed the rest of their collecting equipment: small aquarium-sized nets, a small pale ground cloth, vials of alcohol, and an aspirator—a plastic jar with two rubber hoses coming out of the stopper, used to siphon up tiny insects.

Sue has a thick head of dark hair, gold-rimmed glasses, and a disposition which equals that of her husband, whom she's known since they were fourteen and in 4-H together. An efficiency expert trained in home economics, Sue has spent most of her married life as a de facto entomologist and botanist, working side by side with Charlie in the field (the study of leaf beetles also necessitates the study of their host plants, hence her help with botany). For the past twenty years, she's also been trying to organize the Smithsonian Institution's collection of about a million leaf beetle specimens—"The work of a lifetime," she says happily—something she does as a volunteer.

Our quarry on this bright, gorgeous morning were not leaf beetles, however, but their aquatic cousins. The Staineses, who live in Maryland, had received a small grant to conduct fieldwork in the Smokies from a nonprofit group called Discover Life in America, which, with the National Park Service, is trying to do something that's never been done before: catalog every single species of plant, animal, fungus, and microbe in Great Smoky Mountains National Park. This undertaking is known as the All Taxa Biodiversity Inventory, or ATBI; "taxa" refers to the pigeonholes into which we stick organisms: species, genus, family, order, and so forth. There have been many less-exhaustive attempts to inventory smaller tracts of land, like city parks—they're often called bioblitzes, in which a team of dozens of scientists and volunteers scour the place in a frenzied twenty-four-hour period—but never something on this scale, with the goal of delineating every single kind of organism in a place as large, wild, and rugged as the Smokies.

Or as diverse. Even at the height of the last ice age, the southern Blue Ridge was never glaciated, and the Smokies served as a refuge for many species of plants and animals. Their complicated topography, rich soils, and abundant rainfall make them a natural garden, with such varied habitats as cove hardwood forests with as many as sixty species of trees and shrubs per hectare, high-altitude grass- or heath-dominated "balds," Table Mountain pine stands maintained by fire, northern hardwood forests akin to New England, and spruce/fir forests on the highest mountains. The Smokies are the salamander capital of the world, their waters are full of rare freshwater mussels and fish, and they have more than fifteen hundred species of vascular plants.

Just as the Human Genome Project allowed us, for the first time, to read the genetic blueprint of our species, when the ATBI is completed, it will provide the first opportunity for ecologists to see the entire biotic makeup of a large ecosystem—the genome of the Smokies, in effect—and then use that knowledge to better understand how the system functions and to decide how best to preserve and manage it. E. O. Wilson, the great Harvard ecologist (who is on Discover Life in America's advisory panel), has said that if successful, the ATBI would stand as one of the most important achievements of twenty-first-century science.

It will also be one of the most challenging. About ten thousand species have been recorded in the park, but most estimates put the actual total at more than ten times that number. A few taxa, like birds and vascular plants, are fairly well known, while others—spiders, moths, and fungi, for instance—are largely a mystery. Which brings us back to Charlie and Sue Staines, who are trying to get a handle on the park's aquatic beetles, as well as those that lay their eggs in various kinds of animal dung, some of which specialize in the scat of particular species of mammals or birds. "Bear dung is especially good for beetles," Sue told me, "but any kind of dung works. When we're traveling, we'll stop at roadside rests and go to the pet-walking areas. You get some great stuff, but you also get a lot of funny looks." While I find that kind of biological specialization fascinating, I have to confess I was a little relieved that we'd only be looking for water beetles today.

I'd packed a pair of chest waders, figuring we'd be sloshing around one of the park's picturesque, fast-flowing streams, but our first stop was a parking lot near the Cades Cove ranger station, beside which was—to my eyes—a wholly unpromising rut about two inches deep and as big as a table, formed by the slow seep of water from some hidden spring, and through which a large truck or tractor had obviously driven not long before. Such small wet spots, it turns out, are the preferred home for a raft of small predaceous and scavenging water beetles, most only a few millimeters long. The Staineses use a couple of techniques to catch them. After dark, they will deploy UV lights, attracting the flying adults to a sheet suspended between trees, but a simple aquarium dip net works, too, and that's what Charlie started poking around with. Sue, meanwhile, pulled on yellow rubber dishwashing gloves ("These kinds of places often have lots of glass and junk," she explained) and began methodically kneading the mucky bottom, then pushed the dirt and dead leaves down, watching for any tiny movement. "We're just eight-year-olds who never grew up," Charlie admitted with a smile.

There was a lot of wiggling—the seep was full of toad tadpoles and mosquito larvae—but finally a slim, streamlined beetle the size of a sunflower kernel zoomed with surprising speed and agility across the yellow backdrop of Sue's gloves, and Charlie nabbed it, popping it into a vial. It was *Laccophilus fasciatus*, he said, a pioneering species that quickly zeroes in on new bodies of water.

We were getting plenty of odd looks of our own, especially when we later pulled in to a crowded parking area near one of the park's old country churches and unloaded our gear. A short distance through the woods was a different habitat, a jewel-like pond only a few yards wide. Whirligig beetles, glossy and black, floated in a small cluster until Charlie waded in, when they started spinning and darting chaotically. The drill here was different; Charlie took a long-handled net and repeatedly dragged it through the emergent sedges, then dumped the resulting muck with a wet plop onto a pale ground cloth, where Sue and I picked through it, again waiting for movement.

We found newts, tadpoles, dragonfly nymphs, midge larvae, and a number of even tinier beetles than the last, smaller than poppy seeds. Sue sucked them up into the aspirator, eventually transferring them to an alcohol vial, where they gyrated wildly. Every time she saw a beetle, she let out a little gasp as she darted for it; when I mentioned this, she thought for a moment and admitted she'd never noticed, but then shrugged and said, "Well, yes—it's exciting!" There were a lot of gasps; this was a productive spot, and it was best to take a number of beetles, because some species can only be identified by dissection of their genitals under a microscope, and then only if the specimen is a male. You need a series to make reasonably sure you have at least one male of whatever you've collected. We also bottled up a number of other aquatic insects, including water striders and a slim bug (I use the word in its scientifically precise sense, referring to an insect with sucking mouthparts in the order Hemiptera) related to the water scorpion. "The great thing about these biotic inventories is that *somebody* will be interested in almost everything you find," Sue said.

Later, back in the old rangers' quarters where the ATBI researchers stay, Charlie checked the all-taxa survey's database via computer and confirmed that our catch that morning included a genus of beetle, *Uvarus*, that had never been recorded in the park, while the first beetle we'd caught, the pioneering *Laccophilus*, had never been recorded on the Tennessee side of the Smokies.

Such discoveries are the daily stuff of the all-taxa survey in the park, Jeanie Hilten explained later in the day as she and I sat on rocking chairs on the porch of the park's natural resource center near Gatlinburg. Hilten, a former environmental educator at the Great Smoky Mountains

Institute, runs the Discover Life in America effort, which since its first field season in 1998 has added 3,314 new species to the 10,000 originally known in the park, including many that were new to the state, or even to the Western Hemisphere. More remarkably, ATBI field teams have uncovered more than five hundred species that were completely unknown to science, including twenty-eight crustaceans and seventy-two butterflies and moths; she had to double-check the latest numbers on her computer for me, because they change so frequently.

DLIA and the park take several approaches to the survey, Hilten said, involving more than 140 researchers, park staff, and a large team of trained volunteers. Researchers can, like the Staineses, apply for small grants to underwrite work they conduct on their own timetable, or they can participate in focused efforts that bring together a number of experts for a few days or a week of intensive work on one taxon, like butterflies or bats, or on particular microhabitats, like leaf litter. Because some European visitors who remember World War II objected to the name "blitz," these surveys are now called bioquests, but by whatever name they can be extremely productive; in one twenty-four-hour period a few years ago, scientists documented 860 species of moths and butterflies.

The organizers try to direct the scarce money to habitats that need special attention—high-elevation sites, species like Fraser fir and hemlock under attack by alien pests, Table Mountain pine forests that are disappearing due to pine bark beetles and changes in historic fire patterns, old-growth stands, bogs, and places with unusual soils. (One result of the survey has been the discovery of twenty-one previously unknown soil types in the park, each of which likely has its own unique association of plants and animals.)

Nor is the goal only to count noses; the ATBI is gathering information on distribution, abundance, trends, and much else. For example, Hilten said, volunteers have walked more than a hundred miles of trails, stopping every two hundred meters to set up a survey plot, identify and count all the ferns in it, and delineate the habitat; when they're done, the result will be a fairly complete picture of fern distribution throughout the Smokies. Everything goes into computer databases, where it can be overlaid with topographic, climatic, and geological information, as well as the exquisitely detailed maps she showed me indicating forest communities, a pointillistic swirl of colors—here Carolina silverbells and basswood, there maples and butternuts. In fact, pointillism works not just as a visual description but also as a metaphor for the project as a whole. Each data point—each water beetle record, each fern location, each breeding-bird sighting—blends into the tapestry as you move up and down in scale. What emerges, just as Seurat's colored dots merge into a coherent painting, is a picture of the Smokies in unprecedented, almost incomprehensible detail.

Already the survey has pointed to clear dangers. Great Smoky Mountains has the worst ozone pollution of any of the national parks, and thirty species of plants are showing signs of decline from its effects, including tuliptree and black cherry. By marrying maps showing the distribution of breeding birds with results of ozone-pollution monitoring in the park, even I could see that some species are living in a habitat that's under the gun; the worst levels of ozone occur at precisely the same high elevations required by birds like the black-throated blue warbler.

Of course, all this is just a beginning; only 10 or 12 percent of the park's biodiversity has even been inventoried, and initial estimates that the ATBI could be wrapped up in ten or fifteen years may prove optimistic, Hilten said. But there is a sense of urgency moving everything along. "The park is under attack," she said. "It's a race against time with acid deposition, balsam woolly adelgid, hemlock woolly adelgid, butternut canker, beech bark disease, ozone—the list of threats goes on and on."

As I was sitting on the porch, rocking quietly with Hilten, these worries seemed far away, but then I drove back to the park through Gatlinburg, and it occurred to me that she'd left one threat off the list—the peril of ugliness. I'm not being flip. With the Smokies, you have one of the most stunning natural landscapes on the planet, rimmed by a lot of unfettered visual garbage, in the form of rampant commercial development clogging its main entrances. And if ugliness doesn't reach deep into the heart of a wild place, it surely colors its character from the perimeter, and, more important, it colors the attitudes and perceptions of those who visit it and are forced to run a gauntlet of dross to reach the natural core.

Many parks are grappling with the growth of so-called gateway communities, but no town better illustrates the perils of runaway development than Gatlinburg. A visitor driving north across the Smokies passes through breathtaking vistas and deep green forest for hours, then comes to a curve in the road through which plastic in harsh primary colors peeks from behind the leaves. Rounding the bend, you pass the sign that marks the park boundary, and within a matter of yards you're plunged into the frenetic urban strip of fast-food franchises, condos, tacky attractions, and Swiss-themed motels that is downtown Gatlinburg. There is no in-between, just the park here and, a stone's throw away, the first burger joint. If your eyes don't roll back in your head at this, you have a stronger constitution than I.

A few miles beyond Gatlinburg lies Pigeon Forge, once known for its pottery, now a conglomeration of hotels, shopping outlets, and assorted tourist tawdries, with the Dollywood theme park as its anchor—not to mention an indoor skydiving simulator, an Elvis museum, laser-tag emporiums, and no fewer than eleven go-kart and automobile racetracks. Not hemmed in by the park and surrounding hills like Gatlinburg, Pigeon Forge and neighboring Sevierville have been free to balloon along the state highway. I'm not against some good, clean fun, but the sheer excess of the Gatlinburg–Pigeon Forge strip defies all common sense and taste. I mean, how many Ripley's Believe It or Not! attractions does one community need? In Gatlinburg at least, the answer appears to be four.

Cherokee, North Carolina, over on the south side of the park, is smaller but doing its best to catch up. As I drove through, a new fifteen-story Harrah's was going up at the tribal casino, big-box hotels squatted cheek by jowl with kitschy old 1960s-era units, and outlet malls jockeyed for space with the Bear and Reptile Farm. More power to the Cherokee for bilking stupid whites out of their pension money, but why is it that extraordinary natural beauty always seems to attract its polar opposite, like some seedy suburb? And what can be done about it? A buffer zone around national parks would have been nice, but buffers simply push the detritus farther back from the core—not a bad thing, but it doesn't address the larger problem of the proliferation of unrelated commercial excess beside natural beauty.

I was always inclined to blame this on human nature and the overriding urge to cater to the lowest common denominator—until I visited Stonehenge in England some years ago. I was traveling with a friend who wanted to see the old stone circle, but much as I love ancient history, I assumed it had been tarted up, Gatlinburg-style, with the Stonehenge Gem Shop and the Druid Reptile Farm and Lord knows what else. So imagine my surprise when I found the henge simply standing in the middle of the wide, empty Salisbury Plain, the horizon dimpled with old burial mounds but no sign of commercial development. The car park was a grassy field that held a few hundred vehicles; tourists entered a small visitors center built into the ground, passed through a tunnel beneath the road, and emerged to walk counterclockwise around the circle. There was almost complete silence except for the wind, since everyone was listening to interpretive recordings on handheld wands pressed to their ears. Except for the unfortunate noise of traffic, the ancient monument was simply allowed to exist in its own space and time.*

Why must we North Americans wall up our parks with so many ugly, impossibly congested gateway communities, which, like those around the Smokies, soon grow to have no real connection with the landscape that gave them birth? Is it because for us the thing itself is never enough—not enough to walk among the blooming phacelia and trillium, look for warblers, climb a hill, catch a brookie? It's no shock that Americans in particular want their experiences spoon-fed to them; we've made virtual reality more compelling than reality. In Tusayan, Arizona, you can watch an IMAX movie about the Grand Canyon, and hordes do—even though the real thing is only a fifteen-minute drive away.

What often happens is that gateway communities get their start because of park visitors, but eventually take on an economic life quite independent of the natural attraction, and often in direct conflict with the park's well-being, particularly by forming a developed barricade to the flow of wildlife in and out of the protected core. The explosive growth of Gatlinburg, Pigeon Forge, Sevierville, and Townsend, with road expansions along much of the Tennessee border of Great Smoky Mountains National Park, will, for example, form a barrier to the movement of black bears, which now move into and out of the park to find food; the growth of Cherokee, Maggie Valley, and other communities on the North Carolina side threatens the same, as does a highway—first proposed in the 1940s, and lately gaining momentum—that would run through the park

*My reaction aside, Parliament has branded the current facilities at Stonehenge a "national disgrace." English Heritage, which controls the site, has proposed a huge renovation that would move the visitors center miles away from the monument, create a light-rail system for visitors, remove one road and reroute another through a two-kilometer tunnel, remove fences, and restore the original chalk grassland and its wildlife to bring the surrounding landscape closer to its historical character.

along its now-roadless southwestern edge. Mountain parks and their wildlife are at special risk as ecologically important lowlands become developed and inaccessible.

But that needn't always be the result. Some gateway communities, just seeing the beginnings of creeping Gatlinburgism, are trying to opt out by slowing development and making sure that commercial interests complement natural values, instead of overshadowing or endangering them. Banff, the small town outside the park of that name in the Canadian Rockies, is doing just that, even going so far as to close an airstrip and some attractions and convincing a golf course to restore some of its land to wildlife habitat. The goal is to preserve the historic, small-town feel of the village, as well as critical valley habitat for wolves, grizzlies, and other wildlife.

Gatlinburg and Pigeon Forge are probably a lost cause, except as an object lesson in the worst way to build a gateway. But there is still hope for the rest of the land surrounding the Smokies—though the time to reverse the trend is now and the hour is late.

Five or six times a day on this long trip, I find myself thumbing through my battered copy of *Wild America*, checking my coordinates, so to speak, across half a century: where I am versus where Peterson and Fisher were, what I see versus what they saw. This is an odd, at times disconcerting feeling for me, as a naturalist and an author. I'm used to traveling mostly by myself, but now I feel as though I have these two guys in the backseat, nudging me in the ribs and whispering directions. But the biggest issue is the obvious conceit of trying to follow in the footsteps of two such icons, the leading naturalists of their generation.

I think they'd understand, though, in part because they were, at times, consciously doing the same thing. In a section deleted from *Wild America* before publication, Fisher compared their trip through the southern Appalachians with one made sixty-five years earlier by the pioneering ornithologist William Brewster. Fisher concluded by noting that once allowances were made for the differences in season, their predecessor, though traveling by wagon and without modern optics, managed to see more species of birds in the area than Peterson. Brewster, he wrote, "reached heights of field craft that have never since, and may never again be reached." All I can do is suggest that Fisher was being far too modest on his and Peterson's behalf, and pay his own compliment forward.

There are some species that no one will find in the Blue Ridge Mountains, no matter how sharp their eyes. "Even though the cutover forests are growing up again we shall never again in the eastern highlands know the Wild America that Daniel Boone saw," Peterson wrote. "Two centuries have seen the complete extermination of the timber wolf, the mountain lion, the eastern bison and the eastern elk." But for other animals, time has been kinder. Crow-sized pileated woodpeckers, which were still rare enough in the 1950s to warrant a mention when Peterson and Fisher saw one in Shenandoah, are now common the breadth of the East. Black bears, which were similarly uncommon, have also increased remarkably.

And even some of the vanished species have returned after long absences. Four kinds of threatened or endangered fish—the smoky and yellowfin madtoms, spotfin chub, and duskytail darter—have been reintroduced to the Smokies, while river otters, which were trapped out of the region before the park was established, were successfully returned in the 1980s, along with peregrine falcons. There have even been reports—generally scoffed at by experts—of mountain lions in the central and southern Appalachians. If they are true, the lions either came back on their own or were never completely killed off in the first place, a prospect that seems at least a little more plausible now that the big cats have been confirmed in the mountains of New England.

Both the otter and the falcon reintroductions used techniques that had been honed in other parts of the country, but bringing back larger mammals is a trickier proposition. The red wolf, an enigmatic canine that is intermediate in size and color between a coyote and a gray wolf, once roamed the Southeast, including the Smokies. Red wolves were reintroduced to the park beginning in 1991, but almost none of the thirty-three pups that were born over the next seven years survived past a couple of months of age, probably due to a lack of prey and competition from non-native coyotes, which moved into the park in the 1980s. The wolves also tended to roam beyond the park boundaries, causing conflicts. In 1998, the park management ended the experiment and recaptured the survivors.

But the hope of restoring the Great Smokies' lost wildlife hasn't died. The park's recent effort to bring back elk, once the monarch of the eastern hardwood forests, is part of a larger regional effort—and one that has proven unexpectedly contentious, illuminating the difficulties inherent in reintroducing big, sometimes troublesome wildlife.

Elk once ranged the Appalachians from northern Georgia to the Allegheny Plateau and into southern New York and were in many areas once more common than white-tailed deer. William Penn, writing from his new colony of Pennsylvania, boasted of "elke as big as a small ox" near Philadelphia. But the demand for hides and venison drove the huge deer, which can weigh nine hundred pounds, into a spiral of local extinction. The eastern subspecies had been eliminated from South Carolina by 1737, Vermont by 1800, New York by 1847, and Kentucky—where in 1810 John James Audubon lamented that they were already rare—by 1850. The last eastern elk were killed in Pennsylvania around 1870, wiping out the race.

Game managers and private individuals schemed to restore elk in the early twentieth century, but without much success. Rocky Mountain elk were released in the Adirondacks from about 1900 through the 1930s, but the herd never thrived, and the elk disappeared once more around World War II. Likewise, two small herds of stocked elk lasted about fifty years in Virginia before dying off in the 1970s. Half a dozen other attempts also failed. The only eastern herds that managed to survive were one in northern Michigan and one in the rugged Alleghenies of northern Pennsylvania, where Yellowstone elk were released starting in 1913. At one point, in the 1940s, the Pennsylvania herd dropped to probably no more than two dozen animals, but since the 1980s it has grown substantially and now numbers more than six hundred.

Today, however, elk are enjoying a renaissance in much of their old eastern range, with a number of states either reintroducing them or examining the practicality of such a project. Some wildlife agencies have taken a cautious approach to elk restoration, partly out of worry over the economic impact elk might have on farmers and partly out of concerns over disease; wild elk sometimes carry tuberculosis, brucellosis, and other ailments that might pass to livestock. White-tailed deer spread a parasite known as brainworm, which is harmless to them but can cause a loss of motor control and eventual death in elk, which have limited resistance to it. Furthermore, some western elk herds are infected with chronic wasting disease (CWD), a mad-cow-like ailment that's impossible to detect in live ungulates and might spread to eastern deer if infected elk are imported. (No one knows what risk CWD poses to humans.)

Pennsylvania chose to trap and move animals from its own herd to expand the elk's range in the Alleghenies, reasoning that after nine decades the animals had developed a resistance to brainworm and were known to be free from CWD and other diseases. Tennessee, which has a limited elk restoration project in a four-county area east of the Blue Ridge Mountains, decided to stock several hundred animals from a certified CWD-free herd in Alberta.

Kentucky, on the other hand, jumped into elk restoration with an abandon that has alarmed some of its neighbors. Starting in 1997, the state imported eighteen hundred elk from as far away as Arizona and Oregon. I had a chance to help one of the crews, back in the winter of 1999, that was catching elk in the Wasatch Mountains of Utah, where half of Kentucky's transplants originated. It was bitterly cold, well below zero, and the snow squeaked beneath the tires of our little convoy as we drove back into the Castle Rock Ranch, just a few miles from the Wyoming border. A pair of golden eagles flew off the rimrock, and we scattered small bands of elk and mule deer. After several days of buzzard's luck, we finally struck it rich. The big corral trap, forty feet wide and ten feet high, that we'd baited the day before with alfalfa hay held sixteen elk, mostly cows and young calves, but also a "raghorn" bull with five points on each side of his thin rack.

A cattle trailer was backed up to the chute on one side of the trap, and two of the crew slipped into the corral and tried to herd the cows and calves into the truck while keeping the bull—who might kill the other confined elk with his antlers—out in the open, to be released once the rest were caged. But the bull had other ideas, bolting around the two men and dashing down the chute. A University of Kentucky researcher named Jeff Larkin and I were standing by the mouth of the trailer, and without thinking, we grabbed a piece of burlap from the side of the trap, holding it between us across the narrow chute.

The elk slammed to a halt, six or seven hundred pounds of lathered, angry animal crowding the burlap, his head only two feet from mine, his antlers clattering against the metal trap frame as he tossed his head like a nervous horse. Steam poured in white jets from his nostrils. He was so big, so vital, so potentially lethal, yet Jeff and I held him back with a piece of old cloth while someone clambered into the truck, eased the trailer away from the chute, and closed the door on the other elk. Now the bull could see daylight, stamping his feet and banging his rack; we dropped the burlap, and he exploded out and away, spraying us with snow.

Five days later, forty of the elk I helped trap in Utah rolled into the hill country of southeastern Kentucky, after passing a barrage of health tests and a thirty-one-hour nonstop ride. On a windy hillside damp from predawn rain, the cattle truck stopped near a line of dozens of cars; word had spread, and even though it was daybreak on a Sunday morning in a remote reclaimed strip mine, hundreds of people wanted to be here for the moment. When the doors on the trailer opened, nothing happened at first; then one cow cautiously stepped out, followed by another, and then, like water gushing from a jug, elk poured out the door, their sides brown against the dead grass, their pale rumps like spotlights in the dawn. The herd formed a compact body, running down a draw, then up to a higher knoll a quarter mile away, where they stood in a long line, silhouetted against a bright slit of morning sky. There was muted applause from a few people, but most stood quietly, their eyes watering—and not, I think, from the wind.

Although brainworm has taken a toll on the Kentucky elk, the herd has exceeded even the most optimistic predictions. By autumn 2004, an estimated forty-five hundred animals were ranging over a sixteen-county elk restoration zone—and well beyond. Some of the elk have wandered farther west than managers would like, into farming country off the Cumberland Plateau, and upwards of a hundred migrated into the mountains of southwestern Virginia. This did not sit well with Virginia's Department of Game and Inland Fisheries, which said that while it was considering an elk reintroduction project of its own, it didn't want to have one thrust upon it by a neighboring state, thank you very much. Virginia then essentially declared an open season on any elk that a deer hunter felt like shooting, although this barely seemed to slow the movement of elk into the state. Like pancake batter on a griddle, they seem destined to flow well beyond Kentucky's borders.

Even though North Carolina had twice tried, and twice failed, to reestablish elk in the early twentieth century, wildlife biologists felt that Great Smoky Mountains National Park had high potential for elk restoration, now that so much more has been learned about how best to do it. Starting in 2001, almost sixty elk were released in Cataloochee, a quiet and little-visited valley in the southeast of the park. While Cades Cove had been jammed with bumper-to-bumper traffic, I passed only one other vehicle on the sixteen-mile dirt road, sloppy with overnight rain, from Big Creek to Cataloochee, and when I got to the campground there, it was almost deserted.

I found the elk easily enough; indeed, it would have been impossible to miss them. There were two groups, a band of ten bulls grazing beside a historic barn and the ranger station, and about fifteen cows a mile or so down the road, in one of the long grassy meadows that are all that remain of the farm fields that dominated this valley a hundred years ago. I got out of the car fifty yards from the bulls, which ignored me as well as the handful of cars that passed, slowed, stopped for photos, and then left over the course of the next hour.

Most of the bulls were mature animals, their coats a bit ragged now that the spring molt was starting, their new antlers already branching—heavy blackish growths that looked smooth and taut, like sausages. I knew they'd only just dropped their huge, old racks a few weeks before; the growth rate for deer antlers of any sort is astonishing, and a bull elk will pour as much physical energy into growing a set of thirty-pound, twelve-tined antlers as a cow will into producing a thirty-pound calf. Among the old bulls were two younger males, one probably just getting his first rack, the other a raghorn still sporting his hard fall antlers, three tines on one side and four on the other, dark and curved, though lacking the heft they'd achieve in another year or two. The raghorn's coat was rustier than the others', too, making his pale rump stand out with even greater contrast.

I set up my tripod and took some pictures of my own, with the elk feeding in front of the weathered, century-old barn, but to be honest, I didn't find them very engrossing subjects. All the elk I saw, cows and bulls alike, wore canvas collars fitted with radio transmitters, which dangled loosely. They also wore bright plastic tags on both ears, all of which gave them the look of livestock, which their placid manner did nothing to contradict. Having often chased wild elk through the mountains of northern Pennsylvania, working hard half the day just for a glimpse, I thought this left something to be desired.

I understand exactly why the park does all this manipulation, the tagging and collaring and such; it is the only way to monitor a nascent population, to see how they're faring, to track their movements to determine how far they stray from the park, to discover how they're interacting with other wildlife, like the black bears that have shown an unexpected talent for snagging young elk calves. It's a necessary, unavoidable step, but also, I find, a fairly dispiriting one. I want elk back in the southern Appalachians, filling, even if with a western proxy, that tragic hole in the character of the mountains. But I want *wild* elk, not these manhandled intermediates.

And I will get them, if I'm patient, and if fate smiles on the Smokies. I knew that even as I watched the elk grazing in the fine drizzle of evening, there were cows off in secluded hollows, preparing to give birth to big chestnut calves with gangly legs and creamy spots, which will hide among the trilliums and waterleaf blossoms, their chins tucked to the ground, melting into the play of sunlight and shadow until they are old enough to run with their mothers. The biologists can't catch them all, hard as they try. So each year, there will be more and more elk like them in the hills, elk with no tags, no collars, no computer files or veterinary certificates, elk that have from birth known only the cove forests and the swift streams, the song of scarlet tanagers, the yap of gray foxes, the echo of thunder rolling off the high peaks.

I hope that in fifteen or twenty years the descendants of the elk I was watching will have taken back the Smokies and spilled beyond them, down into northern Georgia, merging in the other direction with the Cumberland Plateau animals in Kentucky and Tennessee, seeping inexorably across western Virginia, maybe someday even bridging the mountains of West Virginia to meet Pennsylvania's well-established population—and the result will be a pan-Appalachian elk herd, for the first time in nearly two centuries.

Or, if the elk don't stay put, if there are too many complaints from landowners outside the park, if there are conflicts between aggressive bulls and visitors ignorant of how to act around big, dangerous wildlife, then the National Park Service may decide to pull the plug on this experiment, as it did with red wolf restoration. I almost wrote that the elk's actions will determine this—whether they will remain well-behaved wards of the federal government, obeying a line visible only on maps, or whether they will follow the siren song beyond the boundary, into those hill-country vegetable patches and orchards, the big grassy yards and dairy pastures where they may not be welcome. But that's not actually the point, is it? It's not, in the end, the elk's actions that will make or break this experiment; it's the choices we make, how far we've come, whether we've learned to make room for something big and occasionally inconvenient. That will tell us whether, as with the wolves, even eight hundred square miles is too small a box for an animal that used to own this world—or at least whether our hearts are still too small to embrace an America as complete as we can make it.

The young bull, the one that still had his antlers, was skittish; I suspect the old guys in the herd beat up on him, and he hung back along the edge of the woods. He tried to join the group, but the biggest male snorted and stamped, and the youngster spun and ran into the woods, climbing the steep slope among the pale gray trunks of the tuliptrees, screened by the gauze of fresh leaves. All I could see were bits of him as he moved—patches of his coat, the flash of his straw-colored rump, the curve of an antler, a haunch bunched and muscular. And suddenly the domesticity of these animals vanished for me, at least for a moment. In those glimpses of color and movement, I saw into the past, and—I said a quiet prayer—maybe the future. The phantom deer arise, and with them, wild America.

FOUR

Replumbing Florida

There may be no more abrupt division between wilderness and civilization in North America than the one you see from the seat of a jet climbing out of Miami International Airport.

It's something I've seen countless times but can never really get used to. Miami, and the Gold Coast to the north, are so utterly urban that although I've seen photos from the 1920s of tangled old-growth mangrove forests near Cocoplum Beach, it's almost impossible for me to imagine this as ever being wild—this place that has been so relentlessly harnessed, a once-horizontal seascape rendered vertical by phalanxes of high-rise condos and hotels spilling inland along the freeways, the housing developments, the malls, and the warehouses.

The plane angles up, banking to the south, and there lies the line—the Miami-Dade metro area to the east, the Everglades to the west, the boundary between wetland and dry marked by the linear blue of long, straight-edged canals and levees: the L-30, the L-31. It is an odd paradox that from a few thousand feet up, it is the developments that have an organic look to them, the seemingly random curves of the cul-de-sacs and ponds and golf courses, while in contrast the Everglades has a stark, flat geometry, broken only by the crisscrossing lines of old airboat trails. But as the plane rises, the westering sun begins to catch on the water, revealing millions of glimmering facets between the sawgrass and the hammocks that breathe visual life into what had, moments before, been featureless.

South of Miami now, and still the line holds, though now the McMansions have fallen away, replaced by rectangles of farm fields in endless shades of brown and green—a mosaic into which pieces flecked with white and terra-cotta have been set, cheaper tract housing that doesn't merit the flowing landscape designs of the trendier developments closer to the city. There are white borrow pits, where machines gouge up the limestone and crush it for cement. And always that line, that knife-cut north-south division between modern Florida and its wild heart.

Higher still; the Glades is green-brown from fifteen thousand feet, with a fuzzy nap like suede, over which white clouds drag soft bluish shadows that lag behind like wayward children. The Everglades runs to the west as far as the hazy air carries my eye, while now below, the sea of grass dissolves into globular coves and inkblot islands in the shallows of Florida Bay. Here the water that spilled out of the rim of Lake Okeechobee, and once flowed in an unbroken sheet through the sawgrass and cypress eternity of the Everglades, reaches the sea. Sunshine gleams off the ocean to the south, the keys stitched together by the twin hair-thin strands of Highway 1. Looking back, I see pillars of cumulus clouds guarding the land like battlements.

Of course, it is illusory to think that the silver lines of the South Florida Water Management District's canals mark a real division between altered and natural; perhaps it is an act of self-deception to see any dichotomy at all. The truth is that the Everglades, as immense as it may appear, has been chained and gutted over the past century and is at best a ghost of its former self—a fundamentally artificial ecosystem whose rhythms and pulse are largely gone. The forests of the Smokies were logged and stripped and burned, but what grew back still functions as Appalachian forest. The Everglades today is not the Everglades in much besides name.

That may sound like a harsh assessment, especially to anyone who has gotten off the road and into the heart of the Glades. It's hard to think that way if you're drifting your canoe with the breeze on a small lake deep in the mangroves, your paddle across your knees, eyes closed, leaning back over the stern, hearing only the slap of waves on the hull and smelling the rich oxygen-charged air with its ripe bouquet of old mud. Or when the sun is going down over the sawgrass, with small flocks of ibises and egrets flapping majestically overhead, their feathers stained with the orange light. It's so much the very picture of the Everglades—the embodiment of Marjory Stoneman Douglas's "River of Grass"—that it seems disloyal to believe anything could be wrong with it, much less to think that this is an

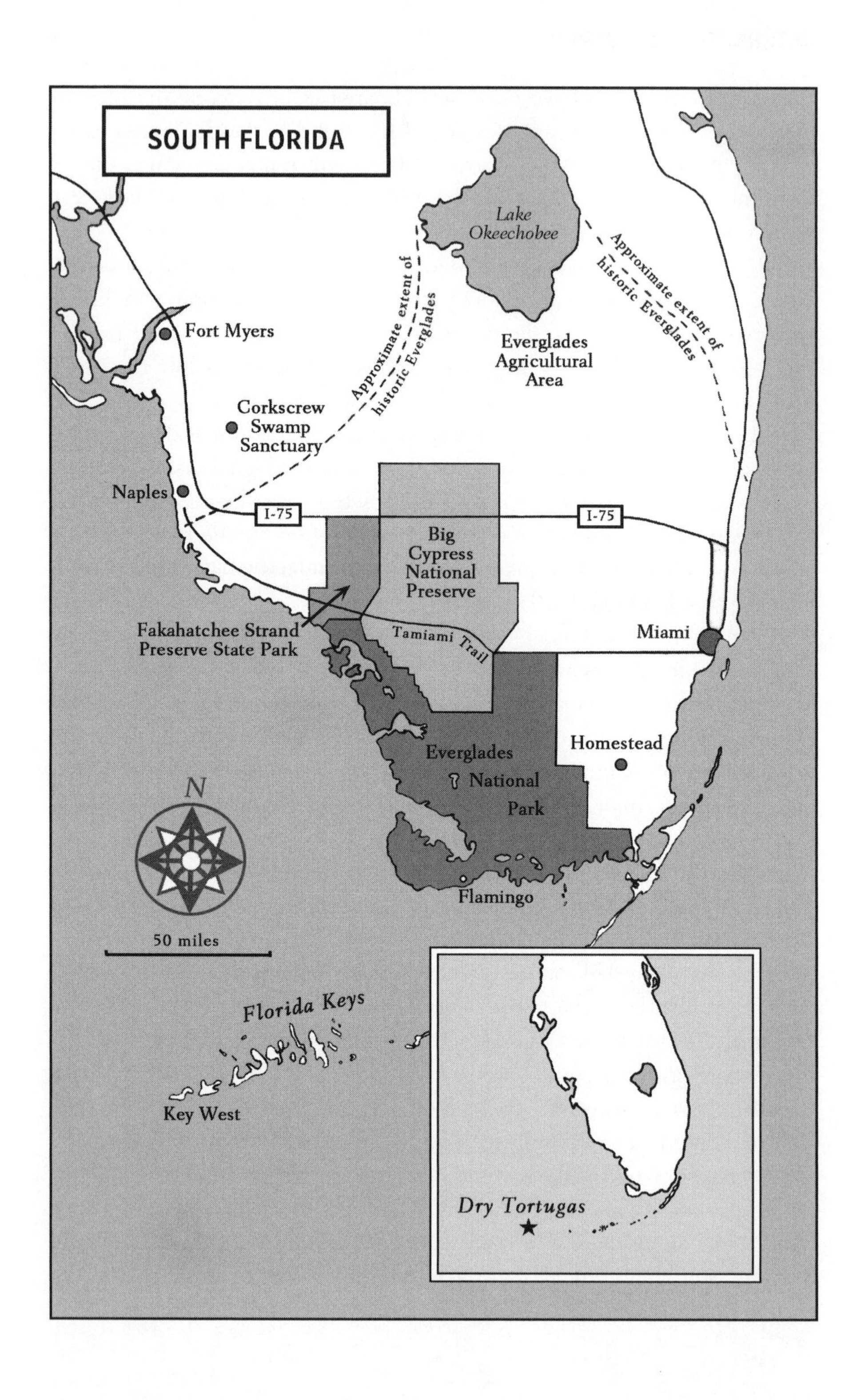
SOUTH FLORIDA
Lake Okeechobee
Approximate extent of historic Everglades
Approximate extent of historic Everglades
Fort Myers
Everglades Agricultural Area
Corkscrew Swamp Sanctuary
Naples
I-75
I-75
Big Cypress National Preserve
Fakahatchee Strand Preserve State Park
Tamiami Trail
Miami
Homestead
Everglades National Park
N
Flamingo
50 miles
Florida Keys
Key West
Dry Tortugas

ecosystem whose major organs are on life support. The eye and the heart rebel at the very notion.

But that's because the Everglades was and is a place of extraordinary subtlety, where a few inches' difference in elevation means the difference between sawgrass and hardwood jungle, between cypress dome and pine key, where the water flows almost imperceptibly, and where the smallest change in the hydrology ripples through every level of the food chain. The insult visited on the Everglades in the past fifty years was anything but subtle, but the effects of undoing southern Florida's natural plumbing sometimes take a trained eye to notice and interpret. And sometimes not. The sky-filling clouds of wading birds are a memory, the wild lions are all but gone, and the once-rich marine nurseries along Florida Bay are giving out; the alligators are back, but exotic fish crowd the waters, and the native species that remain are so full of mercury you shouldn't eat them.

The irony is that even in the remotest corner of the national park that bears the name Everglades (but that holds only a fifth of the historic Everglades), in places where you would stand, waist-deep in tepid water, and without hesitation pronounce it wilderness, the fate of this miraculous ecosystem rests in cane fields far to the north and on the computer screens of bureaucrats in water control offices in Homestead or Clewiston. Open, empty, yes; the kind of place where you can get lost for days if you're not careful. Beautiful, beyond doubt. But even where humans rarely come, this is one of the most manipulated landscapes in America. And for all that, still one of the most compelling, which is why we keep trying to save it.

Peterson and Fisher spent eleven days in Florida, longer than in any other state on their itinerary except Alaska, zigzagging back and forth from the Panhandle to the Everglades to the keys and up to the Panhandle again. And no wonder; Florida is a microcosm of many unusual habitats, full of rare species. Any attempt to see wild America had to include a lot of time in this biologically diverse state, and so it would be for me.

Although they spent most of the time in the boondocks, the guys also hit a few of the tourist attractions like Marineland and Cypress Gardens. "Many tourists see only these show places," Peterson said. "But if they would only abandon the crowded highways and drive down the long lonely roads which cross the Kissimmee Prairie, the Big Cypress, or the ''Glades'—as James and I did—they would really see Florida. Between these slender threads of roadway are great blocks of wilderness, hundreds, even thousands of square miles in extent, wild areas seldom entered except for the occasional cowboy, trapper or Seminole."

Ironically, the "crowded highways" to which he referred are today's quieter byways, like the Tamiami Trail; the real traffic today is on the big interstates like I-75. But I was going to take his advice and stick to the back roads and waterways, trying to find the real Florida. I would, however, be taking a slightly different route from my predecessors, hitting most of the same spots they visited as well as a couple of places that, for one reason or another, they didn't see. I would visit some of the wildest remaining land in both northern and southern Florida (including one place, ironically, that's much wilder today than it was in the 1950s) before heading out into the lapis ocean to the farthest outpost of the keys. But first I wanted to dip a canoe paddle into the Everglades and consider the replumbing of Florida.

Everglades National Park was only five years old in 1953, when Peterson and Fisher came, but Floridians had banded together as early as the 1920s to try to protect at least part of the vast wetlands that covered more than eleven thousand square miles. It was, simply, unique; nowhere else on earth did water move in the famous "sheetflow" of the Everglades, down from the Kissimmee River during the summer rains, overflowing Lake Okeechobee, spilling out across land that sloped gently south at about six inches a mile, forming a shallow river fifty miles wide, eighty miles from north to south, and rarely much deeper than a man's head.

Legislation authorizing a park was passed in 1934, but there was little money and less progress over the next decade, until the Florida Legislature appropriated $2 million to buy up private landholdings, which, along with an existing state park, became Everglades National Park in 1947. This marked a significant departure for the national park system, because for the first time a park had been created, not because of its dramatic scenery, but to preserve a unique and wildlife-laden ecosystem. In the late 1930s, Daniel B. Beard, at the time an assistant wildlife technician for the National Park Service and later the first superintendent of the Everglades, was sent to survey the region's potential for a park. His report urged quick action, but noted the Glades's quieter, more intimate appeal:

> The southern Florida wilderness scenery is a study in halftones, not bright, broad strokes of a full brush as is the case of most of our other national parks. There are no knife-edged mountains protruding up into the sky. There are no valleys of any kind. No glaciers exist, no gaudy canyons, no geysers . . . none of the things we are used to seeing in our parks. Instead, there are lonely distances, intricate and monotonous waterways, birds, sky and water. To put it crudely, there is nothing (and we include the bird rookeries) in the Everglades that will make Mr. Jonnie Q. Public suck in his breath. This is not an indictment against the Everglades as a national park, because "breath sucking" is still not the thing we are striving for in preserving wilderness areas.

Of course, today the Everglades gets more than a million visitors a year, and while few people ever get beyond the visitors centers or the tram tour along Shark River Slough, that's enough to show them that this is a place unlike anything else in the world, breath sucking or no. For most park visitors, though, Anhinga Trail, just inside the eastern boundary close to Homestead, *is* the Everglades. Fisher was impressed with the alligators, egrets, and purple gallinules that ignored the stream of people on the rickety boardwalk, a trait he noticed throughout his American travels. "It is true that in many other parts of the world protected wildlife has become ridiculously tame," he wrote. "But only in North America, so far, has it been possible to combine the national park (where man is encouraged) with the nature reserve (from which, at least in the old days, man was supposed to be kept away)."

Fifty years on, things haven't changed much at Anhinga Trail, except that the boardwalk is much sturdier and the crowds are bigger. It was a clear, upper-seventies day, and the parking lot was full as I ambled out the half-mile loop with Amy, who was able to join me for a couple of days of backcountry camping. The trail starts in Royal Palm Hammock, a place of special significance in Everglades history. In southern Florida parlance, a hammock is an island in the marsh, slightly higher ground covered with hardwood forest; this one was famous for its biological richness, which prompted the Florida Federation of Women's Clubs to agitate for its protection as a state park in 1916, thus forming a four-thousand-acre core of the later national park. Royal Palm Hammock was resplendent, with a stand of majestic royal palms more than a hundred feet tall, and full of Caribbean plants unique to southern Florida; as late as 1917, it was claimed that ivory-billed woodpeckers still nested here.

The Ingraham Highway, an early road linking Homestead with Flamingo down on Florida Bay, opened through the hammock in the 1920s. One of the borrow pits for it, chiseled out of the limestone bedrock by laborers, is today a large pond at the Anhinga Trail entrance, and it was surrounded by clots of people, jostling to see the alligators hauled up in the sun or floating quietly in the tannic water. The pond was full of fish—gars, black bass, African blue tilapia that escaped from fish farms years ago and now inhabit most of the Everglades. Soft-shelled turtles the size of garbage-can lids glided under the surface, their outlines rippling like the reflections in fun-house mirrors. One rose to the surface, its eyes protruding, its long, funnel-shaped nose poking up, its shell like a big, leathery flapjack. I've always been fond of turtles, but there is something—I dunno—so *reptilian* about a big softshell, something not entirely friendly, or at least not trustworthy, in that cold gaze.

A red-shouldered hawk was perched in a tree no more than ten feet off the trail, but hardly anyone noticed it sitting quietly; most folks were watching the double-crested cormorants fishing in the channel by the path, fifteen or twenty feet from the walkers. And I have to admit it was quite a show. The birds arched and plunged like seals, and you could see them going down, down into the clear brownish water, little streams of bubbles rising from their plumage, then coming back up as the water sheeted off their backs. A little farther on, a wood stork was fishing in a small stream that flowed out of the thicket, moving with deliberate, cautious steps, bent over with its big, horny beak held open, just the tips beneath the water—a tactile trap, sensitive to the touch of a fish and able to clamp shut with blinding speed. Wood storks are white, except for the black flight feathers of the wings, which in this light shimmered with purple and green iridescence, and the naked, almost scabrous head. The stork took another step, stirring its pink foot in the weedy muck, then another. Then it bolted, throwing a wing wide for balance, and snapped up a plump reddish fish nine or ten inches long—an oscar, another aquarium import that's gone native in the Everglades.

Overheard on the way back from the boardwalk, a little girl of perhaps six says to her father: "I wanna see another alligator!" His reply, a middle-of-the-vacation growl: "Why? You've seen four already." And from a woman leaning over the railing with a friend to peer at a six- or seven-foot alligator with one half that size sprawled on its back (and speaking, so far as I could tell, in dead earnest): "They must have them sedated!" Perhaps, I thought, it's time to go.

The road to Flamingo is classic Everglades—the great dome of the sky, the thin line of pines on the far horizon or the bulge of a cypress stand, and, around it all, the immense sawgrass flats. It looks so primal, so endless, so untouched and untouchable, but it's hollowed out, a rind of pretty scenery. Even so, I suspect most visitors to the Everglades would find that idea bewildering, given the bird-and-gator show at Anhinga Trail or the dozens of wading birds we saw crowding Mrazek Pond or Eco Pond (actually the sewage-treatment discharge) that evening at Flamingo. As we cooked our dinner in the overflow camping area, a large grassy field a few hundred yards from the water, small flocks of roseate spoonbills flew over, flaming pink in the sunset, and folks in neighboring campsites stopped what they were doing and pointed.

In truth, though, the Everglades—in the broadest sense, encompassing not just the park but all of southern Florida from the Kissimmee and Lake Okeechobee to the ocean, the Big Cypress Swamp as well as the open sawgrass marshes and mangrove forests, the estuaries and keys of Florida Bay—has been eroding its biological wealth for decades. About half of it has been destroyed altogether, drained for farmland or housing. Its arterial flow of water has been so mishandled and misdirected, for so long, to benefit agriculture and development that the legendary concentrations of wildlife, especially wading birds that once stunned observers into awed silence, are almost entirely gone. Now, after years of wrangling, a controversial and breathtakingly expensive plan to restore the Everglades may finally be ready for implementation. Whether we know enough to mend the damage, and whether the feds and the state (its primary architects) have the resources and, more important, the will to actually carry through on their promise to remake the Everglades—well, that remains an open question, as we'll see.

The next morning Amy and I slid our canoe, loaded down with water jugs, food, and gear for a couple of days, into the water at the entrance to the Hells Bay canoe trail and pushed off beneath a canopy of red mangroves. The narrow creek twisted tortuously, the water shallow and dark, with a perceptible current pushing south against us. The mangroves reached above our heads, their gnarled branches intertwined, the air still and punctuated with the whine of a few mosquitoes. Time and time again, we pivoted the canoe around hairpin corners, our paddles digging deep into the peaty mud, past banks where the vegetation had been smashed flat by basking alligators, although we saw only one smallish gator hiding back in the deep weeds.

Eventually the creek opened into wider ponds, where we could stretch a little and feel the breeze. The water below us was choked with bladderwort, its purple flowers rising above fine, tangled stems on which were multitudes of tiny globes—in reality clever traps, for this carnivorous plant seines minute aquatic organisms from the water. White ibises rose from the sawgrass or thickets of mangroves, complaining in nasal tones as they swished away, their wings making the sound of someone shaking out a tablecloth. Ospreys and turkey vultures were always in sight somewhere in the sky, and once, a short-tailed hawk—an ebony-colored southern Florida specialty, and a life bird for Amy—wheeled above us.

In Florida's pioneer days, everyone who visited the Everglades came away with two impressions—that it was a vast, miasmic, impenetrable hellhole through which to travel, and that it was a bountiful land spilling over with wildlife, especially colossal flocks of wading birds, an estimated 1.5 million egrets, herons, ibises, spoonbills, and storks that jammed into vast rookeries. The birds suffered first, and most directly, from the market gunners who slaughtered them for the feather trade until the 1920s; even in 1917, the National Audubon Society's future president, T. Gilbert Pearson, said it was "debatable" whether some species, like the great egret, could be saved at all.

They were, but conservationists lamented the emptiness of the Everglades's skies and commented on the few occasions when they caught a glimpse of the older, better days. White ibises, which were hunted for food but not feathers, and thus fared somewhat better in the plume-trade days, nested in colonies of up to a quarter million. The ornithologist Alexander Sprunt, Jr., who saw one such rookery in 1940, said, "It was as if humanity had ceased to exist, and that we were intruders in a world which was peopled entirely by birds."

Although the Everglades wasn't what it had been in the nineteenth century, Peterson was dazzled by it when he first visited there in 1937, mesmerized by the huge numbers of nesting waders and enduring miserable conditions to photograph them. "I had spent a day slapping mosquitoes in my burlap blind in the midst of a great nesting of white ibises, estimated to exceed 100,000. Long snaky ribbons of these white birds with red faces, and long curved red bills, twisted out of the sky, broke ranks, and fluttered like snowflakes into the mangroves." When he came back with Fisher in tow, the numbers of birds just weren't there; Peterson wrote apologetically that there were only "hundreds, perhaps thousands" of ibises, herons, and wood storks in the colony. Fisher, though, wasn't disappointed; he said the island looked "like an overloaded Christmas tree, as if someone had tried to make snow out of candle wax. All over the mangroves was a frosting of white birds."

In fact, each generation finds the Everglades enchanting, even though each generation sees less and less of the raw abundance that originally made the place so remarkable. The reason lies principally in the history of southern Florida, which can be summed up in two words: "Drain it." In the early nineteenth century, speculators dreamed about pulling the plug on all that swampland and creating an agricultural paradise, but it wasn't until the twentieth century that Americans got down to business. They tinkered with the edges, turning mangrove forests into the city of Miami and draining the margins of Lake Okeechobee for farmland and small communities. When storms flooded the lake in the 1920s, killing thousands, they decided to do it up right and repipe the whole of southern Florida. Deadly hurricanes in the 1940s added urgency.

By the 1950s, the Everglades had suffered almost half a century of dredging and draining, but it wasn't until 1948, a year after the park was established, that the most extravagant plumbing began. That year Congress authorized the Central and Southern Florida Project, a massive scheme undertaken by the U.S. Army Corps of Engineers that was designed to take what had been wetland and convert it to crop fields, and to shunt the original sheetflow of water out of Lake Okeechobee and reroute it for irrigation, municipal use, and flood control through more than a thousand miles of canals, levees, and pump stations. It turned the seasonal ebb and rise of water on its head—too much in some places, not enough in others. The system collapsed. The Everglades, in all its incarnations—not just the famous sawgrass marshes, but the cypress forests, hardwood hammocks, marl prairies, mangrove islands, estuaries, and pond apple swamps—was under attack. Half the wetland was drained, much of it for agriculture, while urban development moved in from the eastern coast. Most of the northern Glades became sugarcane, tomato fields, or permanent sumps, instead of the seasonal marshes that once existed there. Even within the park, which covers only 20 percent of the historic Everglades, the natural system fell derelict.

In the Big Cypress, the twenty-four-hundred-square-mile wetland on the western border of the park, land boondoggles in the 1960s preyed on out-of-staters who wanted a piece of the Sunshine State and led to the draining of huge areas; at the same time, local boosters were pushing for the construction of a thirty-nine-square-mile airport less than six miles from the park's northern boundary. They didn't succeed, thank heavens, and the battle to scrap the jetport was a major victory for the still-young environmental movement. The Big Cypress is now a national preserve, but the airport idea didn't die; rather, the focus has shifted to the former Homestead Air Force Base a few miles from the eastern border, where some folks want a large new commercial jetport with six hundred flights a day. So far they've been stopped, but that's unlikely to be the last word. People look at all that space and just itch to fill it up with, well, *something*. Anything.

All the fiddling with southern Florida's hydrology has had devastating effects on what's left of the Everglades. No longer do the marshes swell predictably during the summer rains, allowing fish populations to explode, then shrink and dry in the winter, concentrating the schools into shallow ponds where birds, which have hungry mouths in the nest, can catch them. There is too much or too little water, at the wrong times and in the wrong places; in rainy years the Everglades is deluged, much of the water contaminated with agricultural runoff, and in dry years what flow there is winds up diverted for humans. Alligators nest, as they always do, in midsummer, laying their eggs in mounds of rotting vegetation just above the waterline, but the canals open, the water rises, and the eggs drown. Wood storks, which breed in the winter, when the fish are easy to catch, find floods instead, while the snail kites, which depend on deeper water, where their food, the apple snail, can breed, find unexpected drought. The kites and storks are both endangered, while wading-bird populations in the Everglades have declined a staggering 93 percent. Even the water isn't the same; loaded with nutrients running off the cane fields to the north, it favors invasive cattails instead of the native sawgrass that defines the central Everglades. Nor was all the damage done in the unenlightened 1950s or '60s. When the C-111 canal system west and south of Homestead was completed in the 1980s, it shanghaied so much water from the park's Taylor Slough that Florida Bay's salinity spiked, fish died, algae bloomed, and nesting waders starved.

Amy rested her paddle across the thwarts of the canoe, laced her fingers together, and stretched her arms behind her, twisting from side to side to ease a crick in her back; her long braid was stuck to the sweat between her shoulder blades. We'd been paddling for several hours, moving from the dense mangroves to open ponds and now into wider lakes, where the breeze was picking up into a tougher headwind. The sky was full of popcorn clouds. We poked among some of the mangrove islands, landing at Lard Can Camp for a bite to eat, the ground squishy underfoot, the vegetation crowded and heavy. Anhingas sat on dead snags, their serpentine necks crooked into question marks.

Another hour passed as we pushed into the wind, moving across a wide lake as the waves cracked over the bow in little explosions of spray. The canoe trail was marked with tall, numbered PVC pipes that leaned this way or that, showing the way through the endless shuffle of low, humped islands of mangroves. About the middle of the afternoon, we came to our camp for the night—a "chickee," a wooden shelter over the water named for the old thatched structures the Seminole and Miccosukee made, though this one had a solid roof and recycled plastic decking. It was, in fact, two shelters, each about ten feet square, joined by a V-shaped walkway, with a chemical toilet at the apex, though the other site was empty. As we unloaded the canoe and set up our tent, a six-foot alligator drifted silently in to wait beneath us, motionless and hopeful for scraps.

As the sun dropped, we cooked dinner, mindful that after dark the mosquitoes might be fierce, though for now there was enough of a breeze to make the tent billow and to stir up little wavelets along the alligator's corrugated back. A lone kayaker paddled into the cove, glided up to the other side of the chickee, and hauled himself out, ignoring our hellos and erecting a tiny tent with practiced ease. He was a lean, unshaven guy given to muttering to himself, short on the social graces but long on strange noises and odd outbursts; we fell to whispering as we cleaned up the meal, watching small flocks of egrets, spoonbills, and cormorants silhouetted against a fiery sunset. The breeze guttered and died; the mosquitoes made an appearance, and we zipped up in the tent to read, the alligator still holding its vigil.

I woke up about midnight, looking out the tent window as the stars broke through large sheets of clouds; dozed off; and woke fully about 4:00 a.m., the tent shaking gently in a strong breeze. I carefully unzipped the tent so as not to wake Amy and sat on the edge of the chickee. There was a bright yellow false dawn to the northeast, the lights of Miami smearing the horizon, but elsewhere the sky was deep black and starry, with Orion low in the west, Sirius just behind him, and the Big Dipper upside down. I heard frogs and insects, and other noises I couldn't identify, and then I heard what I took to be the rumble of thunder, though the sky was cloudless. It came again, clearer, more like a roar, and I realized it was the bellow of a male alligator, somewhere deep in the night, and I shivered with both the chilly air and an ancient fear.

If you want a conservation success story, look at the alligator. In life, the skin is hard, plated, armored—except for the soft white belly, which makes superb leather. The slaughter of alligators has been compared to that of the bison, some 10 million, at a guess, killed between 1800 and 1940. No one spared them much worry, though, until the 1950s and '60s, when it was obvious to pretty much anyone that the alligator was in trouble in most of its range. The gator was on the first two federal endangered-species lists in the 1960s, but those laws lacked teeth, and the hunting, legal and illegal, continued; the 1973 Endangered Species Act, on the other hand, carried stiff penalties, and it finally curbed the trafficking of crocodilian skins, including those of gators, as did the 1975 Convention on International Trade in Endangered Species.

That's all it took. Today the alligator population, which stretches from Texas to North Carolina, is estimated at more than a million, and by 1987 the U.S. Fish and Wildlife Service had pronounced it completely recovered. Interestingly, alligator populations within Everglades National Park are not as healthy as those outside the park; experts blame unnatural water fluctuations and a lack of prey in the form of wading birds. But in most of Florida, alligators are more likely to be the cause of nuisance complaints, in housing development ponds or golf course water hazards, than of conservation concern. But gators may have traded one threat for another, more insidious one, which may affect aquatic life across the country—waterborne toxins that mimic the body's hormones. Some years ago, scientists noticed that alligators in Lake Apopka, Florida's third-largest body of water, were declining drastically; researchers found that breakdown products of banned pesticides like DDT and newer chemicals widely used in agriculture were short-circuiting the gators' endocrine systems and causing reproductive failure.

While Lake Apopka remains an especially dramatic case, such chemicals, known as endocrine disruptors, have proven to be widespread in waterways across the country, and their effects on both wildlife and humans are poorly known but frightening. In Apopka, for instance, male gators were suffering from small penises, but sex-based deformities are proving to be common among aquatic animals. Just recently, researchers have found disturbing reproductive abnormalities in many fish, including male smallmouth bass and suckers in the Potomac River growing eggs in their bodies, that have been linked to human birth-control hormones flushed unchecked through sewage-treatment plants.

The alligator bellowed again. He, at least, was in no doubt about his masculinity.

The kayaker was gone before daybreak, and Amy and I had the rest of the day to ourselves. Paddling back, we ran into a couple of canoes full of birders, chatty and friendly ("Why couldn't *they* have been in our chickee last night?" Amy asked under her breath). We spent another night camping at Flamingo, then drove back east, out of the park, across the startling line of the C-111 canal, which separates the open, wet Everglades from the vegetable fields near Florida City, where irrigation trucks fired long geysers of white water across the fields, spilling rainbows over peppers, tomatoes, and squash with bright yellow flowers. The weedy pastures near town held a few horses and big flocks of white egrets, which Amy and I scarcely noticed—even though fifty years ago these birds were front-page news among ornithologists and figured prominently in the original *Wild America*.

They were cattle egrets, chicken-sized herons with long yellow legs, yellow dagger bills, and a taste for the open road. In the space of a century, this Old World wading bird moved out of its native haunts in Africa, the Asian tropics, and a small corner of the Iberian Peninsula to blanket the globe, and unlike most avian carpetbaggers, such as starlings or house sparrows, which hopscotched the planet with human help, the egrets did it on their own.

Named for their habit of shadowing livestock, cattle egrets spend much of their time in pastures, meadows, and other uplands, often miles from water. Although not large, they are buoyant fliers, capable of crossing immense distances. It appears they jumped the Atlantic from Africa to northeastern South America in the 1870s, aided by prevailing winds, then, as their numbers grew explosively in the early twentieth century, rapidly moved across the Caribbean in a couple of big, distance-eating leaps, rather than flying from island to island. By the 1940s they were in the Greater Antilles, and wayward cattle egrets started showing up in North America proper around the same time—one in Florida in 1941, more in 1948, another in New Jersey in 1951. By 1952 a mini-invasion was under way, with egrets reported as far north as Massachusetts and Chicago. One even wound up on a freighter off Newfoundland.

Peterson, who in 1952 had photographed the birds in Spain, followed this drama keenly and predicted that an American nesting was simply a matter of time. It was a stroke of good fortune, though, that news of the long-expected event came as he and Fisher were traveling through Florida. Arriving near Lake Okeechobee on the night of May 5, they received word that a nest had been found the previous day at King's Bar, not far away. The next morning, armed with head nets against the ferocious wet-season mosquitoes, they followed Audubon warden Glenn Chandler and local ornithologist Sam Grimes into the midst of a huge heronry. Although they found that the historic nest had apparently collapsed overnight, they did see plenty of cattle egrets. The following summer, ornithologists estimated there were at least two thousand of the small herons in Florida, with more apparently moving in from Latin America. "The cattle egret, beautiful and beneficial, is a fine addition to American avifauna. The day might well come when it will be as familiar to many Americans as the starling or ring-necked pheasant," Peterson predicted.

As optimistic as he sounded, Roger Peterson in fact lowballed the cattle egret's prospects in *Wild America,* suggesting it would do well in Florida, in the coastal prairies of Texas and up the Rio Grande valley, and in the Central Valley of California, if it ever reached that far. "But somehow I would not expect it to breed in the northeastern states, nor in the upper Midwest, even though there are plenty of cows," he wrote.

The egret has proven far more adaptable than he or anyone else expected. Although its core range still centers on the Southeast, it is a common breeding bird far up the Mississippi valley and the East Coast. By 1970 it was breeding in Minnesota, and within forty years of that first historic nesting in Florida it had raised chicks in all but six states. (Nor was its only progress to the north—the egret had made it to Tierra del Fuego, at the tip of South America, by 1977.) The species is still marching north, and it may be a sign of things to come that cattle egrets have been spotted in southern Alaska, northern British Columbia, and the northern tip of Newfoundland.

Interestingly, a remarkably similar situation is playing out today across the Southeast—another alien bird that came to Florida at least partially on its own and is spreading with stunning speed. On the telephone wires around Florida City and Homestead we saw big, plump doves, beefier than mourning doves, with squared-off tails edged with blocks of white. In every direction, the air was full of hollow trisyllabic hoots.

The noisy birds were Eurasian collared-doves, or ECDs in birding shorthand. The collared-dove looks a lot like a grayish version of the ringed turtle-dove, the kind a magician pulls from his hat, right down to the black half circle around the back of the neck. Turtle-doves are a mélange of several species, with a domestic pedigree that goes back to the ancient Egyptians. The ECD, on the other hand, is a completely wild species, and one that combines the territorial aspirations of Genghis Khan with an adaptability shown by few animals besides humans and Norway rats.

Originally native to India and surrounding countries, the ECD began moving west early—into the Middle East by the sixteenth century, and into Europe (in the Balkans) around 1900. From there, it moved rapidly across the continent, reaching Norway in 1954. It was seen in England in 1952, and—as Fisher, who studied the species closely, later documented—by 1964 it was nesting in every English county, as well as in Scotland, Wales, and Ireland. A few years later, it had even set up shop in Iceland—not bad for a bird that evolved in the arid heat of Pakistan.

Domestic turtle-doves have escaped many times in North America, and sometimes form small, temporary colonies in urban areas, but in the

1980s a few people realized that the "turtle-doves" they were seeing in Florida were really ECDs and that the number was growing rapidly. Investigation later revealed the species was accidentally imported to the Bahamas in the 1970s by a pet dealer who thought he was getting nice, tractable turtle-doves instead of the scrappy collared-doves, and when he discovered his mistake, he let them go. As far as anyone can tell, the birds made it to southern Florida on their own—and they've been going like gangbusters ever since. Like an overflowing vase, Florida swelled with ECDs, which are pouring north and west from the state. They are a common sight all along the coast west to Texas and north into the Carolinas, as well as inland across Georgia, Alabama, Mississippi, and Louisiana. But like cattle egrets, they disperse far from their core range, and ECDs have been reported in Minnesota and Montana—the vanguard, few doubt, of a continent-wide occupation. A few years ago, I found one in my backyard in Pennsylvania, but it was gone the next day—whether into the crop of a Cooper's hawk or simply over the ridge to greener pastures, I couldn't say. But the smart money's on the dove.

While we saw hundreds of ECDs in the towns, around the scattered housing tracts, and among the farm fields outside the park, inside the Everglades proper we saw only the smaller, grayer mourning doves. So far, collared-doves seem unable to do well in places where there isn't a strong human presence—but then, until recently, such places were also good habitat for mourning doves, ground-doves, and other native species. No one knows what impact the ECD incursion will have on the continent's avifauna, including the white-winged dove, a southwestern species that is moving north and *east* from its old range, expanding around the Gulf even as the collared-dove moves into the neighborhood from the opposite direction, a pincer movement of columbiformes.

Few of the alien species that make it to Florida do so on their own, however. Fire ants, the bane of anyone who has stepped on one of their conical mounds, came to the Gulf coast on ships from South America and stormed out from there, fundamentally altering the ground-level ecology of the South. The state's urban areas have long been hotbeds of escaped pets; the only reason Fisher and Peterson stopped in Miami was to hunt up the spot-breasted oriole, a Central American species that had become established there after caged birds had escaped. It has lots of company these days. The last time I drove out of Miami's airport, a small flock of greenish parakeets buzzed my car and disappeared into a roadside palm tree. Traffic was heavy and I couldn't identify them, but they could have been any one of eight or nine kinds, which along with a dozen or more bigger species of parrots, macaws, and lovebirds now live feral in Florida—not to mention such other exotic birds as mynas, bulbuls, mannikins, and Java sparrows.

Miami ranks, with Los Angeles and New York, as one of the leading ports of entry for exotic animals, and an awful lot of them get away or are released by bored owners. The climate is perfect for many of these foreigners, from capuchin monkeys to geckos. If the thought of a sweetly cooing dove on your windowsill appeals to you, how about a six-foot-long green iguana in the shrubbery, or a troop of macaque monkeys tearing up the garden? Or a twenty-two-foot reticulated python, like the one an animal control specialist was asked to remove—very insistently, one assumes—from under a porch in Fort Lauderdale, where it was sleeping off the raccoon it had just eaten?

The state is bung-full of creatures that don't belong here, from Asian walking catfish and Central American cichlids in the waterways, where they compete with native sunfish and bass, to Nile monitors, predatory lizards that can reach seven feet and feed on animals the size of house cats (the lizards are now common near Fort Myers, where, tellingly, feral cats are said to be rare). Fisheries managers are nervously tracking the progress across the state of the Asian swamp eel, a three-foot-long predator that can slither overland between waterways and has a taste for almost anything that moves. Naturalists in Everglades National Park, where alligators used to be the reptile everyone fretted about, now routinely capture so many pythons and boas that they have a special hotline for those reporting them. Many of the snakes are thought to be released pets, but Burmese pythons, which reach lengths of twenty-six feet, are believed to be reproducing in the park.

And animals are only half of it—or less than half, because from an ecosystem perspective the alien plants may be worse still. By one estimate, more than fifty *thousand* species of exotic plants have been introduced to Florida, more than any state except Hawaii, and 126 of them are considered invasive weeds. Drive anywhere in southern Florida, and you'll see the results of bad gardening decisions run amok, including Brazilian pepper with its cheerful red berries, which covers a million acres of the state, and feathery Australian "pine," which is actually a strange eucalyptus with needlelike leaves. But the granddaddy of invasive plants is another eucalyptus, known as melaleuca—"clearly one of the biggest threats to the ecological integrity of the south Florida ecosystem," in the words of one expert panel.

Brought to Florida as an ornamental in 1906, melaleuca is a botanical sponge, sucking up so much groundwater that it was soon drafted into schemes to drain the Everglades; in the 1930s, its seeds were broadcast by airplane over the marshes. Like many alien species, it found a paradise half a world away from its pests, predators, and parasites and has taken over nearly half a million acres of the state, frequently forming solid, monolithic stands that crowd out all native vegetation. The stuff is devilishly hard to kill; fire only encourages it, releasing the millions of seeds in its fruit capsules. Chop it down and it resprouts; grub it up and you'll leave dozens of rootlets in the ground, each of which, Hydra-like, grows into a new melaleuca. The only effective method is to whack the trunk with a machete and apply a systemic herbicide, tree by tree, which is obviously a slow and tedious process. Yet with great effort and expense, the state is making progress; in the decade before 1998, the acreage impacted by melaleuca dropped by about a third, with the greatest progress in the Everglades and the Big Cypress. That translates into 78 million melaleucas killed, one tree at a time, at a cost of $25 million.

Nature can render such attempts moot, though. Hurricane Andrew in 1992 blew zillions of melaleuca seeds right across the peninsula, into places deep in the Everglades where the pest hadn't appeared in the past. And it also carried other aliens, including Old World climbing fern and air potato, which can blanket an entire hardwood hammock under a suffocating mat of vegetation. Because its spores weigh barely more than air, the fern can pop up almost anywhere, and biologists have found it deep in such relatively pristine areas as Corkscrew Swamp Sanctuary and Fakahatchee Strand Preserve State Park.

Despite the magnitude of the threat, invasive exotics still receive insufficient attention, and efforts at control are often trumped by concerns over international trade. Interdiction and prevention are crucial, but ports are understaffed and underfunded. Inspectors have a hard enough time keeping up with legitimate imports—Miami's international airport has just nine animal inspectors, dealing with up to seventy shipments a day—but the really scary stuff sometimes comes in under the radar, like a bug buried in the heart of an exotic plant, ready to emerge into a virgin and unprotected ecosystem. (The losses aren't always ecological; sometimes they're architectural, like New Orleans's French Quarter, whose historic wooden houses are being eaten right to the wallpaper by voracious Formosan termites.)

Once the invasive is established, the ideal solution is to find a suite of parasites, diseases, or predators from its native land and release them here to reestablish some degree of control. This is more easily said than done; in order to prevent the cure from being worse than the disease, the proposed control species must be tested on virtually every native plant or animal that it might impact—a slow and laborious process, obviously, and one with no absolute guarantees of either success or safety. An Australian moth may offer hope against the Old World climbing fern, but no one knows if it will harm native plants, and even after years of research only a handful of fire ant enemies—protozoans, fungi, parasitic flies, wasps, and nematodes—have graduated to field testing. No one's yet looking at what might eat Burmese pythons, except alligators; not long ago, tourists from the Midwest got video of an Everglades gator killing and eating a python in the park. Score one for the home team.

We skirted Miami, driving up Highway 997 through thickets of melaleuca, then turned west on Route 41, the old Tamiami Trail, which cuts right across the central Everglades and is one of the worst blockages in the region's constipated water flow, ponding it up in the north and leaving the southern Glades short. We passed billboard promises of alligator wrestling, Indian camps, and Buffalo Tiger's Authentic Miccosukee Airboat Rides. The north side of the road was bounded by a huge canal and levee, punctuated every so often by massive water gates; when we could see north or south beyond the margins of the road, the horizon was dominated by cattails until we got close to Shark River.

The canal was L-29, which feeds the Shark River Slough through the center of the park—one cog in the enormous system by which the South Florida Water Management District controls almost every drop of water flowing through the lower peninsula. The SFWMD covers sixteen counties, has an annual budget of $792 million, and maintains, according to its own tally, "approximately 1,800 miles of canals and levees, 25 major pumping stations and about 200 larger and 2,000 smaller water control structures." Every time you cross a canal down here, you see a sign with the agency's symbol, a happy cartoon alligator, identifying that particular link in the SFWMD's vast network.

The system starts at Upper Chain of Lakes and rolls south along what used to be the Kissimmee River, a lovely, meandering, wildlife-soaked waterway with a two-mile-wide floodplain dominated by pickerelweed marsh, willow and buttonbush, hordes of wintering ducks, and one of the greatest largemouth bass fisheries in the world. I say "used to be," because starting in 1962, the 103-mile Kissimmee was channelized by the U.S. Army Corps of Engineers into a ditch fifty-six miles long, thirty feet deep, and three hundred feet wide, walled off into five stagnant pools, whose sole purpose was flood control. Thirty-five thousand acres of wetlands turned into pasture and scrub; the ducks left; the fish, deprived of oxygen, died. Ag runoff, no longer filtered out by the marshes, worsened Lake Okeechobee's already failing health. They even took away the Kissimmee's name and designated it C-38, though now it has its own happy-alligator signs. This process took nine years and cost $35 million in 1960s money—and it became apparent almost immediately that it was a whopper of a mistake. Within five years of the ditch's completion, the state was already studying ways to restore the Kissimmee, an idea the Corps dismissed out of hand.

Congress finally pushed the Corps into accepting restoration—not unchaining the whole river, but at least part of it. The owners of ninety thousand acres of once-and-future floodplain having been bought out, the restoration got under way in 2000, when seven miles of channel were unloosed; dams blew up, levees came down, and the Kissimmee quickly looped itself back into fourteen miles of wandering, oxbow river that stunned almost everyone with how rapidly it began to heal. Pasture reverted into lush wetlands, the fish reappeared, wading birds moved in. In all, twenty-two of the fifty-six miles of channel are to be filled in, and forty square miles of river and floodplain brought back to life. When it's done in 2010, it will have cost twenty thousand dollars per acre to restore what should never have been destroyed in the first place, but by every measure it appears to be working.

So if it's possible to restore at least part of the Kissimmee, shouldn't it be possible to work on a much broader canvas and restore the wider Everglades as well? That's been the holy grail of environmental policy in Florida for generations, the subject of successive bouts of legislation and spending, to little effect. For example, those huge water gates we saw along the Tamiami Trail were built in the 1990s as part of a project known as Modified Water Deliveries, or Mod Waters, intended to create a more natural flow south into the park; instead, after fifteen years and more than $191 million, the project remains bound up in litigation and little closer to actually delivering significant amounts of water to Shark River Slough than it was before it was built.

Finally, after decades of talk, a massive blueprint—the Comprehensive Everglades Restoration Plan, or CERP—was passed by Congress in 2000. It laid out a grand vision of almost stupefying complexity, with sixty-eight major projects under the CERP umbrella and the estimated price tag an equally eye-popping $8 billion. If completed as promised over the next thirty years, it would be the largest, most ambitious, and most expensive ecosystem restoration project in the world.

The stated goal is to re-create, if not the original sheetflow of the Everglades, at least its effects, by mimicking the original seasonal fluctuations. As much as 1.7 billion gallons of water a day, much of which is now flushed down rivers to the sea, will be stored underground in wells and in new reservoirs, then meted out as needed. A couple hundred miles of levees and canals would be eliminated, but almost twice as many miles of new water control structures would be built; unlike with the Kissimmee, where the idea was to let the river go where it naturally would, CERP would make water flow in the Everglades more artificial, not less so.

Ironically, some of the scientists responsible for the Kissimmee success have been among the harshest critics of CERP, which they believe will fail on several counts. It will not, they contend, restore the sheetflow but merely shuttle water from storage area to storage area, neglecting the connectivity that made the Everglades function in the first place; it's being run by engineers, not scientists, and they're trying to have it both ways, providing lots of water for continued human growth while shortchanging what the Everglades needs to live again. As evidence of that, critics point to the CERP road map that the Corps of Engineers released in 2003, which gives equal priority to water supply and ecosystem restoration, instead of directing most of the water to the Everglades as Congress intended. And the parts of the plan that would most directly benefit the Everglades are being pushed back later in the process.

There's also the question of whether agriculture, especially the politically powerful sugar industry, which has badly polluted the water entering the system in the north, can be forced to clean up its act. One reason for a strong role for the federal government in the restoration was to keep a short leash on state politicians, who—even though voters overwhelmingly support saving the Glades—are notorious for caving in to business and agricultural interests. This concern seemed justified when, in 2003, Governor Jeb Bush approved a measure giving the sugar industry an extra seven years to clean up its mess, pushing the deadline back to 2013.

By 2004, the four-year-old project was already two years behind schedule, although just days before that fall's presidential election, Governor Bush announced that the state would pony up $1.5 billion to design and build eight of the CERP projects, advancing the deadline for that section of the plan by ten years. While this promised to jump-start the moribund restoration, it also worried conservationists, who wondered if Florida was staging a checkbook coup, taking control of the restoration project and ultimately shortchanging the Glades once again.

We pulled off on the roadside, where the cattails and sawgrass whispered in the wind, to watch a snail kite soaring overhead, its wings bowed in gentle arcs, its flight airy and dreamlike. It was a male, dark gray but for a white tail tipped in black, its legs carmine against its flanks. We saw another at Shark River Slough as we sat in a long line of cars waiting to get into the visitors center parking lot, but when we did, the tramway was so crowded with people that we did not linger. We pushed on west, through the Big Cypress, where the roadside canal was full of thousands of wading birds and alligators in safari-park numbers—two here, three there, here another two, basking in the sun every hundred feet or so along the miles and miles of road.

At last, late in the day, we turned off the Tamiami Trail for Marco Island on the Gulf coast, where we were to join Amy's family. After days in the Everglades, it was a shock to find ourselves driving through high-rise condos sandwiched against one another, the gated communities, strip malls, and neon that have devoured this once-wild place. Yet this is the modern face of Florida, especially here on the southwestern coast, which is growing faster than almost anywhere else in the country.

How, against such an onslaught, can the state's natural core hope to survive? It seemed impossible, and yet I knew there are still places where Florida's wilderness runs deep and strong. That's where I was heading next.

FIVE

The South's Wild Soul

It was early morning, and there was a damp, empty hush to Wakulla Springs, one of Florida's oldest tourist attractions, a sprawling 1930s lodge in Spanish-revival style, white walls and red-tiled roof, overlooking the huge natural spring that gushes a half-million gallons an hour to form the Wakulla River. Foot-long mullet leaped across the water like skipped stones, and two little blue herons squabbled over a morsel. An osprey, its voice creaky, flew among the moss-draped cypresses.

A squat tower rose from the edge of the three-acre spring like a wader tiptoeing in—a two-story diving platform for those willing to cannonball among the alligators. The water below was clear, with just a faint emerald tinge where it was twenty or thirty feet deep, and in fact it took a couple seconds to realize just how deep the pool really was. Farther out it dropped off to black, over the 185-foot-deep cavern from which the spring emerges. The bones of mastodons and giant ground sloths, relics of the ice age, still lie down there. The strong current rippled the dark mats of submerged plants, against which schools of fish stood out with perfect clarity—catfish, gar, largemouth bass, sunfish, all seen from my osprey's-eye view.

Wakulla is among the world's largest and deepest freshwater springs, and there were efforts in the 1920s to make it a national park. Instead, the financier Edward Ball, who bought up huge tracts of Florida Panhandle land in the 1920s, purchased the springs and its surrounding forest in 1934, turning it into a de facto wildlife sanctuary; he built the lodge and created what was, for decades, one of the signature tourist destinations in northern Florida, with its glass-bottomed boat tours and "Henry, the pole-vaulting fish." Peterson and Fisher stopped here on a hot spring day; they took the boat ride, watched "Henry" (a catfish finning the current near a sunken pole), and photographed a family of limpkins. Peterson referred to places like Wakulla Springs as "wildlife cheesecake," his term for beautiful natural sites with a sideshow atmosphere.

If they came back to Wakulla Springs today, they would notice few differences, even though in 1986 it became a state park; the lodge is still open, the glass-bottomed boats still run every day of the year, and the current incarnation of Henry the fish still does his little routine with the pole. The brochure I got when I paid my small admission fee carried the official state park slogan, "The *Real* Florida." That seems incongruous here, at the very symbol of Florida's kitschy past, even if that past has been completely eclipsed by the theme-park excesses that it presaged.

Peterson and Fisher saw the South at a pivotal moment, as it changed from a winter getaway to a year-round vortex, sucking in population from around the country and across the world, growing at a rate that staggers the imagination. Since their visit in 1953, Florida alone has lost 8 million acres of forest, wetland, and farmland to development—a pace of more than twenty acres an hour nowadays. Can any wild place survive in the face of such rampant destruction, or will Florida soon be rendered nothing more than a Disney version of itself, with a few pockets of carefully stage-managed pseudo wilderness?

But the real South—the truly wild South—is still out there, if you know where to look for it, often surprisingly close to the schlock and profligacy of the modern age. It can be subtle, though, easy to zip past on a highway, with barely a hint of the riches that lie beyond the roads and the palm-lined developments.

I started looking for the South's wild soul an hour west of Wakulla Springs, where Fisher and Peterson hunted for its most vivid icon, the ivory-billed woodpecker. This huge, flamboyant bird—a woodpecker the size of a large duck, with black-and-white wings and a crimson crest—was at once symbol of and slave to the great virgin swamp forests that originally blanketed the southeastern lowlands. Ranging over enormous territories, seeking out freshly dead trees, the ivorybill used its massive bone-white beak to pry off the bark and remove wood-boring beetle larvae the size of a man's thumb. Without ancient trees, there would be no giant beetle grubs; without trees or grubs, no ivorybills.

By the end of the nineteenth century, unchecked logging had so shattered the ivorybill's habitat that it was known (or suspected) in only a few places from the southeastern coastal plain and Florida, west to the Big Thicket of eastern Texas. It was given up for extinct several times, only to be rediscovered, most famously in the 1930s in the Singer Tract of northern Louisiana, where in 1942 Peterson saw the last two females to survive there.

The Singer Tract birds disappeared when the timber was cut during World War II, and the same fate befell those in Florida's Big Cypress, which was also leveled in the 1940s and early '50s. Ornithologists pinned some of their remaining hope on the near wilderness of the Florida Panhandle, and in 1950 they got their wish. Ivorybills were reported by several observers in the deep cypress swamps along the lower Apalachicola River and its tributary, the Chipola, about forty-five miles southwest of Tallahassee. The area was closed to hunting, declared a sanctuary, and one of the men who had spotted the birds, a local guide and snake collector named Muriel Kelso, was appointed warden.

There was no way Peterson was going to pass up another chance to see an ivorybill, so he arranged for himself and Fisher to meet Kelso and a small group of distinguished southern naturalists during their *Wild America* tour. At daybreak on May 14, 1953, they set off in a flotilla of skiffs and canoes into the Chipola River swamps—a spooky region gothically named the Dead Lakes. In fact, it was a single natural impoundment of about seven thousand acres, formed when sandbars on the Apalachicola backed up the smaller Chipola, a place full of dead or dying cypresses and live trees thickly cloaked in Spanish moss.

"We drifted, motors cut off and paddles across our knees, listening intently for the nasal tooting of the ivory-bill," Fisher wrote. "Flicker, pileated, red-bellied, hairy, downy woodpeckers we heard, or saw, but no ivory-bills." They waited near a suspected ivorybill roost hole, but it had been almost two years since the chiseled cavity in a big cypress had been used, and they never saw or heard a hint of the big birds. "No sound. The last Florida ivory-bill—perhaps the last ivory-bill in the world—had gone from its last known roosting place," he concluded.

For decades most ornithologists agreed with him, assuming that the ivory-billed was extinct, until its startling rediscovery in 2004 in eastern Arkansas, perhaps the most dramatic resurrection ever of a species long consigned to extinction. The Arkansas find has renewed interest in the possibility that the great woodpecker might survive elsewhere in the remote swamp forests of the South—including those along the Apalachicola, where unconfirmed rumors of its presence have been circulating for years.

The "primeval oak and tupelo forest" through which Peterson, Fisher, and their colleagues slogged in 1953 is largely gone, though there is plenty of picturesque second-growth cypress along Dead Lakes, where on a drizzly afternoon I watched a trio of wood ducks split the perfect reflection of gray-boled trees, and nodded pleasantly to fishermen winching their bass-boats up onto trailers at the state park boat landing. Not that long ago, there wasn't much to catch here, and Dead Lakes was living up to its name. In 1960, in a misguided attempt to control the lake's fluctuating water level and improve fishing, the Corps of Engineers and the state of Florida built an eighteen-foot-high dam across the mouth of the Chipola. Instead, it interrupted the natural high/low water cycle and cut off migratory fish like striped bass; the lake became choked with vegetation and silt and lost much of its game fish. The situation became so bad that in 1987 the local community prevailed on the state to remove the dam; while it will take years for all the accumulated muck to flush out of the lake, the dam removal has already gone a long way toward restoring fish populations.

Once you drive south from the interstate and the congestion in the hilly land around Tallahassee, this part of the Florida Panhandle is lovely and empty, though the uplands are mostly controlled by timber companies, which have converted the old longleaf pine forests and wetter broadleaf tangles into monotonous expanses of slash pine plantations. Driving here is like passing through ceaseless picket fences, as the rows of spindly trees flicker past your car windows in a hypnotizing blur.

But if the great virgin forests are gone, there are still places where you can find a whiff of the wild South. One is Bradwell Bay, not far from the Ochlockonee River, the next major drainage east of the Apalachicola. In 1952 an ivorybill was reported from this area, and while there may be no more giant woodpeckers here, it's still a big blank spot on the map; one guidebook describes it as "suited only for the most adventurous of naturalists," a twenty-five-thousand-acre swamp in which it is easy to become very seriously lost. Or, as a biologist warned me, "It's the kind of place where if you're hunting bears and your dogs cross the road into Bradwell Bay, you do your damnedest to get them back so you don't have to go in there. It's deep water, deep swamp—just a great big roadless chunk of landscape."

Bradwell Bay is part of the 600,000-acre Apalachicola National Forest and is one of the oldest and largest national wilderness areas in the East. Congress, which passed the original Wilderness Act in 1964, soon realized that a different approach was needed in the East, where little of even the wildest land was truly pristine. In 1975, it passed the Eastern Wilderness Areas Act, and among the sixteen sites it designated was Bradwell Bay, where the dense swamps had stymied even the logging crews, preserving a little of the old-growth slash pine, oak, pond cypress, and tupelo. The logging railroads, known as trams, penetrated the fringe of the bay (a local term for a bowl of lowland amid higher ground), but not its soggy heart.

From the hamlet of Bloxham, it was a good thirty miles south on a little two-lane hardtop that parallels the Ochlockonee through the national forest until I turned off onto a series of sand roads, each one wetter than the last, that led to Bradwell Bay. It had been a rainy year, and the water lay deep in ditches along the road; around one bend I flushed a pair of hooded mergansers, which buzzed into flight and almost instantly disappeared against the dark vegetation—all except the male's white head patch, which on this overcast day glowed like a taillight through the trees.

To say that Bradwell Bay is surrounded by higher ground probably conjures a false impression of the topography; we're talking about the difference of a few feet between the pine flatwoods, where I followed the orange blazes that mark the Florida National Scenic Trail through waist-high saw-palmettos, and the thirty square miles of swamp at the center of the wilderness tract. The trail markers were haphazard guides at best; the trail was only slightly more visible than the many game paths that crossed it, and I had to be aware, as I passed each orange splotch on a tree, where the next was in the distance, because it was easy to make a wrong turn and suddenly find myself confronting a wall of cypress and titi forest.

The lower trunks of the pines were all scorched and blackened, the result of low-intensity fires set regularly by the U.S. Forest Service, but I also passed through large areas in which there were only dead snags, the victims of a ferocious wildfire that hit in 1998, an exceptionally dry year when many parts of Florida burned. Such infernos aside, fire is essential to the health of the longleaf and slash pine ecosystem, including the wiregrass savanna that grows beneath the large, widely spaced trees—one of the most diverse grassland ecosystems in the world, and one that includes an unusually high number of rare and endemic plants. In presettlement days, the pinelands stretched from southern Virginia to eastern Texas, and lightning (along with fires set by Indians) burned up to a third of it every year, keeping encroaching hardwoods at bay.

Between timbering, fire suppression, and grazing (cattle love the tall, bushy, carbohydrate-rich longleaf seedlings), the original longleaf forest all but vanished, and today the regal trees grow on less than 15 percent of their original range, mostly in small fragments and patches. The loss of the longleaf also spelled trouble for the plants and animals that depended on them; the pinewoods harbor more than 125 species that are federally listed as threatened or endangered, or are candidates for federal protection, among them red-cockaded woodpeckers, Bachman's sparrows, and gopher tortoises.

Forest managers in the Apalachicola conduct prescribed burns to keep the fuel load down and prevent cataclysmic fires. This includes the wilderness area, though the swamp forest itself will not burn and stopped even the heaviest of the 1998 blazes cold. I could see why. Even among the pines, I was squishing through wet sphagnum moss and trying not to step on clumps of hooded pitcher plant, a rare carnivorous species whose reddish pitchers were long and slender, capped with rounded hoods like monks' cowls. The trail was soon nothing more than a sinuous trough of water, a roundabout silver line through tangles of titi (a low, gnarly hardwood also known as swamp cyrilla) and small pond cypress. I made no pretense of staying dry. Greenbrier thorns snatched at my clothing, and the only way to move was to twist and squirm under low-hanging branches, trying not to trip on the cypress knees poking from the dark water, which was now up to my shins.

There was a gray-green light in the swamp, and very little sound. There were no bugs—it was, in fact, a chilly day, and my feet were already cold—but I saw lots of animal sign. Deer and turkey tracks pocked the mud, I found coyote scat filled with the orangish shells of crayfish, and I passed a number of trees where bears had scratched the bark; the wilderness provides a much-needed refuge for the threatened Florida subspecies. The water grew deeper, and the trail became more confusing. I also knew, from my map and from what I'd been told, that if it was this wet near the outskirts, I'd be waist-deep before long and swimming by the time I really got into the depths of the swamp. The water moved languidly, seductively; the smooth reflections looked innocent, but I kept finding deeper and deeper holes that I needed to edge past, probing with one foot while holding on to a titi branch for balance.

This was ridiculous—any day on which you can almost see your breath is too cold for wading through a nearly trackless swamp and tempting hypothermia. I stopped, wriggling my numb toes while I debated what to do, and eventually turned to backtrack the couple of miles to the car. But giving up was harder than you might think; there was something pulling me deeper, not only the beauty of the gray cypress trunks and twisted titi trees, which looked like something from an enchanted forest, but the potent call of an unknown place—the knowledge that deeper in the swamp was truly big timber, old tupelos and cypresses that have been there for four centuries, trees that once knew ivorybills and Carolina parakeets, that sheltered panthers and red wolves. A place, guarded fast by mud and inaccessibility, where the twenty-first century is only a rumor. More than once on the hike back I stopped, water puddling in my damp sneakers, and looked over my shoulder.

Florida's image and its reality are at surprisingly sharp variance. The sun-palms-and-gators ideal is at odds with the fact that an awful lot of the state north of Lake Okeechobee is farmland and pasture, quiet and pretty but in no way exotic. In the north, much of it is a monotony of pine plantations. The so-called Nature Coast, along Apalachee Bay at the base of the Panhandle, is particularly empty; Peterson's observation that this was "one of the least settled parts of the state, and except for the Everglades, the wildest," is still essentially true. You can drive almost forty miles from Newport, near Wakulla Springs, east to Perry, and pass only two or three little villages worthy of the name; from there south to the Suwannee River is another sixty-five miles on old Route 98, a divided four-lane whose traffic has been siphoned off by I-75 to the east, leaving down-at-the-heel motels gasping for customers and half a dozen small towns among the interminable pines.

In a lot of other places, of course, Florida is growing like a barn on fire. The state's population jumped from 9 million to 16 million in just twenty years, with a new resident moving in every two minutes these days and a projected population, by 2025, of more than 23 million. Development on the southeastern and southwestern coasts is occurring at such a ripsaw rate that many of the state's municipalities rank among the fastest-growing of their size in the United States—and an outsider, hearing all this, can be forgiven for assuming that, other than the Everglades, Florida is merely days away from becoming solidly partitioned, paved, and drained. (In some places, that's not far from the truth. I once asked an official from Pinellas County, which includes St. Petersburg and surrounding communities on the west side of Tampa Bay, what their growth rate was. "Pretty much zero," he said. "Everything that can be built on has been built on. The only open space we have left is what was protected years ago.")

Fisher and Peterson saw Florida at a crossroads, and a lot of the subsequent change has been due to a single invention. Coming out of the panting heat into a host's air-conditioned home, they marveled at "air as refreshing as a mint julep" and predicted the infant technology would prove to be "a method by which a new vitality would permeate the hot and humid South." Talk about understatements—though of course AC hasn't been the only change. James Fisher found Orlando "a charming inland town" with "block upon block of citrus trees . . . thousands of acres of orderly groves." The town had about 52,000 residents in those days, but in 1955 the aerospace industry moved in, followed in 1965 by Walt Disney and a talking mouse in red shorts. Today, Orlando has 1.7 million residents in the metro area, more theme parks than even the most hardened vacationers (45 million of them annually) can hope to visit, and expectations that half a million more folks will move there in the next ten years.

Such extravagance has long been Florida's stock-in-trade. Fisher, being driven through Miami, wondered, "Was there ever such conspicuous consumption, so many luxury hotels, row upon row, between boulevard and beach, each with its crescent approach? The place was a riot of parthenonian, byzantine, banker's gothic, broker's perpendicular, glasshouse, gashouse, bauhaus, bathhouse, and madhouse." To which a modern visitor can only think, How quaint. In those days, the city (which as an incorporated entity was barely more than fifty years old) had just 250,000 residents; in 2000, there were 2.2 million people in the Miami-Dade metro area, an increase of 16 percent over just the previous decade, and that does not include the 11 million tourists who come each year.

Virtually all of those folks crowd into the Atlantic coastal ridge, which is the noodle-thin rim of slightly higher, considerably drier land that separates the Everglades from the Atlantic and continues north beneath the seamless, linear sprawl of Hialeah, Fort Lauderdale, Pompano Beach, Boca Raton, West Palm Beach, and other communities. Since the 1950s, growth, hemmed in by water on both sides, has mushroomed in two dimensions—density and verticality. The Atlantic is walled off by batteries of high-rise beachfront hotels and condos, so hulking and continuous a barricade that I wonder how the afternoon sea breeze ever finds a chink to sneak through.

The same thing is happening now on the southwestern coast. Ten million people—three-quarters of Florida's population—live within 150 miles of Fort Myers, and the region's population rose by almost 40 percent between the 1990 and the 2000 censuses. Collier County, to the south of Fort Myers, has seen its population skyrocket from about 16,000 in 1960 to 275,000 full-time residents today, with that figure jumping by an additional third each winter when the snowbirds arrive. Naples, its biggest city, is the heart of the second-fastest-growing metropolitan area in the nation.

And right in its path is Corkscrew Swamp, the largest stand of virgin bald cypress forest left on the planet.

The bald cypress is that botanical oddity, a deciduous conifer; each winter, its feathery bright green needles turn orangish and fall, leaving the tree, as its name suggests, bald. It is not a true cypress, but a relative of the giant sequoia and redwood, and shares with them a presence, a majesty that usually forces anyone visiting a grove of old cypresses into quick and reverent silence.

Unless you live in Anchorage or Phoenix, you can probably raise bald cypresses in your backyard, but they really prefer wet soil and thrive best in the deep, mucky river bottoms and swamp forests of the South, where they grow quickly and can reach heights of more than one hundred feet. Their charm is due in equal parts to their size and their shape—gray lightly striated trunks that splay out in wide buttresses just above the water, like curtsying ladies holding wide their skirts, and the famously weird "knees," of uncertain function, which rise up from the roots to poke above the swamp. Once the trees top out above the canopy, hurricanes and thunderstorm winds prune them flat, and they add girth through the centuries until they are true behemoths.

Some of the largest and finest cypresses were in southern Florida, especially in the Big Cypress Swamp, which occupied almost twenty-four hundred square miles to the north and west of the Everglades proper. Despite its name, Big Cypress was not a monolithic block of forest but a mosaic of mostly wetland habitats, from pinewoods and wet prairies to sawgrass marsh and cypress. Much of this last is what's known as pond or dwarf cypress, a small, shrubby variety that grows in tightly packed thickets where the soil is shallow and less fertile, but along the deepest sloughs, where the water and accumulated peat soil were deep enough for trees to gain a real purchase, there were bald cypresses of monstrous size. The big trees grew in what locals called strands—ribbons of forest stretching for miles, which marked a slightly deeper watercourse through the swamp.

One particularly magnificent cypress strand extended from the Okaloacoochee Slough, north of Immokalee, and down into the Fakahatchee Slough, which drains into the Ten Thousand Islands along the Gulf. Another big cypress strand started on the headwaters of a drainage known as Corkscrew Creek, about fifteen miles east of Naples, and ran some twenty or twenty-five miles south to the Tamiami Trail.

Lumbering a cypress swamp was horrible, hot, labor-intensive work. The only way to manage the task was to cut a path into the swamp, heap up the wet soil to create solid ground, and lay a network of railroad tracks along it. "Solid ground" is misleading; the soil was wet and spongy, so ill suited to the task that the locomotive was always laboring uphill, no matter in what direction it was moving—pushing itself along a perpetual sag in the spiderweb of tracks. Because it cost so much to log there, the Big Cypress was mostly ignored until the rest of the South's old-growth trees were cut. World War II changed that; the military needed rot-resistant cypress for the hulls of its minesweepers and PT boats and for the decks of its aircraft carriers. The accessible stands had fallen, so the timber cutters moved into the Big Cypress and kept cutting even after the war ended, marketing the wood for shingles, barrels, even coffins.

At the time of Peterson and Fisher's visit, the old-growth cypresses in the Fakahatchee were pretty much gone, replaced by 150 miles of abandoned tram lines and great slash piles drying in the sun. Much of the southern Corkscrew strand had fallen as well. The Lee Tidewater Cypress Company, which had been cutting in the Fakahatchee, had plans to take out the last of the ancient bald cypresses in Corkscrew, including trees that were more than five or six hundred years old and among which nested the largest breeding colony of wood storks and egrets in the country.

The National Audubon Society, which had posted armed wardens in Corkscrew as early as 1912 to stop the slaughter of birds for the feather trade, stepped into what by 1954 had become a ferocious effort to save the last of the Corkscrew strand. Working through an umbrella group known as the Corkscrew Cypress Rookery Association, with help from a variety of state and national organizations, prominent donors like John D. Rockefeller, Jr., and even schoolchildren around the country who chipped in their dimes and wrote letters of support, by December 1954 they had raised $170,000 to buy 2,240 acres of virgin forest. (The president of Lee Tidewater threw in an extra 640 acres, calling it a Christmas present.) Another lumber concern, the Collier Company, leased an additional 3,200 acres to Audubon, allowing the group to purchase the tract over the course of subsequent years.

The association tried to give Corkscrew to the state as a park, but Florida declined, presumably because a big swamp didn't fit the 1950s ideal of picnic tables and a swimming beach. So Audubon retained control of it, adding adjacent parcels as they became available. Today, Corkscrew Swamp Sanctuary is Audubon's eleven-thousand-acre crown jewel, attracting more than a hundred thousand visitors a year, almost all of whom walk the sanctuary's justly famous two-and-a-quarter-mile boardwalk.

The boardwalk leads past the old plume hunters' camp, which was the closest dry ground to the big wading-bird rookeries, then along a stretch of wet prairie, where in the spring swallow-tailed kites soar. It was very early morning when I arrived, and except for a few volunteers near the big, new visitors center, I appeared to be the only one out on the boardwalk. At the edge of the cypress woods, I found myself surrounded by small birds—common yellowthroats moving like mice in the sawgrass, yellow-rumped and palm warblers flitting through the bare tree branches. Among the trees, the light was still dim, and it was obvious that the night shift had only just retired; there was a fresh pile of otter scat, a puddle of urine, and a set of damp footprints—round, bigger than half-dollars—leading back to the edge of the boardwalk and down into the knee-deep water. I stopped for a long while, watching, but the otter was gone. The surface of the water, though, twinkled with movement—small fish, insects, frogs, who knows what else, were shimmering beneath the layer of duckweed and floating ferns. I was so focused on the animate that it was some minutes before I realized the dark shape a few feet from the boardwalk, speckled with green duckweed, was an eight-foot alligator.

Somewhere deeper in the swamp, a barred owl whooped its slightly lunatic hoot, and a pileated woodpecker drummed, the staccato beat rolling through the still, muggy woods. There were other visitors on the boardwalk now, mostly serious birders, to judge by their expensive optics. The trail led us into the heart of the oldest trees in the sanctuary, massive cypresses 130 feet tall, bushy with bromeliads and resurrection ferns. I noticed that the first of the truly big trees that the boardwalk passes had a peculiar red-brown mark on its otherwise gray, lichen-speckled trunk. I stopped, leaning against the boardwalk railing to think, and almost immediately realized that this was where thousands of people had lightly, almost worshipfully, stroked the tree. As I waited, a pair of birders came by, binoculars ready, but at the tree they paused, reached out, touched, looked up into the high, weighty branches. Then three young people, maybe college kids; the same thing. An older couple with fanny packs and sunglasses; pause, touch, almost without being able to help themselves.

I've been to Corkscrew a number of times over the years, and every visit has its highlight—watching two large alligators slashing at each other in a territorial frenzy, having a red-shouldered hawk snag a lizard just a few feet from my head, or seeing a kettle of a hundred wood storks circling in a rising thermal over the central marsh, ungainly birds transfigured by flight into something at once prehistoric and ravishingly lovely, backlit so that their white plumage was edged with fire. One of the beauties of the place is its ability to lift me out of the modern age and place me, however briefly, in a wilder past.

But this time I couldn't shake a sense of disquiet. When I'd first visited Corkscrew more than fifteen years earlier, the drive out Route 846, the Immokalee Road from Naples, was a long, sleepy stretch of wet pine forest and swamp, with a few homes and pastures and the Big Corkscrew Island firehouse at the turnoff a mile or so away. Electricity service had only arrived in the 1960s.

Today, Naples is lunging toward Corkscrew. On my drive out that morning, construction crews were busily widening the two-lane road to four lanes, and I passed one chichi housing development after another—places with names like Fairway Preserve at Olde Cypress, Pebblebrooke Lakes, Valencia Lakes, and Waterways of Naples, with ornate gated entrances landscaped with palm trees so newly planted they still needed two-by-four braces to stand upright. There was so much traffic on the road, piling up at new stoplights to shop at new strip malls and freshly minted grocery stores, that I could hear the hum of tires on asphalt even in the middle of the old-growth trees out on the boardwalk.

And so Corkscrew, which was saved from the plume hunters' guns almost a century ago and the lumbermen's saws fifty years later, faces an even more implacable enemy today. How do you preserve a piece of wilderness in the face of such an inundation? What will Corkscrew be like in another ten or fifteen years, much less in fifty?

"That's easy for me to talk about, because I spend a *lot* of time thinking about it," Ed Carlson said, after I met him back at the visitors center. We climbed into his dusty pickup truck in the Corkscrew parking lot and headed down a grassy road into the sanctuary's backcountry, where we finally left the noise of the distant road behind.

Carlson has spent almost his entire adult life at Corkscrew. He was born in Jamestown, New York—Roger Peterson's hometown—but his parents moved to Miami when he was four. They never even took him to the Everglades as a kid, but when he got his driver's license at sixteen, he started roaming southern Florida—camping in the Big Cypress, falling in love with wild places. He worked the summer between high school and college at Corkscrew, building the boardwalk, then went to college, got a degree, and came back to the sanctuary as a researcher in 1974. In 1983 the manager's job opened up, and he's been doing that ever since. "It's awfully nice to have studied a place for ten years before you have to manage it," he told me.

Carlson's a big guy, a cypress on legs—tall and broad, with meaty hands that engulf yours when he shakes hello. He has a blondish beard shot through with gray, and a good deal less hair under his cap than he had when he arrived at Corkscrew thirty years ago. He's an articulate, gymnastically verbal man who is very good at conveying his considerable feelings for the place.

We drove through the parts of Corkscrew the public never gets to see—out through pine flatwoods that are so healthy, Carlson said, that when government agencies want to show their land managers what a classic slash pine forest is supposed to look like, they bring them here. Then down by the central marsh, a wet prairie that, like most of the rest of the sanctuary, he and his staff try to burn off every three to five years, maintaining the balance between competing plant communities.

We stopped at a small wooden platform at the edge of the marsh, flushing a pair of red-shouldered hawks, and Carlson leaned on the rail, looking out over the heart of the sanctuary. It was, he said, a view that hadn't changed all that much in the past few centuries, except for the lack of free-range cattle, which the so-called cracker cowboys long ago rounded up to drive to Fort Myers, to sell to the Spanish for gold.

"The really neat thing about Corkscrew is that it escaped what happened almost everywhere else in South Florida," he told me. "It escaped the canals, the levees, the drainage. All our water comes from rainfall, so except for a little discharge from citrus groves over there"—he waved his arm toward a line of distant pines—"we really don't have any pollution problems. We've found out from our research that the relationship between rainfall and Corkscrew's water level hasn't changed at all. Unlike most of the rest of Florida, we don't have to restore Corkscrew; we just have to protect it."

That said, Corkscrew isn't quite the place it was fifty years ago. The sanctuary's wood storks, which numbered nearly five thousand pairs in 1960, have suffered so badly across their range that they are now federally listed as an endangered species, their population having fallen by close to 90 percent. The reason was habitat loss—the misguided drainage in the Everglades, and huge land-development scams like the failed Golden Gate Estates to the southeast of Corkscrew, which drained fifty-seven thousand acres of the Big Cypress and drastically altered the local hydrology. While the legions of storks always nested in Corkscrew, the sanctuary couldn't feed ten thousand adults and fifteen or sixteen thousand chicks. So the adults routinely flew twenty or thirty miles to hunt, and suddenly all that food in the surrounding countryside was gone.

Corkscrew is fecund, but it can feed only a fraction of the storks it once had, and only when conditions are just right. Wood storks are at the mercy of the annual rise and fall of the water table, hitting the jackpot only every few years. That requires a heavy wet season to bring the water up, connecting all the isolated wetlands to form one vast fish breeding factory. The fish spawn like crazy, but because the water is too deep and the fish too dispersed, the storks can't catch them. So the birds head north, to the Okefenokee or the Apalachicola, where the dry season is in full swing and the fishing is better. Then you need a very dry winter in peninsular Florida, corralling the tremendous number of fish in small, shallow ponds. The storks come back south to a feast, with plenty of food for their chicks. That happy chain of events occurs, Carlson said, maybe once every six years.

"Two years ago, we had the best stork-nesting season in thirty years. We had an extreme wet season, and just the minimum amount of rain in the dry season, and that created as much of a feeding opportunity for storks as this ecosystem can generate. We had twelve hundred successful pairs, and about three thousand young fledged. And when they were leaving the nest and learning to fly, up there soaring around with the adults, you could sit there on the porch and see kettles of *thousands* of storks. It was what it must have been like in the old days."

But those sights now come along once a generation. Last year the storks brought off just seven hundred chicks, and so far this year, he told me, it didn't look good; the wet season had been wet, all right, but the winter had been weirdly warm and rainy, and there was water everywhere. Carlson suspected the stork nests would go bust. (They did, bringing off only about 450 chicks, well below the modern average, much less the numbers from the 1960s, when Corkscrew sometimes produced 17,000 young storks in a season.)

But the sword cuts both ways, and if the storks are a shadow of their former glory, Corkscrew is in some respects a wilder place today. "When I started, this land was lawless," Carlson said. The refuge was circled with hunting camps, many of them illegal, and the staff took turns patrolling the swamp through the night, running off armed men trying to jacklight deer or gators. Only in the last decade did the deer poaching end, when much of the surrounding land moved into public ownership and the old hunting camps disappeared.

That's meant a lot more big-game sightings, including bears; the afternoon before we met, one had walked up to the back door of the visitors center and looked in, and the smudge of its nose was still on the glass when I arrived the next morning. The same goes for Florida panthers, Carlson said. "It used to be really easy when people asked me how many panthers I'd seen—three in thirty years. Well, I've had so many panther sightings in the past two years I've stopped counting."

Ed Carlson knows there's no stopping the development tide rising in eastern Collier County, but he also knows it has its limits. Earlier, back at headquarters, he'd shown me a map of the area, his finger tapping along the boundaries. "Here's I-75 and Route 41 running along the west side. Immokalee Road on the south side, which is how you came in—just tremendous growth all along here, as you saw. And with Corkscrew Road on the north, it's the same thing—they put a new college up there, Florida Gulf Coast University, and this is just all explosive growth. And there's Immokalee on the east. But in between"—he tapped his finger again, slowly, this time in the middle—"in between is this great, big swamp."

Big, indeed. While Corkscrew is no pint-sized reserve, it is only a small part of a 130-square-mile swamp and marsh system known bureaucratically as the Corkscrew Regional Ecosystem Watershed, or CREW. The South Florida Water Management District has acquired almost twenty-seven thousand acres upstream and down from the sanctuary and is aiming to buy much more. Most of the remaining land in the watershed is wetland, and so has at least some legal protection even while in private ownership.

"At this point in Collier County, seventy percent of the land is in some sort of preserve—state, federal, or private holdings like Corkscrew—and when all the CREW projects now in the pipeline are done, it'll be eighty percent," Carlson said. "CREW is going to be the Central Park of a developed southwest Florida," he said with a kind of clenched nonchalance. I can only imagine how it must hurt to see such a wild place, where you've lived for three decades, being swallowed alive by 7-Elevens, gas stations, and Publix grocery stores.

There are still plenty of uncertainties; preserving the Corkscrew ecosystem depends on preventing cockamamie drainage schemes, which sprout in the Florida swamplands like mushrooms, and bolstering wetlands protections, as well as maintaining bans on the commercial use of surface water. But Ed Carlson seems optimistic. "We will never see an *expansion* of wood storks," he said. "But I think a hundred years from now, people will walk the boardwalk at Corkscrew and they will still see nesting wood storks. It may be a hundred pairs, it may be five hundred; I don't know. But for the other wildlife—the gators, the otters, the hawks—I'm not worried."

Later I mentioned his Central Park analogy, and he jumped on it, speaking with the most passion I'd heard from him yet. "Yes, but a *living* Central Park. A *natural* Central Park. It will have bears. It will have gators. It will have panthers. It will." I wondered if he was trying to convince me, or himself.

Corkscrew was one of the major forest strands in the Big Cypress, but the real gem was the Fakahatchee, in the heart of the swamp. It's not that the ancient bald cypresses were bigger here, though some of them were immense—up to 120 feet high and 8 feet in diameter—or even that the Fakahatchee was the largest strand forest in the world, stretching almost twenty miles from north to south and up to five miles wide, with panthers and even a few ivorybills.

No, the real beauty of the Fakahatchee lay in its botanical richness. More varieties of native orchids, forty-two of them, have been recorded here than anywhere else in the continental United States, as well as fourteen species of bromeliads, or air plants—also a record. What's more, many of those orchids and bromeliads were found nowhere else in North America except the Fakahatchee, where conditions were exactly right for a tropical flora on the edge of the temperate zone—warmth, water, high humidity, and a deep forest canopy.

Ever since the swamp forests of southern Florida formed, about five thousand years ago as the grassy savannas that once covered the peninsula were overtaken by wetlands, new plants have arrived by air. Hurricanes sweeping in from the Caribbean have carried with them the minute seeds of tropical plants or brought birds that carried the seeds on their feet, in their plumage, or in their digestive tracts. Occasionally, by extraordinary good luck, one of those seeds would wind up in the fertile garden of Fakahatchee and start to grow.

"The Fakahatchee is an island of habitat—the most northerly island in the Caribbean, in a way," state park biologist Mike Owen told me as we waded waist-deep in water the color of strong iced tea. "You have the cypress super-canopy, which traps air beneath it, then the mid-canopy for more insulation, and down here you have all this water and thick layers of peat, keeping the dew point and temperature up and keeping everything down here from freezing." He's ticking these factors off now, swinging his right hand each time in a big arc, slapping one forefinger against the other for emphasis. "Canopy. Peat. Hydrology. That's what allows tropical plants to survive here in brutal subtropical South Florida."

This was my second day wading through the Fakahatchee with Mike Owen, and it had been a remarkable experience—and not only, or even primarily, because of the place. Owen, I'd found, was a force of nature in his own right—manic, gangly, limber as Gumby, a man who machine-guns his speech with exclamation points. He has a rubber-band smile and an over-the-top enthusiasm that would be grating after about thirty seconds were it not for its utter sincerity. This guy *loves* the Fakahatchee, and it's impossible not to be infected with the same giddiness, which he calls his "Fakahabit." He loves it the way little boys love mud puddles—only his mud puddle covers eighty-five thousand acres of the most diverse swamp forest in America.

The Fakahatchee has always been recognized as a place apart. Botanists combed it for unknown species of plants, and as early as the 1920s Henry Ford (who wintered not far away in Naples) offered to buy it for the state as a park. As with Corkscrew, Florida refused, saying it couldn't afford to staff it, and whether that would really have made a difference, once World War II provided a need for cypress wood, is an open question. In any event, the timbering didn't start until 1944, a year before the war ended, and by the time the shooting stopped, half the strand forest still stood. "But they kept logging it, right up until 1953, because there was a killing to be made. It was mining, not forestry. Shingles, coffins, stadium seats, pickle barrels—that's what they cut the Fakahatchee to make," Owen said, unmistakable bitterness in his voice. Only one small area of old growth, at Big Cypress Bend off the Tamiami Trail, survived.

In 1966, the Lee Tidewater Cypress Company, having wrung what profit it could from the swamp, sold seventy-five thousand acres of it to Gulf American Corporation, the same outfit that perpetrated the Golden Gate Estates scam near Corkscrew; here the lots were sold as part of the "Remuda Ranch" development, but fortunately it was never drained the way Golden Gate was. The state, having changed its mind about the importance of the Fakahatchee, eventually stepped in to buy it up and create the preserve—the largest state park in Florida—in 1974. However, in the years between logging and protection, the strand became a no-holds-barred playground for locals and outsiders alike—a place to hunt and shoot, to crash through the woods on swamp buggies and off-road vehicles, to collect rare plants (which survived the logging, however precariously), and to engage in all manner of activities for which official oversight is not desired. There was plenty of resentment—"hatred" is the word Mike Owen uses—toward the reserve, especially when the state assigned a warden in 1975 to start patrolling. Some of that bitterness lingers.

Today the borders of the preserve encompass eighty-five thousand acres, south of Alligator Alley and down to Gullivan Bay in the Ten Thousand Islands, of which about seventy thousand are in state hands; the rest exist in small, mostly acre-and-a-quarter private inholdings left over from the Remuda Ranch days, which Florida buys back from willing sellers as they become available. (The northern portion of the strand, above Alligator Alley, lies within the federal Florida Panther National Wildlife Refuge.)

A few of the private parcels have ramshackle cabins, and we'd passed one on our way in, a corrugated-tin building on an old tram line, beside one of the small, incredibly pretty lakes that dot the swamp. The owners, Owen said, had no intention of selling, and I can't say I blame them; it was an idyllic spot. From there we moved into the flooded forest, walking slowly, feeling our way with our feet, using old ski poles for balance, trying to avoid deep holes and submerged logs. It was another chilly day, which kept the mosquitoes to a minimum and restricted the Fakahatchee's many alligators to the open, sunnier lakes. Still, with the cypress canopy bare for the season, light streamed here and there into the lower reaches of the forest, illuminating the dark water and old stumps, which were sprinkled across the slough like a green archipelago, bright with strap ferns and young palms. In the backwaters and eddies, coppery cypress needles lay in solid mats.

The water moving south across lower Florida, through the Everglades and the Big Cypress, is a shallow sheetflow, but in a few places the slightly acidic water has, with time, eroded the limestone bedrock to create the long sloughs that mark strand forests. The Fakahatchee Slough is three feet deep—a river valley, in effect, or, in Owen's phrase, "the Grand Canyon of Florida." When I'd first slipped into the swamp, I'd expected to sink deep into mud but instead found myself mostly walking through a foot or so of peaty soil, beneath which was the firm foundation of limestone. When I jabbed my ski pole down hard, the metal tip clicked against rock.

With us were Dr. Mike Duever of the South Florida Water Management District, once a colleague of Ed Carlson's at Corkscrew, and a man who probably knows more about the hydrology and ecology of the Big Cypress than anyone, and his wife, Jean McCollom, who manages the Okaloacoochee Slough for the state. All three wore high-topped GI boots, and Duever, a weathered man with a gray beard, had a battered machete hanging from his belt along with two military canteens. Mike Owen wanted to show us one of the Fakahatchee's particular treasures, a colony of *Guzmania monostachia*, the West Indian tufted air plant, a bromeliad native to the Caribbean and Latin America that, like so many other species, is found in the United States only within the Fakahatchee.

Rain several days before had brought up the water, and we pushed north against a gentle but persistent current, elbowing our way through palm thickets that rattled drily, past snapped-off cypress snags ignored by the loggers, and into a relatively open grove of pop ash—skinny trees with convoluted branches three times our height that braided among themselves. Their trunks and limbs were covered with dozens of *Guzmania*, each air plant as big as a basketball, backlit with the early-morning light so they glowed with a pale green translucence reflected in the dark slough water. There were thousands of bromeliads in sight, none growing more than fifteen feet above the water—visible evidence of the lens of damp, warm air that hugs the water's surface and creates an unmistakably tropical environment. Even Duever and McCollom, who have spent a lot of time in the Big Cypress, admitted they'd never seen anything quite like it.

Members of the pineapple family, bromeliads like *Guzmania* are not parasites, but they are often epiphytic—that is, they grow on other plants, using them as a platform rather than a meal ticket. In Florida, some of the more common species often grow on utility wires or home TV antennas—anywhere their seeds find a toehold. The most recognizable species, however, is also the least bromeliad-like—Spanish moss, which is ubiquitous around the South.

The day before, McCollom, Duever, and I had tagged along while Owen had taken a group from the Boca Raton Bromeliad Society on a trek into the swamp—about a dozen and a half people mad for epiphytes of all descriptions. We formed a little convoy at the small park headquarters, driving ten or twelve miles up a dirt road that is the only vehicular access to the preserve, passing most of that way through a dense, jungle-like forest of cypress, pop ash, and pond apple, over which rose mature American royal palms a hundred feet tall, the largest colony of this rare and graceful palm anywhere in the United States, numbering three or four thousand trees. We parked, then walked farther up the road to a spot that, to my eyes, looked exactly like every other spot we'd passed in most of the preceding ten miles. Here Mike Owen plunged through the undergrowth and into the deep ditch beside the road, sloshing merrily through the flooded trees as the rest of us, somewhat more gingerly, eased into the cold water to follow him. ("If this water gets much deeper, you're gonna hear me sing opera," one guy joked.)

But Owen was already yelling for everyone to gather around. Within forty yards of the road, he'd spotted five species of bromeliads, two of them endangered, as well as a leafy vanilla orchid whose stem zigzagged up a tree trunk; in midsummer, the orchid would bear greenish white flowers that are among the largest such blossoms in North America.

"Oh, touchdown!" he hollered, both hands in the air, then pointed to a fast-moving scrap of orange flitting through a gap in the branches above us. "That's a ruddy daggerwing, my favorite butterfly! It's a tropical species only found in South Florida, and its larvae just feed on the leaves of strangler figs, another tropical species that's only found down here. Like the orchids and a lot of other stuff, it probably came in originally on hurricanes."

The woods were filled with laughter and Latin; the conversation was generally along the lines of, "Most likely it's a *fasciculata*, but I'm only about eighty-five percent sure," or, "Here's a really nice *Catopsis floribunda*, even though it's been damaged a little by fall webworm." Our progress through the swamp was marked by frequent yells of "Touchdown!" as the group spotted more and more rarities. One everyone was looking for, though—and the plant that is the only reason most people elsewhere in the world have heard of the Fakahatchee—was the ghost orchid, or palm-polly, *Dendrophylax lindenii*. A Caribbean species restricted in the United States to the Fakahatchee, it is a leafless cluster of green photosynthetic roots that grows on the branches of pop ash and pond apple and in summer produces a wildly improbable, ethereally lovely white flower that has been likened to a face with a Fu Manchu mustache or (less poetically) a leaping frog with legs akimbo.

In 1993, an orchid collector named John Laroche and three Seminole Indians were arrested for removing 136 plants, including several ghost orchids, from Fakahatchee Strand Preserve State Park. Laroche and the Seminoles argued in court—unsuccessfully, as it proved—that treaty rights granted to the Seminole nation by the federal government superseded state and federal plant protection. The *New Yorker* writer Susan Orlean covered the case for the magazine, then expanded on it and the obsessive world of orchid collecting in her bestseller *The Orchid Thief*, which in turn was the basis of the 2002 movie *Adaptation*. No one in the bromeliad society seemed impressed with the movie, which, except for one scene near the entrance to the park, was filmed elsewhere. "None of the plants looked *anything* like the Fakahatchee," one of them sniffed. "I mean, really, where'd they shoot it, Louisiana? Please."

Late in the day, we were following Owen's compass bearing back out; the Fakahatchee, though lovely and intriguing, all still looked generally the same to me, and even Mike admitted he gets turned around sometimes. The group was straggling through cypresses and pond apples when someone yelled, "Ghost! Ghost!" There, on a branch just above our heads, was a ghost orchid—a starburst of greenish roots that reminded me of nothing quite so much as overcooked spinach fettuccine.

Mike Owen charged over, splashing like a dog, almost quivering with excitement. "Oh, man, this is a big, big, *big* touchdown—this is a brand-new ghost orchid, one that we didn't know about! And not just that, but one with a spike—this orchid's gonna bloom this summer!" This made 241 ghosts he's documented in the past ten years—a small number, due to overcollecting, to a huge freeze that hit South Florida in 1977, and probably (though no one can prove it) to the effects of logging fifty years before. It takes a ghost about twenty years to reach flowering size, and most that Owen's found have been young plants, their roots growing just half an inch a year. But this one—he checked the root length with a pocket measuring tape—had a fifteen-inch root, suggesting it was as much as thirty years old.

Everyone was scanning the trees now, and a second orchid, then a third, were spotted as Owen jotted down information in his notebook—the GPS location, the height and diameter of the host trees, other variables that may help him learn what the ideal conditions are for the species. People were bubbling, and Karen Relish, Owen's assistant, who just a few weeks before had rediscovered a species of orchid in the Fakahatchee that hadn't been seen in the United States since 1905, was getting a little miffed over all the ghost fever. When someone yelled that they'd found yet another one, then quickly retracted the claim upon realizing it was a similar ribbon orchid, she growled, "It's *only* a ribbon orchid? Come on, you guys, they're just as endangered as the ghost. The only difference is they don't get a huge white flower—they aren't megaflora."

Finally the hubbub died down, and with obvious reluctance the members of the bromeliad society allowed themselves to be nudged back out to the road, where they chatted in quiet, happy, dripping little groups. One woman lingered in the swamp, looking back at a thicket lush with bromeliads that covered the trees like exotic plumage. "I'm just trying to soak it in, because in two years it may all be gone," she said with a sigh. When I looked puzzled, she said, "You know—the evil weevil."

I didn't know about *Metamasius callizona*, and when I learned, I almost wished I hadn't. A Mexican beetle that apparently entered Florida on a shipment of Mexican plants prior to 1989, it has ravaged bromeliad populations along both coasts. The grub-like larvae burrow into the stem of the air plant, killing it, and Mike Owen has already found the telltale brown dying plants left by a weevil infestation in the Fakahatchee. There's really nothing he or anyone can do; chemical control is out of the question for logistical, environmental, and financial reasons, so the only hope is the discovery of some sort of natural control. But even though the weevil's victims include many endangered species, there isn't government money for research, and what little progress has been made so far, including the discovery of a parasitic fly in Honduras that may target the weevil, has been privately funded by the state council of bromeliad societies.

So I was in a somewhat more sober mood, the next day, when Mike Owen took us to see that huge stand of *Guzmania* air plants deep in the Fakahatchee. We discussed the threat of the weevil, and how not even the most remote location is insulated from the growing risks of an ever-smaller world. Yet I found Mike generally upbeat about the Fakahatchee's future. Despite the old Golden Gate drainage on the western boundary, which lowered the water table there by two feet, the hydrology at the core of the preserve is still pretty much what it was historically. The bald cypresses, a fast-growing species, are approaching the height of the vanished old growth, if not their massive girth. Even the old railroad trams are, in his glass-half-full view, a blessing. Breached in countless places, they no longer significantly impede the water flow, and they function as man-made hammocks, allowing endangered plants like the royal palms to flourish.

"It's hard for me to say this, knowing how much damage the logging did, but I think the trams were a net gain, now that nature's reclaiming them. And the cypresses are growing fast. I'm thinking that within another fifty years, the Fakahatchee'll be back to pretty much what it was before the logging."

Between the extremes of the Everglades and Corkscrew—at one end a vast system but one gutted of its natural ebb and flow, requiring massive intervention, and at the other a smaller but intact watershed about to be enveloped, amoeba-like, by a metastasizing world—the Fakahatchee offers something in the middle, yet perhaps more hopeful than either. Given little more than time and reasonable care, this piece of old Florida should reclaim much of its original splendor, at the heart of a wider matrix of other preserved lands that, with restoration, may also see their wilderness luster restored. It will not be a perfect place—the ivorybills are probably gone, and, barring a miracle, the shaggy blanket of bromeliads may soon be a memory—but in this deep and beautiful forest, the South's wild heart still beats.

SIX

Las Tortugas

If there really are ends of the earth, then the Dry Tortugas must surely be one of them. This sprinkle of tiny coral islands at the farthest edge of the Florida Keys, closer to Cuba than to the American mainland, lies just eighty miles north of the Tropic of Cancer. The islands form one of the four geographical anchors that framed the original *Wild America* journey. Two were its northernmost points—the start in Newfoundland, and the end in Alaska's Pribilof Islands. A third was its southernmost extent, deep in the tropical highlands of Mexico. But Fisher's descriptions of the Tortugas—palm trees, swirling seabirds, coral reefs, and the mysterious Civil War–era hulk of Fort Jefferson, isolated among thousands of square miles of empty blue ocean—made the islands one of the most riveting parts of the book when I first read it as a boy.

Almost seventy miles beyond Key West, this remote cluster of tiny desert islands was dubbed "Las Tortugas" by Ponce de León in 1513, named for the incredible number of sea turtles he found swimming in the shallow waters and crawling up on the coral sand beaches to lay their eggs. In later years, the islands became known as the "Dry" Tortugas, because they have no source of freshwater. Lying along the Florida Strait, which provides passage to the Gulf of Mexico, the Caribbean, and the Atlantic, the reefs of the Tortugas were the bane of sailors. In 1622, nine ships of the Spanish treasure fleet sank or ran aground in the area during a hurricane; one, the *Nuestra Señora de Atocha*, was found in 1985, one of the richest shipwrecks ever discovered. Pirates roamed the area in the eighteenth century, while in the nineteenth the waters were plied by wreckers—schooners manned by salvage crews out of Key West, who did not always wait for fate to deliver a prize but were often accused of luring passing vessels into trouble. John James Audubon, who traveled to the Tortugas in 1832, wrote, "With the name of Wreckers there were associated in my mind ideas of piratical depredations, barbarous usage, and even murder."

To his surprise, Audubon found the wreckers he met to be "excellent fellows," who joined him in collecting birds from the Tortugas—magnificent frigatebirds, pelicans, and two species of tropical terns he'd not encountered before, the sooty tern and the brown noddy. The birds nested on two small keys in the Tortugas, in numbers that amazed him. "On landing, I felt for a moment as if the birds would raise me from the ground, so thick were they all around, and so quick the motion of their wings. Their cries were indeed deafening," he wrote. Audubon and his companions killed about a hundred sooty terns and collected several baskets of eggs, but they also encountered Cuban egg collectors from Havana, who told the Americans they had already gathered about eight *tons* of tern eggs.

In Audubon's day there was nothing but a lighthouse on the largest key in the Tortugas, but in 1846 the federal government, with an eye toward protecting one of the most crucial shipping channels in the country, began to build Fort Jefferson on Garden Key. It would become one of the largest masonry structures in the Western Hemisphere, a slightly elongated hexagon with walls forty-five feet high that swallowed almost the entire twenty-two-acre island. Despite three decades of construction requiring something close to 16 million bricks, it was never completely finished; advances in cannon technology rendered it obsolete only years after the first brick was laid, and it later became a Union prison during the Civil War. To most Americans, it is a historical footnote, the place where Dr. Samuel Mudd was imprisoned, along with several conspirators who avoided the gallows after being implicated in Lincoln's assassination.

The fort later served as a quarantine station through the Spanish-American War, and then as a coaling depot for the Navy, but it was abandoned in 1910, falling steadily into decay thereafter. Teddy Roosevelt had by then declared the islands a federal reservation to protect the seabirds, and only just in time; the number of nesting sooty terns had fallen from the "millions" Audubon described to barely seven thousand pairs in 1903, and the brown noddies numbered just two hundred. A seasonal warden was appointed, and while egg poaching continued, even after the National Park Service was given jurisdiction for what became Fort Jefferson National Monument in 1935, the situation finally began to improve for the birds. By 1950, the sooty tern population had climbed to nearly ninety thousand pairs, almost twice its current level.

It must have been a miserable life at times, out on Garden Key; Robert Budlong, the monument superintendent during World War II, wrote of hot, airless nights when the mosquitoes were so thick the window screens were black with them; of humidity and salt spray that rotted or corroded almost everything he owned; of isolation so profound that his wife was unable to leave the island for twenty months straight; of water rationing when the rains did not come; and, when the supply boats failed to appear, of eating nothing for weeks at a time but the spiny lobsters they caught. Writing to his potential replacement, he warned, "Don't try it unless you know *beyond doubt* that you and your family, if any, can stand utter isolation under extremely trying climatic conditions, where an enormous amount of physical endurance is necessary and where one feels that he is far removed from civilization and its benefits." Having said that, he characterized the time he, his wife, and his young daughter had spent on the Dry Tortugas as one of the happiest of his life.

Right from the start, there was no question that the Dry Tortugas would form a major stop on the *Wild America* tour. Peterson had visited the islands in 1941 and was anxious to get back; he felt the sooty tern colony was "the number one ornithological spectacle of the continent," and Fisher, for his part, had always wanted to visit the Florida Keys and see their tropical seabirds. There was no scheduled transportation to the islands, of course, but Peterson, drawing as always on his remarkable network of contacts, had made arrangements for friends to meet them in Key West with a couple of motor yachts in which to carry everyone on the seventy-mile trip to Garden Key. (Sleeping in a wooden yacht was an improvement over his first trip, when he'd been billeted in the fort itself and had to pound his shoes against the wall each morning to make sure no scorpions were hiding inside.)

The Dry Tortugas are usually described as the second most remote unit of the national park system, to which they were added in 1992, after a unit in American Samoa. Even today, getting there is no cakewalk, but visitation has ballooned. You can charter a floatplane, sail your own boat, or take advantage of two new, albeit pricey, ferry services that run daily out of Key West. Most of the tens of thousands of tourists who come to the fort are history buffs, but for naturalists the Tortugas hold a very different appeal. The park contains the most pristine coral reefs in the continental United States, the best habitat for sea turtles in the country, the only American nesting colony of the magnificent frigatebird, and of course the only large sooty tern and brown noddy colonies on the continent. Sitting in the middle of a wide expanse of ocean, the tiny islands also provide a fire escape for songbirds migrating across the Caribbean and the Gulf, birds so weary and hungry they will hop across a person's feet—a paradise for birders in spring and fall. But for historian, naturalist, or someone just looking for a sandy beach, the jumping-off point remains Key West.

A lot has changed in the keys in the last fifty years, including a dramatic growth in population. What had been a sleepy backwater in the 1940s boomed with subdivisions, starting in the 1950s right through 1992, when growth-control ordinances were passed. Of course, by then, most of what could be developed—the drier upland habitat—had been, and today only a small fraction of the remaining privately owned, undeveloped land is suitable for building. The naval base on Key West, the last of the main islands, pulsed with activity during World War II, when the keys were a submarine battleground and the Tortugas served as harbor for minesweepers and torpedo boats; in the 1990s it was downsized to a naval air station, although there has been a recent upgrade in its mission.

The dock area at Key West Bight has been gentrified from shabby shrimping companies into trendy restaurants, oyster bars, a dive center, and berths for row upon row of fabulously expensive yachts, which stand in contrast to a few old military ships, including a decommissioned sub, to be turned (locals hope) into tourist attractions. But the real behemoths are the cruise ships, which dominate the west end of the island and in a sense the entire place; Key West is now one of the top destinations in the world for cruise lines, with up to a million passengers pushing through town each year—a distinction many here would prefer to avoid, along with the pollution, crowding, and other problems that the giant ships bring.

Hours late, the little twin-prop commuter flight from Fort Lauderdale finally delivered Amy and me to Key West, but not our luggage, including the bag with our spare clothes. We had a ferry to catch at daybreak, so at nine-thirty at night we were on a mad scramble to find toothpaste, underwear, and T-shirts enough for three days on the Tortugas before the last stores closed. (Fortunately, the bag with our camping gear and food had arrived.) Our real mistake, though, was to take the advice of a local, who pointed us down a series of dark, quiet streets where jasmine bloomed on wooden fences, suggesting we try a drugstore on Duval Street. I now suspect he was having a little fun with us.

As a remote post with a big naval base, Key West always had a rambunctious nightlife; Peterson commented on how they had to try "not to run down any of the young sailors having an evening on the town" when they arrived. But not even the U.S. Navy can produce a spectacle quite like that modern phenomenon, spring break, and in Key West, Duval Street is ground zero for it.

Three Florida universities were on break this same week, and the street was jammed with thousands of drunken college students. Music roared out of open bar doors where beefy guys in black SECURITY T-shirts made at least a pretense of checking IDs, the sidewalks sticky with spilled beer. It was almost impossible to move. A pickup truck with a sign reading GIRLSFEST inched down the street through the human tide with four or five young women standing in the back wearing tight shorts and bikini tops. A guy in his early twenties, walking with the sloppy, flat-footed gait of the mostly intoxicated, crossed the street toward them at a lurching angle while trying to drink from his big plastic beer mug, but he only succeeded in jamming the thick ribbed straw in his ear over and over again.

"Did you see that?" a heavy guy in his fifties asked incredulously. Apparently a squadron of twenty kilted New York City bagpipers had marched through the melee at full skirl, and I hadn't even noticed the music over the general din. As the man talked, a middle-aged woman in a lovely sequined white dress, with a wreath of dried flowers in her hair and a pair of silver high heels over her shoulder, skipped barefoot across the street holding the hand of a man in a white tuxedo shirt with a big boutonniere. Newlyweds? Her eyes were shining, but immediately behind her, three drunken girls were trying to shinny up a lamppost to steal the Duval Street sign, before collapsing in a heap and beginning to retch. If I were the new bride, I thought to myself, I'd put my shoes back on. We bought what we needed and retreated. I later read in the local newspaper that the cops had made 344 arrests and confiscated nearly a thousand fake IDs, but deemed the mayhem "not as big as in years past."

The majority of the Dry Tortugas' visitors now come out from Key West on one of two high-speed ferries that arrive each day, disgorging up to two hundred people. Since the ferry services started a few years ago, visitation to the national park has quadrupled, from about twenty-three thousand people in 1994 to more than ninety-five thousand just six years later. All but a handful stay only a few hours—enough time to see the fort, eat a picnic lunch, and do some snorkeling—before heading back to town.

We were staying three days, however, and as per our instructions we had our camping gear, our fresh-from-the-package underwear, and our gallons of bottled water (the Dry Tortugas are still dry, and you have to bring everything you need) down at the dock at six-thirty the next morning. The ferry was a big white catamaran, and once loaded with the day's guests and out in the channel heading west, it was skimming along at more than thirty knots; what had been a six-hour passage in Peterson and Fisher's day would take us barely more than two. We passed the last scatterings of the small keys, then the Marquesas, several small islands capped by mangroves framing a placid lagoon, then finally an empty sea for more than an hour. We sat in the bow, enjoying the sunshine against the chill wind, and watched for birds—pelicans and royal terns at first, then some frigatebirds with their long, scythe-shaped wings, and a couple of masked boobies, these last a tropical relative of the gannets we'd seen in Newfoundland. Of the hundred or so passengers on the boat with us, only one other person had binoculars and seemed to show any interest in the birds.

I'm something of a history buff, so I'd been reading about Fort Jefferson for years, yet all the words and photographs didn't prepare me for the scale of the place as it rose out of the sea at our approach to dominate the horizon. Built to house fifteen hundred soldiers, the fort has six walls, each up to 477 feet long, that enclose a ten-acre parade ground, the equivalent of two Roman Colosseums. Huge underground cisterns were excavated to store a million and a half gallons of rainwater, the only source of freshwater on the island (unfortunately, they cracked beneath the weight of the walls, letting in the sea, a grievous engineering error that helped to doom the fort as a military installation). The fort, in turn, is wrapped in a half-mile-long seawall to form a moat, in which sharks once swam to deter prisoners from escape. A black iron lighthouse stands at one corner, and the structure has a curiously two-toned look: The lower two-thirds of the walls are made of pale orange brick that came from Florida, much of it laid by slave laborers hired out by their owners to the federal government. The upper layer is of darker stuff; once the Civil War started and the Union lost its local supply, it had to ship bricks all the way down from Maine. The slaves, however, were forced to keep working even after Lincoln's Emancipation Proclamation went into effect in 1863, since that order specifically exempted the "Florida Reef," as the Tortugas were known. (The slaves only left after the arrival of convict laborers later that year, which made renting slaves unnecessary.)

But the real draw for us, as it had been for Fisher and Peterson, lay a short distance from the fort—Bush Key, a low island that comprises four or five acres of scrubby vegetation, over which milled a screeching, flashing cloud of thousands and thousands of birds. These were mostly sooty terns, a cosmopolitan species found throughout the tropical oceans of the world, but in the continental United States they breed (at least in any numbers) only here, in this one jam-packed colony. As the ferry rounded the key and eased into the channel, more and more terns passed the boat, coming in from fishing grounds that might be as much as fifty miles away—slim, supple-winged birds, white below and coal black above except for their gleaming white foreheads, which looked like headlamps. Up to forty thousand pairs of the terns breed each spring on this tiny key.

Among them, in much smaller numbers, were brown noddies—an exquisite animal, and one that, over the course of the next few days, I started to think may be among the most beautiful of all seabirds. Though fractionally smaller than the sooty tern, the brown noddy has more visual weight because of its color, a rich brown that, in the sun, has an almost suede-like nap to it, with a white forehead that blends to gray on the nape. The tail is long and rounded, with a shallow dimple at the end, and in the air a noddy flies with the authority of a shearwater or a raptor, and little of the airiness of other terns.

Dry Tortugas is a watery park; of its nearly one hundred square miles, only 104 acres are land, sprinkled among seven islands—though the number depends on how you count. Over the years, some of the islands have disappeared; Bird Key, whose huge tern colony Audubon visited in 1832, washed away in a hurricane about a century later. And in the past decade, land bridges have formed between Long Key, Bush Key, and Garden Key, creating in essence one long, meandering island. To keep visitors from disturbing the birds, the park has roped off the narrow neck of sand connecting the fort and Bush Key, though there is little they can do about the black rats, which have infested the fort since it was built and no longer even need to swim over to Bush Key for a meal of tern eggs or chicks. When they are not menacing the nesting birds, the rats are a hassle for the campers; we had been sternly warned about keeping our food in sealed plastic containers so as not to encourage the pests.

Fort Jefferson so completely swallows Garden Key that only a small fringe of the original sand island remains on the south and east sides. We and the handful of other campers waited on the ferry until the rest of the passengers had flooded down the dock to stake out their picnic tables for the day, then we gathered our gear and picked campsites in the shade of tall coconut palms and buttonwood trees near one of the ruined coaling docks. It was a snug, reasonably private spot with afternoon shade, though the weather during our visit proved so cool and breezy we never felt much need for shelter. As we pitched the tent, a few songbirds moved through the vegetation—parula and palm warblers, for the most part, among the early migrants heading north from the Greater Antilles, where they'd spent the winter. But as I worked, an oddly insistent rustle in the dead leaves finally caught my ear; not a bird, my subconscious realized, nor a lizard, but something heavier and more irregular. A couple of hermit crabs, in shells the size of golf balls, were lurching out of the undergrowth toward our small pile of gear, including the plastic box full of food. Obviously, the crabs had sensed something worth investigating, despite the sealed container. When I picked up one of them, it waved its legs in the air, then seemed to think better of such a display and jerked itself back into its adopted shell, its stalked eyes peering above one large purple claw.

A second ferry had docked, disgorging another hundred visitors, and both ferry crews set up buffet tables beside the moat, which in addition to hungry people immediately attracted a couple dozen ruddy turnstones. These piebald shorebirds, with their stocky orange legs and stubby bills, are notoriously catholic in their tastes, and rather than poke among the sea wrack for invertebrates, as they are supposed to do, these pecked for bread crumbs and wayward pieces of roast beef among the sunblock-greased legs of the tourists.

We wandered through the fort, our footsteps echoing in the long, dim casements and archways, two thousand of them running in endless repetition through the massive walls. During the Civil War, eight hundred prisoners, mostly Union deserters, were sent here, guarded by a garrison of nearly twice that many men; these days, Fort Jefferson can absorb a few hundred noisy day-trippers and still feel empty and a little brooding. We climbed spiral staircases of slick stone up to the second tier, where the old gun emplacements yawned, framing little snapshot images of Bush Key as we walked past, and then onto the uncompleted third tier, where construction was stopped when it became clear the fort was settling under its own weight. The sun was blinding on top, and the wind blew hard from the south, with a half-dozen frigatebirds suspended in it, their thin wings crooked and black. The sea around the key was mottled in shades of blue and green, a mosaic of deep channels, reefs, seagrass beds, and pale sand, and at a small beach on the lee side of the island most of the tourists were paddling around the sheltered water with snorkels and masks.

Back down on the parade ground, in a clump of old, gnarled Geiger trees, we plunked ourselves on a bench near the only running water on this otherwise dry archipelago—a small brick fountain with a trickle of somewhat greenish water that puddled up near the top and flowed down the angular sides to be recycled. While the water might not have looked appealing to me, it was ambrosia to a procession of small birds, which dropped from the trees to drink and bathe—many palm warblers, a pair of parula warblers, and a black-and-white warbler that hitched its way around the brick sides of the fountain like a nuthatch. The star, however, was a beautiful yellow-throated warbler, which kept splashing with complete abandon, sending up a half globe of droplets that caught the dappled sunlight and threw it back as diamonds. It dipped and shook, at one point churning across the shallow basin like a paddle wheeler in what seemed like something close to ecstasy.

And it may well have been. That small fountain is a lifesaver for many of the birds that use the Dry Tortugas as a migratory stopover, crossing the Caribbean from the Greater Antilles to Florida and the Gulf coast. While the prospect of flying across hundreds of miles of open water would be a terrifying one for humans, even the smallest songbirds make those kinds of journeys with relative ease—when conditions are right. Each night in the peak migration period from late March through early May, millions of birds will lift off from Jamaica, Hispaniola, and Cuba at dark and head north, usually with a southerly tailwind helping them along, saving both time and energy, and bringing them to the mainland after fifteen or twenty hours.

But frequently the migrants encounter storms over the Caribbean, or fierce headwinds that slow their progress to a crawl and divert them from their intended course. Some may be in the air for thirty, even forty hours. Under such conditions, the Tortugas can be awash in tired, hungry, dehydrated birds—a bonanza for visiting birders, but a real life-or-death drama for the migrants. Many still have energy reserves and can rest briefly, reorient themselves, and head for the mainland coast more than a hundred miles away. But those that have used up most of their fat, or have even started catabolizing their muscle tissue, need freshwater and food in order to recuperate. Unfortunately, both are in scarce supply; except for the small fountain in the fort, there is no open water except for rainwater, which rarely puddles on the sandy islands.* And insects are few and far between in spring. Most of the weary birds depart after as little as a few hours, but some never manage to escape. One day during our visit, the northeast wind brought in a flock of barn swallows, which may have come from as far away as Argentina. The next morning, only three remained, endlessly patrolling the air around the fort but finding few flying insects. Later, we found one dead, floating in the moat; another sat, fluffed and lethargic, on the seawall as I crouched a few feet away; it watched me with dull eyes, too far gone to care.

We waited until the ferries loaded up in mid-afternoon, leaving the island almost deserted, and finally slipped into the water with our snorkeling gear. We had the reef to ourselves. The wind-driven swells had stirred up a lot of silt, making the water a little murky, but we prowled around within shouting distance of the fort, gliding over beds of translucent green seagrass and coral heads that rose up from the bottom. Queen conchs, their gray foot-long shells fuzzy with sponges and gorgonians, plowed slowly through the sand, and schools of jack crevalles and grunts swam in

*Sadly, the fountain and many of the old trees in and around the fort were destroyed six months later by Hurricane Charley, which struck the Tortugas with 130-mile-per-hour winds.

and out of view. At one point, an eagle ray with a five-foot wingspan materialized from the dim haze of the water almost at my elbow—a huge dark shape in the corner of my eye that startled the bejesus out of me until I realized what it was.

Even the old seawall of the fort was full of life, softened by time and biology into a living quilt of corals, sponges, purple sea fans waving in the surge, and sea urchins with their black lethal-looking spines hiding in the dimmer crevices. The fish were mostly familiar faces to me, from time spent on reefs elsewhere in the Caribbean and Central America—sergeant majors like zebra-striped bluegills, yellowtail damsels with ultramarine bodies, butterfish and blue tangs, a big parrotfish with powder blue eyebrows and red lips, nonchalantly chewing away on a coral head, the limy grit expelled in little puffs back through its gills. I followed a needlefish, a slender, living tube about a foot and a half long with a javelin beak, then took a breath and dove down toward a coral head where a cluster of small, bright fish were swimming. A young French angelfish, satiny black with yellow stripes, charged out as if to contest my right of passage, staring at me a few inches from my mask.

Coral reefs are among the most complex ecosystems on earth, rivaling or surpassing rain forests for their diversity of life. A few reefs are found as far north as the Florida Panhandle and near Jacksonville in the Atlantic, but the only true, living barrier reefs in the continental United States hook around South Florida and along the twelve hundred islands of the Florida Keys. And the best of them, ranking among the finest in the entire Caribbean basin, are those in the Dry Tortugas. Here grow more than seventy-five species of both hard and soft corals, including some like elkhorn and fused staghorn that are locally rare, as well as deep reefs that may be among the oldest such communities anywhere in the Western Hemisphere.

They also have a particular importance these days as reef health deteriorates all over the world, and especially in Florida. Reefs may be biologically rich, but they are also uncommonly fragile, susceptible to a variety of human-caused ills. At the ecosystem level, these include sedimentation, chemical pollution, ultraviolet radiation, and rising sea temperatures, some of which may be contributing to the spread of bacterial diseases and to coral bleaching, a poorly understood phenomenon in which stressed coral polyps expel their symbiotic algae, leaving the reef ghostly white and many of the corals dead. At the other end of the scale, a carelessly dropped anchor can shatter coral heads that took centuries to grow, while the brush of a diver's hand—never mind some doofus who tries to stand on living coral—can kill or injure the colony, opening it to disease. By any measure, Florida's reefs are in a bad way, with almost 30 percent of the coral in the keys dying just since 1996.

Given the overall high quality of the park's reefs, managers want to designate forty-six square miles—roughly the western half of Dry Tortugas National Park—as a research natural area, closed to fishing and boat anchoring. Combined with the adjacent ninety-square-mile Tortugas North ecological reserve, part of Florida Keys National Marine Sanctuary, it would be one of the largest no-take marine sanctuaries in the world. (Another sixty-square-mile unit of the federal marine sanctuary, closed even to diving, lies just to the southwest of the park's waters.)

What happens in the Tortugas will have ramifications for the entirety of the keys and Florida Bay, since there's plenty of evidence from no-take marine reserves elsewhere in the world that protecting one relatively small area improves the quality of fisheries in surrounding waters. Because the park already serves as a somewhat protected spawning area for reef fish, shrimp, and spiny lobster, whose larvae are dispersed by currents to distant areas, the rest of the keys would be in even worse shape without the steady influx the park provides.

And protection may come only just in time. Fisheries in the keys have been pushed too hard, for too long; mainstay species like grouper and snapper are way down, and by one estimate 60 percent of the reef fish in the keys are overharvested. This is a potential disaster for a region so economically dependent on commercial fishing and sportfishing. "The multi-species reef fisheries of the Florida Keys are under siege from fleet expansions and increased vessel fishing power that threaten to overexploit, destroy habitat, change marine environments, and reduce biodiversity," the park's management plan notes, and the National Oceanic and Atmospheric Administration has called the keys one of the nation's most significant—and most stressed—marine ecosystems.

Because the park is closed to commercial fishing, you'd think it would be immune to these trends, but sportfishing is permitted in most of the

Tortugas, and while the park is in better shape than other parts of the island chain, there are fish declines here, too. For one thing, as the fishing worsens in open areas, more and more fishermen—equipped with faster boats and armed with sophisticated gadgets like sonar and GPS systems—are able to find more quickly and efficiently the fish that remain in the park. James Fisher, in 1953, commented on the abundance of big fish like barracuda and yellowtail snapper (and caught a lot of the latter to feed himself and his companions), but comparisons of old reports with current conditions have convinced park managers that many of the Tortugas' most popular species have drastically declined in the past century and may be potentially rare or endangered.

Marine habitats are in trouble as well. Although isolated, the Tortugas attract a lot of visitors, and some of the most heavily visited sites are in danger of being loved to death. Divers, snorkelers, fishermen, and boaters have had a serious impact on coral health in much of the park, and a lot of its seagrass beds have been damaged by boat props and dragging anchors. (In the keys as a whole, some thirty thousand acres of seagrass, a critical nursery for marine life, have been scarred by propellers.) As the seagrass erodes, bottom sediments previously locked in place are freed, making the once-clear water much more turbid.

The park is also a critical nesting habitat for several species of sea turtles, a group that has fallen on hard times. Audubon saw multitudes of green, hawksbill, loggerhead, and leatherback turtles in the keys, but professional "turtlers" who caught them on the nesting beaches, or who harpooned the reptiles at sea, took a fearsome toll. The soldiers at Fort Jefferson ate a lot of turtle—almost forty thousand pounds of the delicious meat in just two years—and egg collectors probing the sand with steel ramrods to find the hidden nests further decimated the population. Today all four species of sea turtle found in Florida waters are threatened or endangered, and as mainland beaches have become more and more dangerous—rimmed by brightly lit buildings, for instance, which disorient the hatchlings and lead them inland, to their deaths—the remote sands of the Tortugas have grown in importance. Loggerhead turtles are the most common breeding species, but the park is home to the largest nesting population of green turtles left in the Caribbean.

The sun was lowering by the time we climbed out of the water, shivering a bit in the cool breeze, and toweled off; there was no way to wash off the salt, since our freshwater was limited, and for the rest of our time on the islands we simply built up sticky layers of sunblock, sea salt, sweat, and fine sand, knowing that everyone here pretty much smelled the same. It was supper time, which was going to be a fairly basic meal. Another of the warnings we'd received prior to the trip—this one from the ferry operator—was that under no circumstances were we permitted to bring cooking fuel on board, either propane or white gas. If we wanted to cook on the island, we were told, we should use charcoal in the braziers the park provided. While this makes perfect sense in theory, we discovered that in practice it means that everyone who comes out brings a bag of charcoal—and a nice big can of highly flammable lighter fluid with which to start it. How this is safer than a small propane tank is beyond me, but we'd opted for a more rudimentary solution; we figured that for three or four days we could manage on cold food, and so had packed bagels, PB and J, snack bars, and some canned ravioli and pasta. There was a moment of panic in the planning stage when Amy, a confirmed caffeine addict, realized she'd have to do without a morning cup of coffee, but then she found some hideous canned mocha stuff that I could barely look at but that she swore was fine.

Long before the trip, I'd made arrangements so our visit would coincide with that of one of the National Park Service biologists who covers the Everglades and the Dry Tortugas, in order to help him band nesting terns on Bush Key. But a couple of weeks earlier, he called to say that the terns were ahead of schedule and that he would be leaving the park the day before we arrived. "But be sure you talk to Mike Ryan, the chief interpreter there," he said. "Mike's been on the Tortugas for years, and knows the place as well as anyone."

I found Ryan in his cubbyhole office the next day, tucked into the brick guts of the fort, more or less below what had been the cells for some of the Lincoln conspirators. He was a barrel-chested man of forty in an NPS uniform and shorts, blond hair cropped low and stiff, with a goatee and a quiet, deliberate way of speaking. A specialist in coastal forts, he's worked from Alabama to New York City, but despite the harsh climate and the isolation he and his wife relish their posting on the Dry Tortugas, now in its fourth year. We talked happily about the fort, its history, and whether Samuel Mudd was, as he claimed, innocent of complicity in Lincoln's assassination (Ryan is doubtful), before we turned to the many environmental issues that face the Tortugas.

Mike Ryan repeated what I'd often read: that by its very isolation, the park has been spared some of the problems that plague inshore waters closer to the Florida coast. For example, algal blooms from Florida Bay, which kill many marine organisms and are caused by nutrient-laden runoff from land, have come close but have so far missed the park, while tests show that the park's water is incredibly clean, with just a tiny fraction of the fecal coliform bacteria counts found near Key West. But with ocean temperatures rising and coral bleaching (which appears tied to the warming) increasingly common, even an outpost as remote as the Tortugas may not be remote enough.

Ryan prefers to worry about the things the park can control. "Here at Dry Tortugas, it's all going to depend on transportation. In the last fifty years we've seen exponential rates of change in transportation, but let's face it, those seventy miles from Key West are still a challenge. But who knows—fifty years from now, we may all have personal hovercraft, and we could have ten thousand people who can just fly themselves out here."

It won't take Buck Rogers technology to cause a problem, either; the current visitation rate, facilitated by the two commercial ferries, is already having a deleterious impact on this fragile park, especially the area around the fort and Garden Key. Both the park and the ferry operators (who have onboard naturalists that give a do-not-touch lecture to the passengers) push the message that reefs require a strictly hands-off approach—and feet-off as well, since simply standing on a seagrass bed can damage the plants, break up the root system, and create a fast-growing hole chewed ever deeper and wider by the currents. But despite signs, brochures, and further preparatory lectures from rangers—despite the best intentions in the world—a significant number of those ninety thousand visitors are going to forget, or slip, or stumble. In just the relatively short time he's been in the park, Ryan's seen a noticeable decline in the coral.

Most days, especially at the peak of the winter snowbird season, the ferries are full, and with talk turning to even larger vessels and bigger crowds, the park imposed a moratorium on new commercial activity in 1999. The NPS has since informed the two ferry operators that future contracts will be with only one company, which will be allowed to bring 150 people a day—a boost for whatever firm wins the bidding, but a one-quarter reduction in the number of daily visitors. The park also recognizes that the number of private boats coming there has jumped dramatically, too, but has chosen to defer any restrictions on them for now, Ryan said. That issue will become critical, though, if the federal government ever lifts travel restrictions on visiting Cuba. "If Cuba's ever opened to Americans, this becomes the natural layover for a boat trip between Florida and Havana—the perfect place to stop and break up the trip."

Besides the reefs, the park's other premier natural spectacle is the tern colony on Bush Key. Several evenings we sat by the rope line across the land bridge to the island, in the slanting orange light of sunset, and watched them trading back and forth on nimble wings. Ground-nesting seabirds like terns thrive on the kind of isolation and freedom from mammalian predators that the Tortugas offer. Originally, the terns would have had to contend only with the frigatebirds, known in years past as man-o'-war birds for their piratical nature, and with predatory crabs, but now the intrusions are more profound, like jets from a fighter-training base at Boca Chica Naval Air Station, which fly near the park and rattle the mortar out of the fort walls with their sonic booms.

Our last afternoon in the park—after the ferries had gathered up their passengers like hens calling in their broods of chicks, and Garden Key was quiet again—Mike led Amy and me past the rope barricade that bars the land bridge to Bush Key, and around the edge of the colony. The terns were up in agitated flight, and the air was full of their *wide-a-wake* cries, so loud they made conversation difficult. Most of the nests were toward the middle of the small island, where the sea lavender, prickly pear cactus, and bay cedar stood a couple of feet high, but some of the birds were nesting among the low mats of green fleshy sea-purslane plants, which afforded marginal shelter at best. The nests were just shallow scrapes in the sand, and only half still had speckled, heavily camouflaged eggs; the rest were occupied by chicks, one per nest, some so newly hatched their down was still wet and dark.

Mike stopped at a thin trail that led through the purslane into the middle of the key. Anyone watching from a distance would have been puzzled by our exaggerated careful movements, like a trio of mimes, each of us trying to place our feet exactly where the one ahead had walked and watching the ground for eggs or chicks on which we might otherwise step. The terns were simply everywhere underfoot, sitting tight unless we had to step right over them, in which case they would scuttle away with growling alarm calls, occasionally making a stab or two at our shoes before clambering into the air to join the maelstrom of birds around us. Yet when we stopped moving for even a few moments, they settled right back down again. The chicks, heeding a warning in their parents' squalls, played possum and lay flat on their stomachs, chins pressed against the sand and eyes closed. Only when an adult landed and waddled back to stand protectively over them did I see their small, bright eyes peering out from beneath mom's or dad's sheltering belly.

Some of the birds around me, I knew, could easily be twenty or twenty-five years old; sooty terns are remarkably long-lived, and at least one banded tern was still hale and fertile at age thirty-two. And there are a lot of them. A recent tally of sooty tern colonies in the world's tropical oceans suggests a global population of between 18 million and 23 million breeding pairs, more than half of them on the far-flung islands of the Pacific, where a single colony may contain more than a million pairs. Brown noddies have an equally expansive, pan-tropical range, with the biggest concentrations in the southern Indian Ocean, around Hawaii, and off Australia. They are far less common in the Caribbean and tropical Atlantic than the sooty terns, though, and, even on a global scale, probably number fewer than a million pairs.

The noddies on Bush Key occupied the middle of the island, perched in the taller shrubs, where they were surrounded by suburbs of sooties. Unlike their neighbors, they prefer some height for breeding and usually build grass-and-twig nests instead of scrapes. They seemed to be lagging behind the sooties in their breeding schedule, and many were still bringing in nesting material—dried seagrass and other flotsam plucked from the wrack line along the beach or hauled in from heaven knows where across the blue water. They also seemed much more phlegmatic, sitting quietly on the edges of their messy, guano-stained nests even as I edged in for photographs.

Up close, I was struck again by how dramatically plumaged a bird the noddy is. I struggled for an analogy that fit the warm velvety brown and the stark white cap, and finally decided the birds looked as though they were made of powdered cocoa, with a smudge of confectioners' sugar on their heads that blurred to dust on the neck. In a sense, they were a photographic negative of a typical tern—darkest where others are whitest, and vice versa. Even when they took off, if we came too close, there was none of the shrill histrionics of the sooty terns, just a complaining, crow-like caw as they flew away.

"I guess it's because I'm a quiet, soft-spoken kind of guy," Mike said, "but I've always liked these noddy terns. It's a matter of personality, I suppose."

One of the biggest questions for park managers, of course, is how the Tortugas' most famous avian residents have fared over time. It's hard to judge how many terns there were originally; early visitors viewed them as a food supply or a curiosity and made no attempt to count them with any rigor. Emily Holder, the wife of a post surgeon at the fort during the Civil War, mentioned in her journal that "in June the gulls always came in thousands to lay their eggs on Bird Key, the season being in the nature of a festival and feast for us, as we made up egg-collecting parties." Sgt. Calvin Shedd, a Union soldier from New Hampshire also stationed at the fort, wrote to his wife and children in 1862 about another egging expedition to Bird Key, which at the time was the main breeding colony: "Oh! how I wished you could have been there & see the Thousands on Thousands of Birds Black & white about the sise [*sic*] of wild Pigeons they were very pretty & their cries were almost deafening & so tame that I could have caught hundreds with my hands." Audubon's account speaks of sooties and noddies "by [the] millions," but a closer read shows he was just repeating what a local ship pilot told him; if the artist tried to make any count of the Tortugas colony while he was there, he didn't record the results.

The commercial egg collecting was clearly a disaster for the terns, which like many long-lived species lay only a single egg each season; Fisher and Peterson calculated that based on an average egg weight of thirty grams, the eight tons of eggs that Audubon credited the Cubans with collecting in 1832 represented a quarter-million tern eggs. And given that these species invariably lay a single egg, their calculation also suggests that the colony in the early nineteenth century was probably in excess of a quarter-million pairs—three times the number today. How rapidly the numbers fell isn't clear, because reliable counts of the nesting terns didn't begin until they were almost gone. From a low of about seven thousand pairs of sooties in 1903, the population climbed to fifteen thousand pairs in 1935, to between thirty and fifty thousand pairs in the 1940s,* and to a modern peak of ninety-five thousand pairs in 1950. Fisher's estimate for 1953 was about eighty thousand pairs, almost exactly what the park service found a few weeks later when it conducted its annual census. Since then, the population has declined to its current level of about forty thousand pairs, a trend blamed on island erosion and a reduction in shrubby nesting cover.

Black rats from Garden Key, which used to swim the channel and now can simply walk across, may also be a threat, though Mike Ryan said that trapping on Bush Key doesn't seem to bear that out—the rats may be much happier picking up the scraps left by picnickers. But all that food has attracted another problem—gulls. The gull population on Garden Key has started to grow, a subsidized gang of layabouts that, when the ferry buffet isn't set, raids the tern colony for eggs and chicks. Unlike along the mainland, where food sources for gulls are legion, in the Tortugas the park managers have a chance to nip the problem in the bud. The ferryboat operators were told to keep a tidy table, and Ryan said things have improved, but with hundreds of people a day eating on the island, including boaters who may get no advance warning at all, Garden Key is

*Robert Budlong, the Tortugas' superintendent during the early 1940s, wrote what must rank as the wittiest, most consistently irreverent official reports ever filed by a federal bureaucrat. This was his report from July 1944 on the subject of censusing terns:

> All the terns are doing nicely, both the Noddies and the Sooties . . . though I had a little trouble in my one attempt to count them. I don't know how they do it in the other wildlife regions, where they show their numbers boldly in the usual round numbers. Here our methods show precision; we use the very latest and refined techniques of experts—we don't guess them here, we count them. We row over to the island, and I send My Man Friday [Budlong's lone assistant, Franklin Russell] to ride hard upon the sea-birds, have him round them up together, by their many tens of thousands, all in one big flock he holds them, like a howling herd of dogies; when I make a certain signal My Man Friday runs them past me . . . or he herds them past me flying, and I count them in their passing; one by one I count the sea-birds, for I know our wildlife experts want those figures with exactness. And I almost had them counted, when a ship came in the harbor, passed quite close by the island, and a blonde stood at the railing in a most breath-taking costume, stood and called and waved and beckoned, and I quite regret to state that I stopped counting for a moment. That was fatal, that was tragic . . . I quite forgot my counting, quite forgot to think of sea-birds, though my mind was still on figures yet I did not think of sea-birds. I regret that circumstances interfered with this year's census.

one big scavenger's smorgasbord. And even though it's illegal, commercial fishing boats—which operate just outside the park boundaries, and then moor for the night in the sheltered basin by the fort—sometimes clean their catch in park waters, with a cloud of gulls picking up the scraps.

In the end, this issue—like almost every conservation issue in this lonely archipelago—comes down to visitation, to how many shuffling, swimming, boating, lunch-nibbling human bodies you can add to a delicate ecosystem and not tip it badly out of balance. There's no neat equation to answer that question, either, just a messy trial and error that gets harder and harder as visitation swells. Visitors like me. I crouched on the warm sand, the breeze lightly scented with guano, the high walls of the fort backlit and dark in the distance. The sound of the tern colony covered me in a rough-woven blanket of growls and barks, but through it I could hear the whine of another floatplane on its way in. One more person was coming to look for paradise.

SEVEN

Wilderness Lost (and Found)

If you drive around the Gulf of Mexico—making a great arc from the Florida Panhandle to the west and south that skirts New Orleans and Beaumont, cuts below Houston, around Lavaca Bay and San Antonio Bay to Corpus Christi, then parallel to Padre Island and the gleam of sun off the distant water—eventually you'll come to Brownsville at the tiptoe end of Texas. Not far south of town, through the cotton fields and rows of palm trees, you will run up against the Rio Grande, a turbid river when the rain has been falling, which makes a huge dip to the south here, scribbling across the flat landscape in looping oxbows before winding north again to empty into the Gulf. Nestled inside one of those loops is the most southerly piece of American soil in the state, known appropriately as Southmost Ranch, a place where, because of the Rio Grande's erratic course, you can actually look *north* into Mexico.

In its day, Southmost was a big producer of row crops, citrus, and ornamental plants—palms of several varieties that thrive in the hot, humid South Texas climate, Ruby Red grapefruit sweet enough to die for, all manner of exotics grown in its fields and greenhouses. Of Southmost's twenty-two hundred acres, almost all were under cultivation, except for a few hundred acres of thornscrub forest and native sabal palm groves along the river and the old cutoff oxbow lakes known as *resacas*.

That's pretty much the story for all of the land in this part of Texas, along the lower Rio Grande valley. Once the richest, most biologically diverse ecosystem in North America, home to tropical birds and butterflies, to jaguars and ocelots, it was all but obliterated in the first half of the twentieth century, as mechanized agriculture transformed old cattle ranches, mesquite and huisache thickets, and impenetrable *resacas* into sprawling, bountiful fields or orchards.

By the time Peterson and Fisher arrived in the lower Rio Grande, the alteration was complete. Peterson described it as "a country of recent, rapid, far-reaching changes, which have swept away the jungle-vegetation of the old cutoffs and oxbows of the Rio Grande . . . and put in their place broad highways, shiny new neon-lit towns, great citrus groves, and cotton fields." The valley had become one of the most important agricultural centers in the country. That didn't stop the two naturalists from enjoying a spectacular day of birding, because even in its reduced straits, the lower Rio Grande remained one of the birdiest places in the country. Fisher tallied 132 species, more birds than he'd ever seen in a single day in his life, including 38 life birds that were brand-new to him—a lot of "Tally ho's." But even with all the birds, Peterson and Fisher lamented the extinction of the jaguars, the ocelots, and much more, and their feelings came through loud and clear in the chapter title they chose: "Wilderness Lost."

That paradox remains. The lower Rio Grande valley (or LRGV, as it is widely dubbed) is still astonishingly fertile ground for the naturalist, and nature tourism has, in recent years, become a major economic engine, even as agriculture fades in the face of cross-border commerce and roiling urbanization. There's a boomtown feeling to parts of the LRGV these days, driven by free-trade treaties and the proximity to Mexico; between the last censuses, the valley's population grew by almost 40 percent, and the pace is climbing. From any of the many triple-decker expressway overpasses around McAllen, Harlingen, and Brownsville, you see a horizon-to-horizon skyline of fast-food, hotel, and retail signs poking up above the street-side palms. They've even coined a trendy new name for the urban border region: the Rioplex.

Yet at the same time, there is an unprecedented attempt to stitch up the lingering fragments of the older, wilder Rio Grande. A variety of federal, state, and private entities are working to create a 275-mile-long corridor of wildlife habitat, from the Falcon Dam to the Gulf coast, much of it comprising old farmland being slowly restored to thorn forest. The irony is that while places like San Benito, where the last jaguar in Texas was shot in 1946, are well on their way to becoming urban centers, and privately owned upland brush disappears at an accelerating rate for subdivisions and commercial development, there is a good deal *more* protected, restored wildlife habitat along the river itself than there was in the 1950s.

Every day, the divide becomes sharper as the fuzzy middle ground—the private land like farms and ranches where wildlife is welcomed but not necessarily husbanded—moves into either the protected or the developed column. In a few more decades, most local experts agree, this process of polarization will be nearly complete, and the LRGV may be a sea of people with—and this is the hope—a linked and functional chain of natural land preserving most of the region's biological wealth.

Southmost Ranch is already part of that chain. When the owner of the farm decided to sell in the late 1990s, it wasn't to developers but to the U.S. Fish and Wildlife Service, which incorporated about half the land into Lower Rio Grande Valley National Wildlife Refuge, and to the Nature Conservancy, which bought more than a thousand acres, including most of the agricultural land. While the group continues to farm part of the ranch, using it as a demonstration project for sustainable agriculture techniques, within a few months of the purchase the conservancy had begun the laborious process of revegetating some of the old fields with native plants, fifty-five thousand seedlings, most of which they now grow themselves each year. Out with exotic palms, in with lotebush and Texas ebony, sabal palm and cedar elm—and a little bit of the old wilderness character of the border comes back to life.

And the character of this part of the border is like nowhere else in the United States. Biology doesn't recognize human boundaries, and when the Spaniards first arrived on the lower Rio Grande in the sixteenth century, they found a decidedly tropical ecosystem on both sides of what would eventually become the U.S.-Mexican border. Ecologists refer to this region as the Tamaulipan Biotic Province, which rises from the Mexican state of Tamaulipas to cover the horn of southern Texas—almost 20 million acres of semiarid brushland dominated by drought-resistant trees and shrubs like huisache, mesquite, ebony, and prickly pear cactus. The one universal seems to be an abundance of spines and spikes; an especially well named shrub is simply called allthorn. The old Texas cowboy joke goes, "If it don't bite, sting, scratch, or poke, it ain't from around here."

The densest thorn forest, however, grew on the rich alluvial soils of the lower Rio Grande valley—not a valley so much as a wide delta, periodically flooded by the great, meandering river. Thickets of sabal palms lent a jungle air, so much so that the river was originally called Río de las Palmas, while toward the Gulf, where barrier islands running for hundreds of miles on either side of the delta formed the expansive Laguna Madre, the land was drier and more open, a coastal prairie where grassland, brushland, and wetlands made a rich mosaic. Many parts of North America can claim to be biological melting pots, but the lower Rio Grande is an exceptional example, blending elements of the Chihuahuan Desert to the west, hardwood forests to the north, Gulf coast communities, and, naturally, the tropics just a couple hundred miles south in Mexico.

Most important, the valley provides a corridor north for species that otherwise never make it onto American soil—for wildcats like ocelots and jaguarundis, for butterflies, frogs, and snakes, and for nearly five hundred species of birds, including green jays, hook-billed kites, buff-bellied hummingbirds, great kiskadees, and clay-colored robins. It is the most biologically diverse corner of the United States, and in the past century it was almost completely eradicated.

The Spanish established missions in the valley in the early eighteenth century, and for the next two hundred years cattle ranching was the Rio Grande's lifeblood. Where they could, the ranchers hacked and grubbed out the brush (though because of overgrazing, some mesquite actually expanded into what had been grassland), but with the advent of mechanized farming in the early twentieth century the offensive became a war. Bulldozers smashed brush flat, and herbicides, ever more sophisticated and effective, kept it from growing back. Between 1920 and 1950, more than 95 percent of the native Tamaulipan brushland in the United States was destroyed, with just 1 or 2 percent of the subtropical riparian thorn forest along the river delta left in existence.

That makes the surviving pieces priceless. Few places showcase the valley's original richness as well as Santa Ana National Wildlife Refuge, on a bend in the Rio Grande just south of the town of Alamo, about forty-five miles upriver from Southmost. It is, as when Fisher and Peterson were here, a tiny jewel, only two thousand acres in size, yet it holds more than four hundred species of birds (more than half the total found in all of North America) and more than 250 species of butterflies—so many that if Santa Ana were a state, it would rank fifth in the nation in butterfly diversity, edging out California's 101 million acres.

The refuge visitors center doesn't open until eight in the morning, but you can hit the trails at daybreak, when the air is still cool, full of the low hoots of white-tipped doves. The huisache was in bloom, these native acacias covered with tiny, globular clusters of orange flowers, and there were even a few blooming honey mesquites, their fresh leaves bright green and flickering in the breeze. Migrating broad-winged hawks passed overhead in twos and threes, and there was a lot of movement in the brush, a mix of early migrants heading north, like gnatcatchers and warblers, with resident birds like olive sparrows and long-billed thrashers. Rosebelly lizards skittered across the trail, always sounding bigger than they were—I was on a hopeful edge, knowing that there are ocelots and jaguarundis in the refuge and cherishing the notion, however unlikely, that I might see one.

Something much larger than a lizard moved in the undergrowth, and I froze, but it wasn't a cat. Five birds that looked like small, slim turkeys paraded in single file across the trail, then began hopping, with an alacrity belying their size, into the shrubs and then up the slender branches of a huisache to its crown. They were plain chachalacas—members of the tropical family known as cracids, to which the guans and curassows also belong—greenish brown but for white edging to their long tails and a small patch of red skin at the throat. In the distance, I could hear other flocks of chachalacas sounding off—first a lead bird with a lower call, then the whole lot of them opening up, a braying call that's usually described as a repetition of their name but that to my ears sounds more like *Wake them UP! Wake them UP! Wake them UP!* However you transcribe it, it's an ear-shaking noise when a whole flock starts belting out just a few feet over your head, and one of my favorite sounds of the humid borderlands.

The refuge's *resacas* were full of waterbirds—coots squabbling among themselves, and flocks of black-bellied whistling-ducks, long-legged and goose-like, which splashed and dove for cover when a hungry Harris's hawk made an unsuccessful attack. Around the water, the forest was eerie—tall ebonies, Berlandier ash and cedar elm, old mesquite, all of them wrapped in the thickest blankets of Spanish moss I've ever seen, hanging in drifts that almost reached the ground and covered even the trunks of the trees like heavy cotton batting; it felt like a vast, jumbled room in which all the furniture was wrapped in sheets.

At a small, drying pond rimmed with mud and dimpled with tadpoles and frogs, a great kiskadee with a lemon yellow breast and a head striped like a skunk's kept sneezing out its three-syllable name. Kiskadees are tropical flycatchers, but they have an all-consuming diet that goes way beyond bugs to include basically anything they can jam down their throats; this one leaped from the branch, swooped over the edge of the pond, and came back with a frog almost as big as its blocky head. After giving the amphibian a couple of whacks against the branch to still it, the bird opened wide and swallowed, the frog's legs still kicking as they went down.

Nearby, I could hear the ascending, buzzy call of a parula warbler, a sound I associate most closely with the spruce forests of coastal Maine, but this song was slightly different, with a burr on the final, sharp note. I had been expecting this; the tropical parula that nests in this corner of the refuge is, like many of the birds of Santa Ana, an avian celebrity, and several people had given me very precise directions on exactly where to go along the trail system to find it. A celebrity, but a reclusive one; I spent twenty minutes patiently searching, while the bird sang maddeningly from the treetops, always out of sight, sometimes shifting locations without my seeing it move, a trick that seemed to have more to do with teleportation than flight. When at length I found it, though, the prize was worth the search—a handsome bird about four and a half inches long, with a dark blue-gray back and a sunshine yellow breast and throat, across which stretched a band of deep orange.

With so much to see, birders come from around the world to visit Santa Ana and the rest of the LRGV, and as their numbers have grown over the past fifteen or twenty years, the valley has taken note. Visitation at Santa Ana alone is more than a hundred thousand people a year, and overall the lower valley's national wildlife refuges get half a million visitors annually. And that's just a small slice of the pie. The last time anyone tried to put a figure to it, birding and other forms of nature tourism brought $100 million a year into the local economy, an estimate almost everyone concedes is probably a fraction of the actual total. In most parts of the country, watchable wildlife viewing (the catchall term for such pastimes) is largely invisible to municipalities and the business community, even where it forms a significant part of the economy; unlike, say, golfers, birders don't usually gather in public places where they're easy to spot, and so they buy their restaurant meals, pay for their gas, and book their motel rooms without anyone ever suspecting why they're in the area.

Not in the LRGV. Here, more than almost anywhere else in the United States, local chambers of commerce have embraced birds and other forms of nature tourism and made attracting birders a top priority. There are three major annual events—the Rio Grande Valley Birding Festival in Harlingen each autumn, the Texas Tropics Nature Festival in nearby McAllen in the spring, and the Brownsville International Birding Festival in July, which specializes in Gulf of Mexico seabirds. Up and down the valley, municipalities are doing the same thing, often by appealing to other niche markets. The town of Mission has a successful butterfly festival, capitalizing on the lower Rio Grande's unparalleled diversity of lepidoptera, while Weslaco has also gone the insect route with a dragonfly festival. The region is also home to the new World Birding Center, a joint project of Texas Parks and Wildlife, the U.S. Fish and Wildlife Service, and nine lower Rio Grande communities.

At daybreak on a muggy, overcast morning in McAllen, volunteers with clipboards were sifting through a throng of birders, most of whom had risen at five so they could grab a bite at the Holiday Inn's breakfast buffet before heading out on one of the half-dozen field trips offered that day by the Texas Tropics festival. Some would be going to Bentsen–Rio Grande Valley State Park near Mission for hook-billed kites and Altamira orioles, while others would be canoeing the river below Falcon Dam, hoping for red-billed pigeons, or visiting a ranch bed-and-breakfast that caters to birders and claims a 100 percent success rate with its nesting ferruginous pygmy-owls. As far as I could see, everyone was chatty and alert despite the time—as a rule, we birders are disgustingly chipper at dawn—but for late risers there was a butterfly field trip scheduled for mid-morning, when the sun would stir the insects to activity.

At last count, there were more than 125 annual birding festivals in the United States and Canada, as well as several in Mexico and the Caribbean, and while many of them had their genesis with nature centers or conservation organizations, a surprising number are run by business leaders who see an opportunity. That's the case with the Texas Tropics event, which is presented by the McAllen Chamber of Commerce and usually attracts more than four hundred participants, largely from outside the region. Held over four days at the end of March, it's a mix of daytime field trips, evening lectures and seminars by guest speakers, social events, and a variety of nature-themed vendors. And if four days isn't enough, participants can also sign up for a sister festival held immediately thereafter in Monterrey, Mexico, as well as pre- and post-festival field trips into the northern Mexican mountains.

I found a seat on one of two buses heading to Bentsen, and once we got there, we split up into several groups, finding chachalacas and quail, orioles and golden-fronted woodpeckers. A small herd of javelinas trotted across the road, their hooves clattering on the pavement, but the biggest entertainment was one of our own group, a middle-aged fellow from Tennessee, a very new birder making his first trip to South Texas. On a field trip the day before, he'd seen seventy life birds, and he was anxious to rack up a similar total at Bentsen. His enthusiasm was childlike, and everyone was enjoying finding him still new additions for his list, but some of us tried to rein in his expectations a little; there was no way he could maintain such a pace, even in the lower Rio Grande, and we were afraid he might be setting himself up for a fall. But almost every birder goes through this kid-in-a-candy-store phase, before realizing that nabbing new species is only part of the thrill.

One of the field trip's highlights was a hook-billed kite, a strange tropical raptor that specializes in eating snails, which was discovered nesting north of the border first in 1964 and again in 1976, and now breeds in a number of locations in the LRGV. There have been quite a few species, once thought to be Mexican-only birds, which are now found along the U.S. side of the border. Some, like Altamira orioles, have definitely expanded their range north, probably as a result of habitat changes (and it's worth noting that two other orioles, the Audubon's and the hooded, have declined in the area, probably for the same reason), while others may simply be better detected these days, with so many more birders looking.

At midday, everyone reluctantly boarded the bus for the ride back to town as the tour leader, Red Gambill, pushed back his hat, smiled a huge white smile in his tanned, deeply lined face, and grabbed the microphone for the vehicle's PA system. He started working his way through the checklist of the morning's sightings as people hollered out what they'd seen. By the time they'd finished, the list stood at 132 species, a half-day total that would dazzle birders in most parts of the country but is considered merely respectable for the LRGV. (This was, ironically, exactly the same total James Fisher had posted fifty years earlier.) The guy from Tennessee was visibly disappointed that he'd only gotten forty-some lifers.

Of course, there's more to Texas birding than just the LRGV, and the state has capitalized on that fact. In the mid-1990s, nature tourism expert Ted Lee Eubanks and the state Parks and Wildlife Department, with support from the Texas Department of Transportation, created the Great Texas Coastal Birding Trail—an auto tour route covering twenty-one hundred miles and encompassing forty-three counties, with more than three hundred bird and wildlife viewing sites, each marked with a distinctive, numbered highway sign that corresponds to a printed map and guide. The idea is that birders could follow the trail—for weeks, if they had the time—and never be more than an hour's drive from the next location. The birding trail concept has taken off; Texas has followed it up with new trails in more regions, and other parts of the country have copied the idea, with birding or wildlife trails in twenty-eight states and one province, and more coming every year.

The birding trail through the lower Rio Grande valley connects widely scattered sites, but the only way that the region's biodiversity will survive the twenty-first century is by physically connecting the increasingly isolated pieces like Santa Ana. In 1979, Congress authorized the Lower Rio Grande Valley National Wildlife Refuge, which will eventually encompass more than 132,000 acres in the four counties from Falcon Dam to the Gulf—the heart of what proponents hope will be a wildlife corridor stretching along 275 miles of the river and protecting parts of eleven distinct habitat types. While much of the refuge is immediately adjacent to the Rio Grande, it includes crucial tracts away from the river, like La Sal del Rey, a brine lake that's been mined for salt since the days of the Spaniards and provides migratory and winter habitat for thousands of birds, including a tenth of all the long-billed curlews in the world.

Jeff Rupert, a tall, lanky man with a goatee and glasses, manages the LRGV refuge, which includes Santa Ana; the day I met him to discuss the future of the wildlife corridor, the first stop was, of all places, the headquarters' main hallway—the only place large enough for the detailed maps that show the extent of the project. Refuge land, owned by the U.S. Fish and Wildlife Service, was outlined in blue; the refuge is by no means a contiguous ribbon, and in some places there are large gaps. Even though it's been twenty-five years since the refuge was first approved, and the project consistently ranks near the top of Department of the Interior land acquisition funding, it's taken a long time to assemble the ninety thousand acres—just under 70 percent of the eventual total—that the agency now owns.

That's about a third of the U.S. side of the river, but there are lots of colors on Jeff Rupert's map besides blue. "This is a huge, across-the-board partnership," he said. "I see the Lower Rio Grande Valley National Wildlife Refuge as the backbone, but it's far from all there is to the wildlife corridor." The map also showed parcels of state land like Bentsen–Rio Grande park, as well as the holdings of nonprofit groups like the Sabal Palm Audubon Sanctuary and the Nature Conservancy's Southmost Preserve. Some cities in the corridor are managing portions of their land as wildlife habitat.

Although the refuge includes some relatively pristine areas of native thorn forest, Rupert's staff has to take what it can get, and that usually means farmland, not brush. The idea had always been to buy up agricultural land from willing sellers and restore it to wildlife habitat, but just letting the land go fallow, managers quickly discovered, was a disaster.

"We found that simply retiring ag land guaranteed a monoculture of exotic buffle grass, Bermuda grass, or guinea grass, which chokes out everything else. And then when it gets dry, that stuff is a tinderbox and we'd have huge wildfires, which would further retard brush growth," Rupert said. "All that argued for keeping those fields in production."

What the USFWS now does is lease out the farmland under agreements in which the planter is required to revegetate a certain percentage of it each year with native species, using government-supplied seed and with agency assistance. It's a phaseout program, as Rupert terms it, in which farming will eventually cease because all the land will have been restored, but that will take a while, since they're currently buying about fifteen hundred acres a year and restoring only about half that much each season. It may take fifty years to complete the transition from field to forest on all the refuge property.

Even if the wildlife corridor can be fully assembled and restored, there's no getting around the fact that several fast-growing cities exist smack in the middle of it, essentially severing the linkage in a couple of places. The worst choke points are the old international bridges to Mexico, which occupy both banks of the river with nothing left in the way of wildlife habitat around them (newer bridges span a wider piece of real estate, leaving a little more undisturbed land below them with which to work). A further complication is the question of border security, always a hot topic here but one exacerbated by the 2001 terrorist attacks. With the Department of Homeland Security pushing to tighten border crossings, Rupert has to find a way to squeeze wildlife through the narrowest of keyholes, negotiating with agencies whose primary concern is not how to safeguard wandering bobcats. Below the Hidalgo-Reynosa Bridge, for instance, he will be lucky to restore a ten- or twenty-yard-wide passage, but that will still be an improvement, because right now there's nothing there but concrete and dirt.

"So in a sense, the future has never looked brighter for the corridor. But when you go north of the river to Highway 83, that's where it gets scary," he said.

Even though the nearest town is seven miles away from refuge headquarters at Santa Ana, the Lower Rio Grande complex is already an essentially urban refuge, Rupert said. U.S. Route 83 runs parallel with the Rio Grande and a few miles inland, from Brownsville all the way past Laredo. Ever since the North American Free Trade Agreement in 1994, economic activity—especially along the stretch from Mission to Brownsville—has exploded, and so has rampant development, much of it completely uncontrolled; some of the counties don't even have authority from the state to enforce local zoning. "Most of the land we manage is backed up against subdivisions, retail zones, and warehouses," Rupert said. That will only get worse as Route 83 is incorporated into I-69 and a proposed quarter-mile-wide mega-highway to funnel trade across Texas, and as Brownsville pursues plans for a huge new truck/rail bridge into Mexico and an expansion of its deepwater port.

What's happening, in Rupert's view, is another change as tectonic as the one Peterson wrote about fifty years ago: the inevitable demise of large-scale agriculture in the lower valley as the land shifts from farming to commercial and residential development. The corridor will be the only hope for wildlife, and in many places it will be a north-shore hope only; the human population is about four times greater on the Mexican side, most of it concentrated right along the river. Altogether, the U.S. and Mexican population along the lower Rio Grande is about 2.4 million and climbing fast.

Although the corridor is far from finished, Rupert believes the effects are already evident, especially from near the Falcon Dam, where there are still some large tracts of brushland forest on the Mexican side, down toward Mission, with growing numbers of large mammals like deer, javelinas, and bobcats using the revegetated land to move into new areas.

Nor will it be a resource solely for wildlife. "The corridor has always been recognized as a hotbed of diversity on a national scale," Mike Carlo, who handles recreation planning for Santa Ana, told me. "But when agriculture is diminished and we have this shining ribbon of habitat running through essentially an urban environment, local people will be able to use it for recreation in ways they never thought of before. It'll be 275 miles of opportunities to tell people what it was in the past, and what it could be in the future."

The wildlife corridor has always enjoyed strong support from a variety of conservation groups in Texas, who pushed the concept years ago and continue to shepherd it by lobbying politicians and community leaders who, with some exceptions, have proven largely supportive as well. But in a state where only 4 percent of the land is in any form of public ownership, the real wild card for habitat is what happens on private land, especially the somewhat more arid cattle ranches a bit farther away from the river, which still support brush.

The size of some of these ranches beggars an easterner's imagination; the legendary King Ranch, founded in 1853, sprawls across more than thirteen hundred square miles—more than 800,000 acres—between Brownsville and Corpus Christi. The Kleberg family still runs the Santa Gertrudis cattle that were developed there, but today most of the ranch is devoted to intensive wildlife management and hunting, with the focus being on white-tailed deer, wild turkey, and bobwhite. Hunting for a trophy whitetail buck on the King Ranch can set you back more than five thousand dollars.

James Fisher, struck by the similarity of the brushland ranches of South Texas to the African veld, wrote, "I was so overwhelmed by the novelty by then that a lion or a giraffe or a herd of antelope would scarcely have been beyond belief." I couldn't help recalling his words as I bumped down a dirt track on the La Coma Ranch near Edinburg, a twenty-two-hundred-acre tract owned by the Bentsen family, one of the old and powerful clans in southern Texas. There were no giraffes, but in the late-evening light we passed Indian blackbuck, kob, and lechwe antelope from Africa, red stags from Europe, and axis deer from Asia. Besides the native game, landowners like the Bentsens have for years brought in a bestiary of foreign species like gemsbok, oryx, and even black rhinos, which thrive in the hot climate.

This has had a number of effects on native wildlife. Properties like the King Ranch, managed for big-game and bird hunting, often have excellent habitat for a range of indigenous species, from wildcats and birds on down to insects. On the other hand, some of the animals that have been stocked have proven extraordinarily destructive and adaptable. Feral hogs—a blend of escaped domestic swine and wild European pigs imported for hunting—are a very different creature from the native javelina, even belonging to a separate family; the hogs are now established across much of the lower Rio Grande and the coast, where they churn up soil, destroy vegetation, devour the eggs and chicks of ground-nesting birds, and compete with native species for food. Nor are they the only exotics to have escaped or been released. Texas now has sizable, self-sustaining populations of axis, sika, and fallow deer, blackbuck, Barbary sheep, and nilgai, this last a huge Indian antelope that can weigh up to six hundred pounds. The state even has feral emus, the ostrich-like flightless bird of Australia. No one really knows what impact this new menagerie is having on native ecosystems.

Private ranchers are also catching on to the economic value of native, nongame wildlife, and more and more are permitting birders and naturalists access to their land—for a price. Here again, the King Ranch has been a leader, offering an array of birding, butterfly, dragonfly, and native plant tours, a model that other ranchers have been following in growing numbers.

Building bridges between private landowners, nature enthusiasts, and land conservationists has been helped in the LRGV through a unique project of the Valley Land Fund, a conservancy formed in 1987 that sponsors the South Texas Shootout, now the world's richest photography contest. All the photographs must be taken on private land, with the cooperation of the landowner, who is, with the photographer, required to pay an entry fee. What's more, all the pictures must be shot within a five-month period, giving the contest an immediacy lacking in those where photographers can dig into their files of old material. At the end of the five months, each photographer submits a portfolio of his or her one hundred best images. Prizes as high as thirty thousand dollars are split between the photographer and the landowner.

The contest, which attracts photographers from around the world, is such a huge endeavor that it is only held every other year. The lengths to which the photographers go are equally astounding; most of them practically live on the properties they are shooting, sometimes building watering holes or elaborate feeding stations to lure in the wildlife. One fellow dug a huge pit, covered it with a roof, and put a deer carcass beside it. Then he crawled in with water, food, and a jug for wastes and stayed for a week. He had to ration his food toward the end, but he got prizewinning photos of coyotes and other predators feeding on the carrion.

Karen Hunke is the former president of the Valley Land Fund, and she and her husband, Phil, a pediatric dentist in McAllen, own a four-thousand-acre ranch where more than a few winning photographs have been taken. On a warm, breezy day, Karen—a trim, elegant woman with blond hair and a pair of worn snake boots—climbed into a stripped-down Ford van converted with old bus seats and a roll bar into an open-air safari wagon, and showed me around the place, which abuts the vast King Ranch.

While the initial goal of the photo contest was to raise money for land conservation, Hunke said the partnership that arises between photographer and landowner has some unexpected consequences. "It's really a learning process for the owner, seeing your land through the eyes of a photographer. It's been wonderful to see some of these families who've been here for a million years—people who inherited their land from the old Spanish land grants, but who say they didn't realize what they had on their land until this contest. And there's the competitive spirit, too. People figure that they're not going to let so-and-so at the next ranch beat them, and so they start doing habitat improvements on their own property," she said.

The Hunkes named their ranch El Tecolote for the great horned owls they saw the first time they visited, and managing it for wildlife has been a passion ever since, eclipsing (so far as I could see) their interest in the two hundred or so cattle they keep. As we drove around, roadrunners legged it up the dirt path ahead of us, pacing the van, and we flushed coveys of bobwhite. When the Hunkes built new deer-hunting blinds on the property, they left up the old ones, which are perfect nesting sites for barn owls; we pulled up to one, the engine idling, and as Karen eased open the door of the plywood box, one barn owl shot out a hole in the side, squirting a stream of whitewash, while the other stood its ground over a single egg—the first of a new brood for this pair, which had just fledged five chicks a few weeks before.

Although the ranch is forty-five minutes or more from town, the town is closing in fast. The Hunkes swallowed hard and bought a two-thousand-acre ranch on their northern border when it came on the market the previous year, fearing that otherwise it would be chopped up into ten-acre ranchettes, as is happening all along the roads closer to McAllen. The land boom is gathering strength and economic weight; between the time they bought their first fifteen-hundred-acre parcel and this last, the per-acre price of land jumped from $650 to $2,000.

The Hunkes have done such a fine job with their property that they received a statewide land stewardship award from Texas Parks and Wildlife, and they have turned the spread into something of an informal research station. But Karen Hunke regrets the species that are gone. She's hopeful that the aplomado falcon, which has been reintroduced along the coast an hour or two away, will eventually return, but some of El Tecolote's wildlife is lost for good. Standing by a pond, watching blue-winged teal dabbling for food, Karen recalled a conversation she'd had with an elderly woman who'd grown up near the ranch before World War II, when this was a very much wilder place. The woman told Karen that one hot Christmas Day when she was about ten, she was riding her new bike and stopped to take a rest under a shady tree. Hearing a noise, the girl looked up, "and she said, 'There was the most huge cat, and it was black and it was yellow, and it had the biggest yellow eyes.' And I said, 'My gosh, Thelma, you saw a jaguar!'"

That must have been one of the very last jaguars to venture into Texas, but there are still wildcats around. Mountain lions have been seen on El Tecolote, and bobcats are common. Karen believes she's twice seen a jaguarundi, the strangely elongated, otter-like wildcat of the neotropics, which comes in reddish and gray color morphs. There are even ocelots on a neighboring ranch, but while biologists have set automatic camera traps at El Tecolote, they didn't capture any photographs of ocelots there—to the Hunkes' great disappointment.

The ocelot is the most mysterious and charismatic of all the lower Rio Grande's wildlife, a wildcat of remarkable grace and beauty but so little known that until recently no one was even sure if it still existed in the United States. A tropical species found as far south as Argentina, ocelots once ranged up into southeastern Arizona and across southern Texas to the canebrakes of northern Arkansas and the coastal forests of Louisiana. But demand for their lovely spotted pelts was unrelenting, and predator-control efforts aimed at coyotes, bobcats, and other, more common species took a toll as well. The biggest blow, though, was habitat loss. The lower Rio Grande valley was their last haven in America, but Roger Tory Peterson assumed, like most naturalists in the early 1950s, that they'd been exterminated there as well. It wasn't until 1957 that a trapper caught an ocelot on Laguna Atascosa National Wildlife Refuge, along the coast north of Brownsville, and released it unharmed. The ghostly cat, it seemed, was not gone after all.

Not for another two decades did anyone try to study Laguna's ocelots, and when they did, the population was thought to be just twenty cats—a low guess, it turned out, but not by much. Today, after a great deal more research, scientists think there are only eighty to one hundred ocelots remaining in Texas, in two disjunct groups—one on and around Laguna Atascosa, and the other on extensive private ranchland about twenty miles to the north and west of the refuge, separated by farmland. As a federally endangered species ecologically bound to the brushlands, the ocelot is at once a victim of the changes sweeping the valley and a major player in how land there will be conserved.

"That poor animal," sighed Jeff Rupert at the LRGV refuge, when the subject of ocelots came up. "The weight of the world is resting on that cat's back."

Linda Laack grew up on a dairy farm in Wisconsin and admits that the heat, humidity, and bugs of South Texas took some getting used to. She was working in northern Minnesota in 1985 when she was offered a chance to do her master's thesis on ocelots, and it was early summer when she got to Laguna Atascosa, which was then a forty-five-thousand-acre refuge of coastal prairie, Tamaulipan thornscrub, and tidal flats along the Laguna Madre. She was picking up the threads of a project started a few years before by another graduate student, but there wasn't much for her to go on.

"I'd read that ocelots wouldn't cross open ground, not even a trail, so I spent months dragging these heavy traps way back through this dense brush, crawling on my belly, because if you're any bigger than an ocelot, it's the only way to move around in there," Laack said as she stopped her pickup to unlock a gate to one of the backcountry sections of the refuge. "It was hot, it was humid, the mosquitoes and chiggers were awful, and I wondered what on earth I'd gotten myself into. It took me two months to catch my first ocelot, and longer than that to realize that ocelots are lazy, just like us, if they can be. They'll cross roads and use trails. Now I set my traps out here, where it's a lot easier."

It was a brilliant morning, and Laack—a no-nonsense woman with wavy blond hair and a crisply pressed USFWS uniform—was checking ten traps she had set around the refuge. No longer just a grad student, she has since 1990 been Laguna Atascosa's resident biologist, and probably the most experienced ocelot researcher in the country. Each winter and spring, she tries to trap about ten of the roughly thirty ocelots on the refuge in order to radio-collar them as she continues to unravel the mysteries that surround this slip of a predator.

Ocelots are among the world's most bewitching animals. Weighing fifteen to twenty-five pounds, they are as heavy as a bobcat but more gracile—leggy and tapered, with a long tail, and intricately patterned with black scrawls and russet spots against a background the color of summer-dried grass, which makes them almost invisible in the dusky, dappled light beneath a thorn forest. As Laack discovered, they will cross open ground when they have to, but they prefer a solid roof overhead, which means they are restricted to brushland and other dense cover.

Laguna Atascosa is a captivating landscape, where the thornscrub that the ocelots prefer mixes with coastal prairie that runs down to the miles-wide lagoon, with South Padre Island hazy in the distance. Caracaras and Harris's hawks soar overhead, and around the lakes and tidal flats there are thousands of shorebirds, waders, and waterfowl in season. The prairie is dominated by big yuccas fifteen or more feet high, with crooked, multiple trunks, most of which were in bloom when I was there—shaggy crowns of waxy creamy-white flowers, on which Couch's kingbirds perched. The yuccas give the refuge an otherworldly, Seussian air, as though the Lorax might hop out at any moment.

Some of Laack's traps were tucked in the shade beneath thorn trees right along the dirt road; at other spots Laack hopped out of the truck, grabbed a long stick, a jug of water, and another jug of pigeon food, and trotted back into the prickly pear and mesquite. Swarms of mosquitoes shadowed us; there'd been several inches of rain two weeks before, bringing on a big hatch, but they wound up being the lesser problem, compared with the legions of tiny seed ticks and chiggers I picked up.

Each trap was a large wire cage, with a separate holding cell, reinforced with fine screening against snakes, containing a live pigeon. Laack topped off the lure bird's food and water and moved on, one empty trap after another, and I was resigned to coming up dry until we approached one set and could hear, unmistakably, the rattle of the wire cage and the wings of the panicked pigeon. Laack tensed, moving in slowly, then relaxed. "Just a possum," she said. "Usually they're asleep in the day, and when you hear the cage clatter like that, it means a cat." She opened the trap, propping the door with the stick, and started feeding the pigeon, but it took the better part of a minute for the opossum to figure out it was free to go.

Over the years, Linda Laack's caught everything from javelinas to alligators in her traps, but one animal she has yet to catch is the ocelot's close relative, the jaguarundi. Although a few were seen on the refuge through the 1970s and early '80s, the last record in the area was one killed by a car near Brownsville in 1986, and Laack isn't sure they even still occur on the refuge; she gets lots of jaguarundi reports from the public, but the photos she's seen have all been of house cats. It may be that the jaguarundis weren't able to compete with the far more abundant and adaptable bobcats, while ocelots have been able to coexist with them.

Her research has shown that male ocelots defend a territory of about five hundred acres if they can, though in their cramped haven they must often make due with much less. A male's range usually overlaps those of one or two females, as well as a couple of subadults who skulk around the margins, trying to stay out of the dominant cats' paths until they can find a territory of their own. Even though Laguna Atascosa is big—the refuge has grown to more than eighty-eight thousand acres in recent years—much of it is prairie or island sand dunes that aren't good ocelot habitat, and what land there is for the cats is full to the brim. And because the two Texas populations are isolated from ocelots in Mexico, and from each other, they appear to be highly inbred.

The young ocelots try to move out, not across the open farmland that separates Laguna from the other, more northerly ocelot population, but south through the fragmented brush toward Brownsville, where they immediately bump into the rising tide of new development. A few of the young cats somehow manage to hang out in those small thickets and *resacas* among the new housing developments until they are old enough to come back and take a vacant territory, but it's a rough neighborhood. Roadkills account for a hefty chunk of the mortality among Laack's ocelots, and plans to upgrade two roads through the refuge, and connect one to a proposed new causeway to the resorts on South Padre Island, are unsettling to ocelot supporters.

Although Laguna Atascosa lies to the north of the formal Lower Rio Grande wildlife corridor, there are plans to connect the two via a curving ribbon of land sandwiched between Brownsville and Harlingen. This may give the Laguna ocelots someplace to go, but Laack is troubled by the fact that the corridor will be, well, mostly corridor—a very linear strand of habitat, with few extensive pieces of land like Laguna to anchor it. "For me, I think you start with the big blocks of habitat, and then you connect them, instead of starting with the corridor first," she said. She realizes how much of the impetus for the corridor has come from concern over the ocelot, which ranks as a glamour animal even among the many wild celebrities of the LRGV. But she's unsure how well it will serve her cats. "We sold the corridor on the strength of the ocelot," she said, "but we haven't sold the corridor *to* the ocelot."

Meanwhile, the refuge buys up what land it can or arranges permanent conservation easements on neighboring plots, aiming for the 108,000-acre total it's authorized to purchase on the mainland and South Padre. "You perform triage," Laack said. "You pick and choose what's important. For ocelots, you buy dirt and you plant brush back on it—there just aren't many big tracts of brushland available."

Now even the dirt is climbing out of the refuge's price range. An acre of farmland, being sold as such, goes for one or two thousand dollars—and that's all the refuge is allowed to pay. But as developers buy up neighboring land for subdivisions, the asking price is skyrocketing to three or four times that amount. That morning, on the southern boundary of Laguna, I'd passed a large sign advertising almost eight hundred acres of farmland for sale. The refuge would love to buy it but can't. "They want six thousand dollars an acre for it, but if we tried to buy it at farm prices, the owners would just laugh at us—in fact, they'd be insulted. So wildlife is getting the leftovers. The best ocelot habitat grows on the most fertile soils, which we can't afford."

But even with the refuge expanding and the corridor taking shape, Laack sees all the odd, brushy little corners and jungly old *resacas* to the south of the refuge disappearing into new housing construction. "I'm still not sure there's a net gain," she said. "When they say that all the brush was cleared years ago, that isn't true—all these small towns had little *resaca* systems, and the cats were able to move from place to place. There were ocelots that spent months down in those *resacas*, waiting until they could find an adult territory. That connectivity is going."

Maybe, Laack wondered aloud at one point, we're wrongly putting all our eggs in one basket. And after all, ocelots were once found across a much larger part of the south-central region, far from the brushland, where they will never return without human help. "We've been so focused on trying to save this one little population that I don't think we've looked hard enough at what their historical range was," she said. "Maybe we ought to be thinking about range expansion. Now, in Arkansas they were in canebrakes along the rivers up in the north, and that's all gone, but what about Louisiana, or the Big Thicket in East Texas?" A reintroduction program could use Mexican ocelots of the original subspecies, and because ocelots are well-behaved predators (with no documented cases of them taking livestock), conflicts with humans shouldn't be an issue. Given the slender toehold the cats have in Texas—at risk from one bad hurricane, or a disease outbreak, or some other disaster—spreading them around may make sense. (There is some indication that ocelots may also be pushing north again into Arizona, although hard evidence of this is scarce.)

We checked the last traps, then drove out toward the vast lagoon, where Caspian terns plunged for food and ospreys sat on the mudflats eating fish, surrounded by patient squads of laughing gulls hoping for scraps. The hotels and condos of South Padre made a jagged skyline along the horizon, a reminder of how more and more closely the outside world is hemming in this last bit of the old borderlands.

Soon the siege will be complete. "I think this refuge will be surrounded by a sea of houses, which will make Laguna Atascosa an even more precious place than it already is," Laack said when I asked what the future holds. As is the case up and down the lower Rio Grande, the unanswered question is whether conservationists can save and restore enough land quickly enough to keep all the pieces—even the sensitive, wide-ranging ones like ocelots—in the urbanized world that will be South Texas.

But sitting by the bay, I knew there was another face to the great Laguna Madre, one where there are no clamorous resorts and hordes of beach buggies filling the barrier islands, where there are still huge areas of Tamaulipan brushland along its inner fringe—a place where ocelots and even jaguars still hunt. It was time to cross the border and see how wild North America is faring in Mexico.

EIGHT

The Real Treasure of the Sierra Madre

Well before dawn, the air in Tamazunchale was so humid, so oppressive, that the gentle rain seemed almost redundant. Bob McCready and I climbed sleepily into our truck and drove out of town, passing Huasteca Indians carrying on their backs what were surely enormously heavy loads, supported by wide tumplines across their foreheads—firewood, huge woven plastic sacks of citrus, and, in the case of one man, an old-fashioned milk can.

Tamazunchale, which sits in the foothills of the eastern Sierra Madre about eighty miles from the Gulf of Mexico, was a small, gritty town, and we were happy to see the last of it. Several days earlier, Bob and I had crossed the border from South Texas, driving deeper and deeper into rural Mexico, baking in the heat and humidity of the coastal lowlands. But at last we were turning away from the foothills and into the mountains themselves. The clouds followed us for an hour or so, but as we drove higher and higher up winding roads, they began to break up, as though shattering into gray plates on the mountains. We left them behind entirely, far below us, as we switchbacked up canyons and around great sheer precipices, always climbing, climbing west into the heart of the Sierra Gorda. Ahead of us, heavily forested massifs reared nine or ten thousand feet into the cool blue sky.

Roger Peterson badly wanted to include Mexico in the *Wild America* itinerary but knew that he and Fisher wouldn't have a great deal of time to spend in the country; a taste would have to suffice, somewhere they could see the greatest diversity of species in the smallest area. The answer seemed obvious; a few years earlier, Americans conducting a Christmas Bird Count at Xilitla, in the mountainous state of San Luis Potosí, had tallied 230 species, the highest single-day count on the continent to that point. The rugged mountains around the small village offered everything from lowland jungle to high cloud forest.

Because neither he nor Fisher had been to the tropics before, Peterson enlisted a guide, Bob Newman, a university ornithologist from Louisiana who had worked extensively in Mexico. I was also fortunate, as I prepared to follow their route south, in having a good guide and traveling companion, who was coincidentally also named Bob—Bob McCready, who lived and traveled around Mexico for years, including a stint as northeastern Mexico director for his employer, the Nature Conservancy. Months earlier, I'd met Bob at a conference, and we spent hours discussing the original *Wild America* trip—what Peterson and Fisher saw, what they missed, and the tremendous changes in Mexican conservation since their visit. Bob was full of suggestions, and before either of us knew it, he'd decided to take some vacation time, borrow a company truck, and show me around himself.

We would have the luxury of more time than my predecessors. Fisher and Peterson spent only five days in Mexico, much of it driving as fast as safety allowed on poor roads the four hundred miles each way to Xilitla, high in the eastern Sierra Madre. They really only spent about thirty-six hours doing any serious birding and wrote little of the trip south along the Mexican portion of the Laguna Madre, or back north through the Chihuahuan Desert on the western side of the mountains. We'd have ten days to cover much the same ground, which Bob promised would provide a fascinating look into the challenges and potential for wildlife and wildland conservation south of the border.

Our plan was to cross the Rio Grande—or Río Bravo, as the river is known in Mexico—as Fisher and Peterson had, from Brownsville into Matamoros, and explore the Laguna Madre de Tamaulipas, the southern portion of the great lagoon, which runs for more than 125 miles below the border. Then we would angle inland, intercepting the Sierra Madre Oriental right at the Tropic of Cancer and, climbing into the mountains, stop at a number of birding sites before reaching Xilitla, which has developed a minor tourist industry in recent years.

But Xilitla would just be the beginning. What we knew, and what Peterson probably didn't, is that if they had pushed just a bit deeper into the mountains west of Xilitla, they would have reached one of the most spec-

EASTERN MEXICO
Falcon Dam
TEXAS
Rio Grande R.
McAllen
Harlingen
Reynosa
Brownsville
Matamoros
Monterrey
Saltillo
MEXICO
La Soledad
GULF OF MEXICO
Sierra Madre Oriental
Ciudad Victoria
TROPIC OF CANCER
El Cielo
Ciudad Mante
Tampico
N
Xilitla
Tamazunchale
Jalpan
Sierra Gorda
75 Miles

tacular and wildlife-rich regions of Mexico, the Sierra Gorda range, which today, thanks to a grassroots effort among local communities, is an enormous biosphere reserve that may be the greatest conservation success story in that country. That would be our major focus before we, too, headed north along the western flank of the mountains, through the high, arid grasslands of the Chihuahuan Desert.

I met up with Bob in Harlingen, and we loaded my gear into an old cream-colored four-by-four that he had, years earlier, dubbed La Blanquita, "the Little White One," which bore the scars of many cross-border trips. McCready, for his part, carries his age a good deal better; he's a forty-year-old with a rusty goatee and the build of an enthusiastic tennis player, which is offset by an ever-present pinch of snuff in his lower lip. Bob tends to favor shorts, sneakers, and faded tropical-print shirts, and he punctuates his conversation with expressions like "Geez Louise!"

We crossed the border on a silky, warm evening, the trees along the river (which Fisher, expecting some great torrent, had found disappointing) full of creaking, jangling grackles, as a slow procession of pedestrians walked back into Mexico over the bridge—men in cowboy hats carrying plastic shopping bags, women wheeling collapsible handcarts with small appliances they'd bought, teenage girls laughing among themselves, young boys spinning in orbit around their mothers. Our hotel was an older building of terra-cotta and colorful tile work, and once we'd cleaned up, we ate grilled goat in a restaurant Bob knew just a few blocks away, maps spread around us as we talked through the trip.

In the morning we headed to the coast, past huge new American chain hotels on the outskirts of town, built for businessmen following the flood of maquiladoras, foreign-owned assembly plants in Mexico. The light was just coming up, and Chihuahuan ravens were sitting along the roadsides on fences and old snags, along with Tamaulipas crows, a regional specialty most commonly seen on the American side of the border at the Brownsville dump, which explains that landfill's fame among birders.

We turned off the main road toward the distant village of Puerto El Mezquital, which sits on a barrier peninsula between the Gulf and the Laguna Madre. Much of the land we passed was tilled and planted; Bob explained that in the 1970s, the Mexican government increased subsidies to clear land for farming, which resulted in widespread replacement of thornscrub with corn and sorghum. But the cost in terms of soil has been severe. "It's a real clayey soil, and, man, is it windy down here—you're right on the ocean, there's no relief. This is a place made for wind. When it's dry, you get a mini-dust-bowl effect." The area, he told me, has the second-highest rate of erosion in the country.

We passed enormous fields, hundreds and hundreds of acres each, some privately owned, some controlled by the local *ejidos*, or peasant landholdings set up after the revolution when the vast haciendas—some of them containing millions of acres each—were broken up and given to the people who had once been nothing more than serfs.

Because agrarian reform and the *ejido* system bear so strongly on modern attempts to conserve land in Mexico, it's worth spending a few moments on their history. After the revolution, some of the old landowning families were able to reassemble their holdings, though on a much-reduced scale, and because the government restricted ownership to no more than a thousand acres per person, the pieced-together ranches often had titles spread among dozens of members of an extended family. Titles remain foggy and confusing to this day, which makes trying to purchase land or arrange for a donation extraordinarily difficult.

As for the *ejido* land, Bob told me, title to it remained with the federal government, much as on Indian reservation lands in the United States. Consequently, the *ejidos* had no collateral for loans, nor (often) much education in how to manage a farm. The result had been a steady decline in the fortunes of many *ejidos*, even as private landowners fared much better. In 1992, Mexico changed its constitution to allow for private ownership of the *ejido* land, with local collectives dividing most of their holdings among their members; the thinking was that this would give them collateral against which they could borrow for machinery and other supplies. But predictably, it also gave them the right to sell. "They can sell to conservation groups like us, but there aren't enough of us out there," Bob said. "Or they can sell it to corporate agribusinesses or other large landowners." And the latter is exactly what's been happening—a great irony, since the *ejido* system, which was created to spread landownership among the poor rural population, has now sparked a trend back toward very large landholdings.

As we drove, the land dropped by subtle degrees to the sea, with the newly risen sun peeking through high fog banks; the fields were replaced by occasional thickets of low mesquite, clumps of prickly pear bright with yellow or orange flowers, and wide vistas as the mist burned off. There were now great mudflats to our north, where flocks of shorebirds worked like sewing machines, probing, probing, probing—dowitchers, stilt sandpipers, yellowlegs, avocets, black-necked stilts. These were not, as I first supposed, tidal flats; the "tide" is caused by the wind, for on these extremely wide, extremely shallow lakes, even a moderate breeze will pond up the water on one side while leaving enormous areas of exposed mud on the other. The same phenomenon occurs in the laguna itself, on a much larger scale, when there's a stiff north or south blow—the wind simply empties one end or the other of the lagoon, creating ephemeral feeding bonanzas for the shorebirds, which move nomadically around the entire bay system depending on where the best conditions might be found.

The Laguna Madre is a binational ecosystem, and one of hemispheric importance. Its highly saline waters grow lush beds of seagrass, which are the main diet of the diving ducks known as redheads. Eighty percent of the world's redhead population winters here, moving freely between the American and the Mexican sides, along with two-thirds of the shorebirds that winter in eastern Mexico. It is an incredibly important spawning and nursery area for fish, shrimp, and shellfish, supporting both sporting and commercial fisheries.

A finger of the laguna proper opened to our right, with white pelicans soaring overhead; there is a local colony, the most southerly in the hemisphere. We got out to move a mud turtle across the road and to watch a snowy plover hunting along the water's edge. Out in the lagoon, maybe half a mile away, a man was poling a raft so small that he looked as though he were gliding across the water itself. Just up the road, a flock of thirty or forty reddish egrets—half in typical maroon plumage, half the striking white morph this species exhibits—were working a school of fish in close formation, dancing and lunging, throwing wings to spook the fish in a tight, violent way that looked a great deal like a rugby scrum.

Here and there, we saw the remains of old shrimp weirs, V-shaped lines of stakes extending hundreds of feet in either direction, on which nets would be strung, leading to a small collecting box at the apex. When the adult shrimp come into the lagoon to spawn, they are trapped in the box until the fisherman comes by—a passive, labor-saving means of fishing that takes only shrimp and, because of the coarse netting, only mature ones at that. But now fishermen, many moving north from Veracruz, come up to Tamaulipas with huge fine-mesh seine nets up to two kilometers long, which they tow through the lagoon between heavy skiffs with extra-powerful engines.

"Of course, they get everything," said Bob, who in his earlier life was a social scientist and did research on Mexican fishing communities. "The redfish, the speckled trout, the big shrimp, the small shrimp, everything—the bycatch is outrageous, along with the way they scour the bottom of the Laguna. These heavy engines burn through the gas, and the fishermen burn through engines, so they have to make a lot of money to make it pay. Whereas the weir fishing has virtually no bycatch and can be sustainable—it doesn't even require much gasoline, because the fishermen use these little put-put engines." And the reason the Veracruz fishermen are here in the first place is because they have already stripped clean their home waters. "It's desperate people, doing desperate things," he said.

In some respects, the two sides of the Laguna Madre are polar opposites, each reflecting what's missing in the other. While the Mexican side suffers from the kind of strip-mine approach to fishing that Bob was describing, it is largely undeveloped and free from the dredging conducted in Texas to maintain the Gulf Intracoastal Waterway, sedimentation from which has badly damaged seagrass beds. The American half of the lagoon, in contrast, has fairly healthy fisheries, with a robust sportfishing industry, but the barrier island has been greatly impacted by heavy recreation and the South Padre resort development. And unlike in Texas, there are large areas of intact thornscrub around the Mexican lagoon, habitat that still holds parrots, ocelots, and even jaguars.

Not long ago, the Mexican government included all of the Laguna Madre de Tamaulipas in a 1.5-million-acre protected area, including the barrier islands and some of the inland thorn forest. While this probably scuttles plans to extend the intracoastal waterway to the south—a scheme that never made much economic sense anyway—it doesn't throw a fence around the Laguna Madre or affect private landownership within the protected area. Instead, local communities and fishing cooperatives like Puerto El Mezquital, which were instrumental in getting the designation in the first place, will have a chance to create a zoning system of core reserves, areas for sustainable development, agriculture, and so on. This is the common model used in Mexico for land conservation; the goal is to protect the ecological character of the reserve while allowing residents there to make a better life for themselves.

We drove back inland, looping hours to the south and passing through two military checkpoints, a little show of muscle aimed at drug smugglers; the young men with machine guns politely rifled our bags, asked about our binoculars, but seemed puzzled only by the power pack for my laptop. The land became hillier, more arid, the scrub lower and scruffier. There were roadside vendors selling yucca flowers, the great masses of white blooms hanging like plumes of cotton candy from the roofs of their thatched stands. There were scissor-tailed flycatchers beyond number, pearly gray kingbirds with impossibly long forked tails that would flash deep pink under their wings as they flew, and lots of raptors—Harris's hawks, a gray hawk, harriers, white-tailed hawks, a white-tailed kite, and several Swainson's hawks on their way from Argentina to the Great Plains.

We overnighted in Ciudad Victoria, which sat hard by the base of the Sierra Madre Oriental—big, serious mountains, rising without preamble from the lowlands, the tallest of which was Cerro El Potosí, more than twelve thousand feet high. The next morning we pulled out an hour before dawn, climbing up into low hills; a short while later, we passed a sign reading TRÓPICO DE CÁNCER, and while I, unlike Fisher and Peterson in 1953, had been to the tropics many times before, it was still a thrill to make the transit, as though daybreak would unveil wonders wrought by the passage of this theoretical line.

Oddly, dawn did exactly that—and this shouldn't have been a surprise. Ornithologists have long commented on the remarkable jump in avian diversity just south of the Tropic of Cancer, and that was certainly the case for us. As it grew light, we saw mountains just to our west, the Sierra de Guatemala, rising more than seven thousand feet, thickly covered in forest; flocks of twenty or thirty red-lored parrots zoomed by in hurried flight. We turned off the main highway and zigzagged higher and higher to the trim little village of Gómez Farías, bright with ornamental shrubs and flowers, which sits on the edge of El Cielo Biosphere Reserve, a popular birding destination.

There was a distinctly tropical flavor to the birds here—lots more parrots, social flycatchers, hummingbirds, and orioles. A pair of masked tityras flew in, gray birds with black wings and robber's masks. The tityra impressed Fisher with its improbable grunting vocalization, a call that one authoritative field guide to Mexican birds describes accurately (if indelicately) as "fart-like."

The pavement gave out at Gómez Farías, but we thumped and rumbled up an increasingly bad road of dirt and broken limestone, climbing several thousand feet in a series of hairpins and switchbacks. Mist rose from the valley, along with the crowing of roosters, the barking of dogs, and the choruses of chachalacas. The trees grew in stature, a pleasant change after days in the scrubland, their branches festooned with heavy locks of Spanish moss, among which small birds flickered—tropical parulas like the one I'd seen at Santa Ana, but also migrants heading north for the breeding season. These were birds that would fan out in all directions in a few weeks—Wilson's warblers that might go as far as central Alaska, western tanagers that breed in the northern Rockies, and a large flock of rose-breasted grosbeaks, black and white with a pink triangle on each of their breasts, which would hang a left across the Gulf and go back to the hardwood forests of the East.

The higher we climbed, the heavier the epiphyte load each tree carried—a few pines and lots of large oaks and sweetgums, heavy and widespreading, their branches cloaked in bromeliads, orchids, moss, and a shaggy forest of ferns that hung from them like green fringe. The hills folded in around us, high limestone escarpments overhanging the narrow valley. Many of the trees were in bloom; one species, covered in flowers of an almost neon fuchsia, was common enough to make some mountain slopes look purple, while a few trees had gained unexpected grace in death thanks to a vine cloaked in large yellow blossoms. Brilliant blue morpho butterflies drifted along the road, as big as my hand and looking lazy in the air, but I knew better; Peterson, a butterfly enthusiast, had chased them all over the place during his time in the Mexican mountains and caught only one ratty specimen for his troubles.

The Sierra Madre Oriental is one of the most unusual mountain ecosystems in North America. Because it spans the temperate zone and the tropics, it brings together species that otherwise do not mix—the most southerly population of black bears, for example, with the most northerly spider monkeys. Common ravens and American dippers are found with military macaws and emerald toucanets. The complicated topography, along with rainfall and wind that vary with elevation and distance from the coast, creates dozens of habitat communities ranging from lowland rain forest to Chihuahuan Desert.

A lot of the remote mountain tracks are as bad as they were in the 1950s, but we had the advantage of better roads in the lowlands as we continued south along the eastern side of the Sierra Madre. We drove through the beginnings of true tropical dry forest dominated by tall, disheveled palms, though most of it had been cleared for pasture or sugarcane, then moved into rugged country again, higher and wetter as we headed into southern San Luis Potosí. The villages through which we passed were heavily Indian (though in much of Mexico *indio* is a shunned word, with racist connotations; the preferred term is "indigenous"). This is still an intensely traditional area; a few of the oldest Huasteca women still sported colorful headbands, and many of the women we saw still wore, as in Fisher and Peterson's day, white blouses embroidered with floral designs in primary colors.

We crossed the Río Axtla, shallow but with a good current, a faint blue-green tint to the water, and lined with the tallest trees in the area; one was fruiting, and its branches were full of red-billed pigeons guzzling berries. The ferry that was in operation in the 1950s had been replaced by a bridge; below us, a man in his underwear and a face mask was swimming with infinite care along a series of shelving rock ledges, a short speargun in one hand. The gun fired, and he added a nine- or ten-inch fish to half a dozen others on a stringer tied to one wrist. "*¿Mojarra?*" Bob yelled down. "Bass?" We'd been seeing signs at the little restaurants advertising fried *mojarra*, a catchall term for cichlids in this region. "*Sí*," the man answered with a laugh—anything that swims.

The road rose quickly from the river up into the cramped canyons that lie within this part of the southeastern Sierra Madre, which is known as the Sierra Gorda—the "Fat Mountains." The land around the Axtla had been cleared, or was covered in nondescript secondary growth; there had been huge old-growth trees and thick lianas, which Fisher called "the finest bit of jungle we were to see in Mexico," but that forest was long gone. We snaked up a couple thousand feet, into wraiths of fog and mist, hugging the thickly wooded slopes from which cement gray cliffs loomed. Around us, the Sierra Gorda soared up into the clouds proper, above eight or nine thousand feet. Nestled among them was Xilitla.

Fisher called Xilitla "a primitive village, an Indian village, a poor but happy village," the last bit of which was presumptuous fluff at best, given that he, Peterson, and Newman spent barely more than twenty-four hours there and were hardly in a position to make that kind of judgment. Though still fairly poor, Xilitla (pronounced Hee-LEET-la) is no longer a primitive village, but a town of about ten thousand people, spread over several hills so steep that I doubt there is a square meter of flat ground in the whole of it. Houses, many of unpainted cinder block, stood in tiers above the narrow streets, which were crowded with people waiting for buses or setting up produce stands near the old, fortresslike church, a row of large bells hanging above its roof.

Xilitla today is perhaps best known for El Castillo (the Castle) and Las Pozas (the Springs), a fantastical sculpture garden created by the eccentric English mining heir Edward James—pillars, fountains, spiraling staircases that go nowhere, and open-air structures with names like the House with Three Stories That Might Be Five. Our goal, however, was to find La Cueva del Salitre, a large sinkhole near town that Peterson and Fisher had visited and described as rich in birds. When we finally found someone who could direct us, we learned that Xilitla had moved; the cave, once a mile from town, was now on its very edge, and the entrance was through a three-man cinder-block-making operation, which created a horrific, ear-grinding din. We paid the twenty-peso entry fee to one of the workers, edged through a narrow gap in a barbed-wire fence, and slid down a slick path through a steep pasture below the road until we were, blessedly, out of earshot of the block makers.

The Sierra Madre Oriental, including the Sierra Gorda, is largely a karst landscape, ancient marine limestone that has been uplifted and eroded into high, jagged ridges—and also eroded belowground, into a wealth of caverns. Once, Salitre was a large cave, but half of its roof collapsed in eons past, forming an immense sinkhole with an amphitheater-like wall on one side, facing the valley below. On the lip, rooted among the rubble of the old roof, grew large trees, while the overhanging wall, which was at least 120 feet high, was studded with short, bulbous stalactites where rainwater percolated through the limestone. Edging carefully down the slope into the cave, we could see a muddy pool of water at the bottom, perhaps another hundred feet below us. This was, despite its proximity to town, still a very impressive natural space.

And a loud one, though not because of the block factory. Fisher had called Salitre the noisiest cave he'd ever visited, thanks to dozens of nesting green parakeets, and their great-great-great-progeny were still raising a ruckus, their screeches echoing in the muggy air. Dozens of rough-winged swallows flew in and out, disappearing in crevices in the wall, while from another cleft a canyon wren poured out its sweet, descending song. I heard a gurgling sound, like water glug-glugging from a jug, and saw a Montezuma oropendola perched in one of the overhanging trees—a big, crow-sized bird related to orioles, which makes a pendulous woven nest four or five feet long. Oropendolas are among the arresting birds of the tropical forest—warm russet with yellow tail feathers and long, heavy bills that look like they are made of two-toned plastic, blue-black at the base and pink at the tip. But what's most enchanting about them is the display males like this one perform. As I watched, the bird opened its wings, fanned its tail, and fell forward while hanging on to the limb, belting out its loud, chugging-water call as it flipped completely upside down, around in a loop, and back up again.

With a local guide to lead them, Fisher, Peterson, and Newman had then hiked from Xilitla up into the cloud forest, which, half a century ago, had covered the mountains surrounding the town. This was to be the main attraction of the Mexican swing and, for both Peterson and Fisher, a highlight of the entire *Wild America* odyssey—and no wonder, for cloud forests are among the most diverse ecosystems in the world, with many endemic species. Those of the Sierra Gorda, which owe their existence to the warm, humid northeast trade winds off the Gulf, have more than thirty species of oaks and fourteen of pines, along with sweetgums, white firs, Mexican beeches, and even magnolias; the strongly temperate character here has more in common with, say, the Great Smokies than with the ceiba trees and palms growing at lower altitudes. And the flip side, sometimes lying a mile or less away, is the starkly drier inland slopes starved for rain, scrubby with agave, small palms, and clumps of tall, columnar cactus. As often as I saw this juxtaposition, I was shocked anew by it each time.

Because cloud forest is restricted to a narrow altitudinal band, there was never very much of it to start with—about 1 percent of Mexico's total forest area—but 10 percent of the country's plant diversity is found within this single ecosystem. On a local scale, the numbers are even more lopsided; fully 60 percent of the plants in the state of Querétaro, just south of Xilitla, depend on cloud forest, along with a hefty chunk of the state's animal diversity. That would make the Sierra Gorda's cloud forest a natural candidate for protection, but the history of conservation in much of Latin America, and particularly in Mexico, has not been an especially encouraging one. Parks and reserves tend to be protected in name only—"paper parks" designated with a governmental flourish, but lacking in funding or any effective, on-the-ground enforcement to prevent illegal logging, hunting, farm encroachment, overfishing, or other destruction.

That's why, when I first heard of the Grupo Ecológico Sierra Gorda, I was so intrigued. Founded in the late 1980s by a dynamic husband-and-wife team, Roberto Pedraza Muñoz and Martha "Pati" Ruiz Corzo, this small environmental organization had, in fifteen years, grown into an astonishingly effective social and conservation force in the eastern Sierra Madre—working with local residents to plant more than 3 million native trees on denuded slopes; installing thousands of fuel-efficient cookstoves to reduce firewood consumption and dry composting latrines to reduce pollution and recycle wastes; creating community-based ecotourism programs; and employing dozens of environmental teachers to work with tens of thousands of children in 130 small communities throughout the Sierra Gorda.

Most impressive of all, the GESG managed to hammer out a broad consensus among those mountain communities, which successfully petitioned the federal government to declare almost a million acres—the entire northern third of the state of Querétaro—a biosphere reserve. The Mexican government, in turn, appointed Pati Ruiz the federal director of the reserve, with the GESG as co-manager. As with the new reserve encompassing the Laguna Madre, the Sierra Gorda biosphere designation doesn't alter the fact that most of the land within its boundaries is privately owned, by individuals or *ejidos*; the only way it will succeed is with the cooperation of the hundred thousand residents of the reserve itself, many of whom depend on the forests for their livelihood. The goal isn't to seal off the reserve but to manage its resources and the way they're used in a sustainable fashion.

The success of the GESG, and the bright future of the reserve, owe much to Pati. "She's a real Mother Earth type," Bob had told me months earlier, "a hippie in a place where they didn't have many hippies. She's big, she's loud, she's smart, and she's won more environmental battles than anyone else in Mexico. She doesn't get aggressive, she just doesn't take no for an answer, and she's accomplished more with very few resources in very little time than anyone I've ever seen." Another conservationist who knows her put it more simply: "Pati is the personification of the belief that one person can change the world."

I had hoped to meet Pati Ruiz while I was in Mexico, but she was, ironically, in Washington, D.C., during our visit, lobbying for international assistance for the reserve. But we'd made arrangements to meet her son, Roberto Pedraza Ruiz, a sturdy twenty-nine-year-old known as Beto, whom Bob had gotten to know years earlier. Beto wanted to show us their latest coup: La Joya del Hielo, "Jewel of Ice," a fifteen-hundred-acre tract of cloud forest his family, working with other Mexican conservationists through a new NGO created for the purpose, had just purchased—an unusual event, because the idea of a private group buying land just to protect it is a fairly new concept in Mexico. What's more, the group had made the deal using donations from several corporations, including Grupo Bimbo, a huge Mexican baked-goods firm and one of the country's largest businesses—an important detail, since corporate donations for conservation are also a rarity here.

We drove up a steep and rocky road through juniper and live oak, with a few pines mixed in, and parked in dry oak forest amid lush stands of wildflowers. As rich as the Sierra Gorda is known to be, its botanical diversity is still being cataloged; Beto recently discovered what appears to be an unknown cactus on one of the highest ridges, and an undescribed species of agave was found in this newly purchased parcel of forest. We shouldered packs and started up the mountain toward the cloud forest, several thousand feet higher; the air was bracing and chilly—and, for someone like me used to lower elevations, noticeably thin.

My mental image of cloud forest comes from my experiences in Central America: mysterious, damp woodlands at the highest elevations, gnarled, ancient trees dwarfed to elfin proportions, where absolutely every surface of trunk or branch is covered with mosses and air plants. And so Joya del Hielo was a revelation to me: a cloud forest of majestic oaks, cedars, and firs a hundred feet tall, draped with mosses and spangled with bromeliads—a soaring, humbling place. This is the *real* treasure of the Sierra Madre.

We hiked up for more than two hours, taking our time to bird from one plant community to another—first juniper, then oak-juniper, then predominantly oak with madrone and cedar. Some of the oaks were huge, with limbs from which dangled hundreds of long, narrow green strands, like cooked spaghetti—epiphytic cacti, just opening dozens of pink flowers. There were bromeliads on most of the branches, but one especially attractive species grew on the trunks of the trees, its leaves all tipped in vivid crimson; a male magnificent hummingbird, big and black-green, flew in and checked out the faux flowers but left hungry.

The birding was good, and Bob and I lagged behind Beto, who was expert enough to note the call, flick his binoculars to his eyes to confirm the ID, and move on. For us, it was all wonderful and new, and we lingered over each species. And what wonders there were, like the golden-browed warblers, clear yellow with rusty heads that bore wide canary stripes. There was nothing standoffish about them; they all but parted our hair when we made quiet *spishing* sounds. We stopped time and again to listen to the brown-backed solitaire, a gray thrush with sober, Quakerish plumage, but a song that makes the forest ring—a startling, loud tumble of bell-like notes that cascade together, sounding like two or three birds singing at once.

And we lingered longest over the most spectacular of all, the mountain trogon, a member of a tropical family that ranks among the showiest birds in the world—a foot long, with a bright red belly and an iridescent green back that would put a hummingbird to shame, which deepens to blue-green on its squared-off tail. The males were calling, a *kewp-kewp-kewp* call that came from all quarters, and where we saw one, we invariably saw more: males chasing females, males chasing males, swatches of crimson darting through the woods like embers. I recalled a line of Fisher's, after he'd spent a day in the cloud forest: "We were becoming sated with novelty."

But glitzy as its birds may be, the Sierra Gorda is more important for what we didn't see. It is so big, and so relatively complete, that it still has all the pieces—almost all the wildlife that it started with before the Spanish conquest, including the big, shy predators that require enormous areas over which to roam. Jaguars are here, along with mountain lions, margays, and jaguarundis. Beto himself rescued an ocelot near Xilitla that someone had captured and caged, and released it in the preserve. The Sierra Gorda is critical for the world's most southerly black bears, which vacuum up the cornucopia of acorns each season, and there are white-tailed and red brocket deer, javelinas, kinkajous, coatimundis, hairy porcupines, otters, dozens of species of bats, and much more. (The one loss, unfortunately, was the Sierra Madre's spider monkeys, now gone.)

The birds also include some high-profile species, like huge green military macaws more than two feet long, a globally vulnerable species with a small, disjunct population here. The preserve is perhaps the last, best hope for saving the bearded wood-partridge, possibly the most endangered bird in Mexico, which is restricted to cloud forest only in a slender strand of real estate in the eastern Sierra Madre. A stocky brown bird with red legs and a pert crest, it is probably extinct in much of its old range, but it remains fairly common in the heart of the Sierra Gorda, where a few hardy birders from around the world come to look for it. They must often make repeated trips, though, because this quail, even in the Sierra Gorda, can be impossibly shy and retiring. And so I will advance an apology to all those weary, disappointed birders before admitting that barely an hour into our hike, a big brown bird with red legs and the build of a football buzzed across the trail at waist height. Beto gasped, said "Wood-partridges!" and pointed at two more making hurried tracks away from us; Bob muttered a delighted "Geez Louise!" under his breath as he jockeyed for a glimpse. I really am sorry. Honest.

The higher we went, the lusher the forest became—oak, white cedar, Mexican basswood, a surpassingly rare species of magnolia with eight-inch-tall buds ready to open, like slender pink candle flames. Everything was wrapped in moss—hell, the air even *smelled* green, don't ask me how. It was charged, oxygenated, fresh. Through the trees, we could see fragments of the neighboring mountains, washes of orange and bronze like the start of autumn in New England, but this was the tint of life, not impending death. Beto explained that the new oak leaves first appear in March in a blaze of color, only later gaining the respectable green of chlorophyll; we had missed the peak by a couple of weeks, but it was still breathtaking.

At first, the cloud forest looked pristine, but here and there we noticed moss-covered stumps, and Beto told us that the old-growth cedars had been cut to make roof shingles fifty or sixty years ago. "The best thing that happened to these cloud forests was when they introduced corrugated-metal roofs," he said. In fact, the mountains around Joya del Hielo have been logged time out of mind—sometimes selectively, as for the high-elevation cedars, sometimes much less discriminately. Fires have swept through some areas, especially since the climate appears to have warmed and dried over the past twenty years, making the blazes worse. That's what makes the remaining old growth so valuable.

"Over here are the fat boys," Beto said, leading us off the trail, our feet scuffing a thick mulch of dead leaves. The fat boys were old-growth red oaks, *Quercus laurina*, and they were dandies; some of the biggest trees up here are 150 feet tall and five or six feet across the base. We hiked among them, giant after giant, moving unconsciously from one life zone into the last and highest, a pure stand of white firs stretching to the top of the peak a few hundred feet above us, their branches, too, swaddled in moss. I wandered in a happy daze, but still attentive to my feet. Beto had warned us earlier that there were an unseemly number of hefty black-tailed rattlesnakes in Joya del Hielo, and much as I would have liked to see one, I didn't want a careless step to offer offense.

The topographical and climatic jumble of the Sierra Gorda makes it rich in life, with at least fourteen major habitat types within the preserve. They harbor a Noah's ark of life: more than 325 species of birds, 131 mammals, and 650 kinds of butterflies (and counting). And it is big country. After we bade Beto farewell, promising to meet him in town in a couple of days, we drove for hours down a white-knuckle road into a hot, sun-seared valley to the town of Zoyapilca, where we left the hardtop. Then we started back up again, the truck laboring in the heat and dust of a rutted dirt road that gained grade with the appetite of a mountain goat. Outside Soledad de Guadalupe, a village of a few dozen homes, there were fields walled off with drystone fences, protecting huge agaves, their fleshy gray-green leaves as tall as a man and each spiked with a long, fiercely sharp needled tip; these are grown to make pulque, an alcoholic beverage. Higher, among chaparral and stunted oak, we admired a short-tailed hawk circling on a thermal; and higher still, we moved back into a deep forest of juniper, the air a good ten degrees cooler than it had been in Soledad de Guadalupe.

And this whole time, we were surrounded by great mountains, their crests bare rock, their sides scored with wide limestone cliffs where the *guacamaya verde*, the military macaw, nested. The mountains rode to the horizon, rank upon rank for maybe thirty or forty miles at a glance, fading into the blue distance—seemingly untouched and uninhabited. But this, I knew, was wrong on both counts; throughout the Sierra Gorda there are countless villages like Soledad de Guadalupe, with fields and pastures, with bony cattle that range free even up to the cloud forests, with men who cut *pino* and *cedro blanco* in such impossibly rugged terrain that they mill the logs into rough boards on the spot and pack the lumber out on the backs of mules.

So it's not the land it was before the conquest, but it's still damned close, and the Grupo Ecológico Sierra Gorda has taken giant strides in making sure it stays that way. "I just can't believe how much they've accomplished, even since the last time I was here four or five years ago," Bob said, clearly jazzed by all the encouraging developments Beto had related. "They're kicking ass and taking names."

One of GESG's many goals is an ambitious plan to help communities around the biosphere reserve tap into the growing ecotourism market. We were heading for one such project, a joint venture with the village of San Juan de los Durán. Two years before, with GESG expertise and federal funding, the village built five or six tourist cabins at the edge of a high valley, beneath soaring mountains. For the equivalent of about twelve bucks a night, visitors can stay in the rustic but comfortable cabins—no electricity, a communal bathhouse with hot showers and flush toilets, and wonderful meals for a small fee. We rolled in late in the day, dusty and tired, and dropped our gear in a room with three bunk beds. There was still a fresh-scrubbed feel to it; the cabins had a rich pine scent from the locally milled wood, and for light we had several tall glass candles with pictures of Christ and the pope. The next day, with a local guide, we hiked high into the mountains, exploring one of the many caves that pock this limestone region, following the flickering light of Jacinto's burning pine torch into the cool darkness of the cavern, where colonies of large bats fussed above us and where until a generation ago priests made pilgrimages to hold Mass in the great entryway.

We packed up La Blanquita and headed west, still farther into the Sierra Gorda to Jalpan, one of the old mission towns in this part of Querétaro, founded by Franciscans in the 1750s; today there are about five thousand residents, and the mission church still stands on the square, pale orange, its facade heavily decorated. After dark, people by the hundreds milled about the plaza, buying snacks and long white candles, preparing for an open-air Easter Mass in the courtyard in front of the church, where a big stage had been set up. Like most Mexican squares, it was shaded by old trees full of screaming boat-tailed grackles and had an ornate fountain in the center. Roger Peterson, attracted by the grackles, had watched evening strollers in the square at Ciudad Victoria in 1953—young men walking clockwise around the plaza, young women walking the opposite direction, both genders safely chaperoned by parents or grandparents. That tradition is largely gone from much of modern Mexico, but the plaza is still a see-and-be-seen part of courtship, and many of the teenagers in Jalpan that night with their parents obviously had more on their minds than piety.

The next morning we followed Beto's hand-drawn map to the Río Escanela, an hour west of Jalpan in the more arid region of the biosphere reserve. The hills overlooking the river had been cleared almost to their tops, with small rock-walled terraces in a few places—and also some evidence of reforestation with native pines, a major focus of the GESG. The Escanela was a lively cold-water creek that came down a remarkably deep and narrow canyon, cool beneath the shade of massive old-growth sycamores, some of which must have been six or more feet in diameter; I found it reminiscent of the deep canyons of southeastern Arizona.

The tiny village, clustered near an old silver mine, is popular with weekenders and holiday visitors who camp and swim—and who, sadly, make a real mess of the place in the process, since there are no sanitary facilities. So with help from the Grupo Ecológico Sierra Gorda and the federal tourism agency, the village has taken the first steps toward both controlling and increasing tourism by requiring that all visitors to the canyon employ one of ten local guides, who have been trained in everything from natural history to safety. Soon they hope to have toilets for visitors, and eventually tourist cabins.

We tagged along with one guide, Roberto Vega Banderas, a newly minted photo ID pinned to his shirt, and a family of four from Mexico City, who were camping along the stream and wanted to see El Puente de Dios, "the Bridge of God," which I'd gathered from Beto was some sort of natural limestone arch. There was a decent trail along the creek, and we quickly realized that the Escanela had, obviously, once flowed through a miles-long cavern, the roof of which collapsed long ago to form today's high-sided canyon, its walls hung with weathered stalactites. At one point, the canyon narrowed to a slot, through which the village had just built an ingenious walkway, bracing concrete slabs against one wall about eight or ten feet above the water. The walkway, and the requirement that all visitors have a guide, grew in part from a fatal mishap a year before, in which a tourist who tried to climb around the canyon slipped and fell.

We crossed and recrossed the creek, sometimes on log bridges, sometimes on stepping-stones, and finally we had to unlace our boots, roll up our jeans, and wade knee-deep in the bracing water—very gingerly, for the shards of limestone were thorny underfoot. "Geez Louise, I really am a tenderfoot," Bob muttered, trying to ease his way across the creek. I had to agree, but when we reached El Puente de Dios, our complaining feet were forgotten. The stream poured from the mouth of the remaining portion of the cavern, probably sixty feet wide and twenty high, with curtains of old stalactites hanging like melted candle wax around the opening, draped with ferns, mosses, and small palms. One especially large stalactite, which looked as though a giant had squished it into a flattened tube, was snapped off at the bottom, and from its hollow interior there shot a fifteen-foot-wide sheet of water, hammering straight down with tremendous velocity—the ultimate shower massage for the family we were with, who had already stripped to their bathing suits.

That was our last stop in the Sierra Gorda; the next morning we left the mountains, driving west to Cerritos, then north, this time with the Sierra Madre Oriental on our right. But although the sierra towered above us, we were still quite high, for now we were on the Mexican Plateau, the arid (and, thanks to a recent cold front, very chilly) altiplano, dominated by Chihuahuan scrub and grassland. Peterson, Fisher, and Newman had crossed into this desert farther north on their way back to Texas, but they wrote almost nothing about this bit of the trip, dismissing it as an anticlimax after the cloud forest. But as with the Sierra Gorda, if they'd made a small detour, they would have encountered a place little short of spectacular—a glimpse of the Old West that has vanished virtually everywhere else in North America.

We tend to imagine deserts as rocky, cactus-studded places, but—especially before settlement and overgrazing—many of them were surprisingly grassy, including the Chihuahuan Desert, whose southern reaches we were now crossing. The shortgrass prairie of the arid Texas Plains blended with the Chihuahuan, creating a sparse grassland ecosystem that continued for hundreds of miles into northern Mexico, complete with buffalo herds, vast prairie-dog towns, and endemic birds.

The bison are long gone, but the prairie dogs and birds remain. In the high valleys south of Saltillo, some of the best Chihuahuan shortgrass prairie still survives on the great llanos,* the griddle-flat plains that lie between high, parallel ridges, in an area known as La Soledad. But its survival, and that of the creatures that depend on it, hang by a thread, as more and more of this precious relict is turned into irrigated potato fields. Bob—who coordinates the Prairie Wings program for the Nature Conservancy, a tri-national effort to secure both the breeding and the wintering grounds for a suite of grassland bird species—had been telling me about this region, its incredible wildlife, and the work TNC had been doing with a local conservation organization, Pronatura Noreste. Early on, we had decided to make the high prairies of La Soledad the finale of our Mexican trip.

We drove north beyond Matehuala, up Route 57, the so-called NAFTA highway, carrying trucks filled with goods heading to or from the United States. Much of the way, we passed little shacks and tarp-covered shanties along the road, where people tried to scratch out a living by selling travelers dried rattlesnake skins, two or three feet long, which flapped stiffly in the slipstream of the trucks. Even here on the valley floor, the elevation was more than five thousand feet, but on the southeastern horizon sat huge Cerro El Potosí, which we'd first glimpsed a week and a half earlier from the east, its hulking, rounded dome more than twelve thousand feet high, wreathed in clouds.

Our destination, San Rafael, was an armpit of a town, strung out along the highway and consisting mostly of ragtag stores, mechanic shops, and vendors servicing truckers; the cactus and brush a long way from the road were awash with trash snapping in the cold altiplano wind. Pronatura Noreste rents a small house on a back alley for its researchers, with bunk beds and cots on which we unrolled our sleeping bags. Our host was Armando Jiménez, a young, intent biologist who runs Pronatura's operation

*Some experts contend the true shortgrass prairie did not extend south of central Texas, and prefer the term "arid grassland" to describe this desert plant community.

in the area; the next day we'd be joined there by a couple of other Pronatura staffers, as well as by one of Bob's TNC colleagues.

Daybreak the following morning would have been frosty if there had been enough moisture in the desert air to condense into ice. Clouds of steam hung in front of our noses as we loaded the truck, and as we drove down a narrow lane into the llano, western meadowlarks and curve-billed thrashers sat high on clumps of cholla cactus, facing the newly risen sun for warmth. A scaled quail perched on a fence post and gave its *queesh!* bark, its head snapping back with a flick every time it called. Burrowing owls, staring at us with flat, bored glances, stood beside empty prairie-dog burrows.

Soon we began seeing the prairie dogs themselves in small patches of shortgrass prairie among the potato fields. These were Mexican prairie dogs, one of five species that once dominated mixed- and shortgrass prairies from southern Saskatchewan to northern Mexico, in numbers that simply beggar belief—more than five billion by one estimate. Because they were thought to compete with livestock for grazing, they have been mercilessly persecuted, often with government assistance in the form of trapping, shooting, and poisoning campaigns, with such tenacity and success that today even the most common species, the black-tailed prairie dog, occupies only a small fraction of its original range.

The Mexican prairie dog, isolated in the high grasslands of the Chihuahuan Desert, initially fared better than its American cousins, but as agribusiness has moved into the altiplano, its population has dropped by more than half in just the past decade, and it is one of the most endangered mammals in Mexico. Today, La Soledad is the last stronghold of the species, but it is quickly disappearing under tractor blades as wealthy investors from Monterrey and other large cities lease up *ejido* land for irrigated potato farms. How quickly? We passed prairie dogs milling about the verge of the road and standing aimlessly in freshly furrowed fields that Armando said had been a thriving dog town just a week before; the rodents were refugees, dug up out of their homes. Off in the distance, we could see tractors plowing still more virgin prairie, while machinery was demolishing forests of giant yuccas twenty or thirty feet high nearby. The prairie dogs and the rare yucca forests led the government to declare this a protected area, but it is protected in name only, it seems; the potato growers are politically powerful, and they can often get the state agricultural agency to issue a permit for clearing despite the protected status, without bothering to consult their environmental counterparts.

Overall, the Mexican prairie dog has already lost 70 percent of its range, and even here in La Soledad only half the dog colonies remain—"And the rest are going like that," said Miguel Ángel Cruz Nieto, the director of conservation programs for Pronatura Noreste, snapping his fingers. Bob and I were jammed in a truck with Miguel Ángel and Jeff Weigel, the northeastern Mexico director for the Nature Conservancy, the job Bob used to have. Our visit was an excuse to bring together Mexican and American conservationists, with a common, urgent interest in the grasslands of La Soledad, for a couple of days of brainstorming and discussion.

We rode out into the heart of the largest of the dog towns, and it's hard for me to convey what an exciting experience it was, because it's all a matter of scale. As individuals, prairie dogs (which are, of course, actually large colonial ground squirrels) are immensely appealing. The Mexican species is yellow-brown, hourglass-shaped with a plump belly that droops when it sits upright, and endowed with a textbook's worth of interesting behavior, like its barking alarm chirps, its instinct for digging elaborately architectural burrow systems, or its habit of "kissing" (actually more of a face-to-face nuzzle). But prairie dogs aren't about individuals; they're about numbers. And here at La Soledad, in the last place on the planet where they exist in truly enormous numbers, the scale simply overwhelms you.

The llano covered a colossal U-shaped valley set among juniper- and yucca-covered mountains another two thousand feet high. The valley ran for miles and miles, and while much of its northern end has been converted to potato fields, virtually everything in sight—as much as twenty thousand acres of shortgrass prairie, four or five miles long and a couple miles wide—was a magnificent dog town. The mountains are all but uninhabited, the haunt of mountain lions, jaguarundis, and maybe a few jaguars; Armando told us that not long ago, a local rancher claimed to have seen a Mexican wolf, a species thought to be long extinct down here. It was a slice of an older, wilder North America, and it was a sight I never thought I would see—one that, until just a few years ago, I and much of the world didn't even realize still existed.

The grass was only a couple of inches high, cropped close by the hungry rodents, and the burrows were scattered with almost geometric regularity, fifteen or twenty feet from each other—mounds of dusty whitish soil punctuated by dark holes, beside which sat alert dogs, watching us with caution. Many were females with a couple of babies playing around them; when the mothers yipped, the youngsters would scurry back and stare, their eyes huge, as we slowly drove past. Some of the mounds were several feet high, like old, worn volcanoes, and Miguel Ángel mentioned that researchers can roughly age a colony by the height of the diggings—the higher, the older. Later, when I had a chance to scuff around, I found the bones of long-dead prairie dogs in the dirt, brought up by fresh generations—the unwritten history of La Soledad.

The prairie dogs weren't the only animals competing for our attention; the colony just hummed with life. Miguel Ángel, who has spent years researching La Soledad's burrowing owls, pointed them out by the dozens; I saw more in that day than in the rest of my life combined. There were hundreds of Cassin's and western kingbirds, stopping in their migration, hunting insects or just sitting on the ground. We flushed desert cottontails and black-tailed jackrabbits, watched coyotes trotting through the heat haze, and saw red-tailed hawks, golden eagles, and kestrels hunting. At night, there are kit foxes and badgers prowling, though the black-footed ferret, a prairie-dog specialist and one of the rarest mammals in the world, was never found this far south. We spotted Worthen's sparrow, a species found nowhere else on earth. The high grasslands are also one of the most important wintering sites for ferruginous hawks and long-billed curlews, the latter of which have been seen in flocks of up to six thousand, but these had already migrated north.

Miguel Ángel jammed on the brakes about lunchtime, pointed, and said, "*¡Chorlitos!*" Just ahead of us, two birds materialized out of the brown-white earth as if they'd just been created from it—mountain plovers, among the rarest and least-known shorebirds in North America. They were shaped like killdeer but with snowy breasts, sienna upperparts, and black caps above clean white eye stripes. The name "mountain plover" is a gross misnomer; "bare-ground plover" would be better, for they inhabit high-elevation short grasslands where the soil they match so closely is exposed. It's thought that they evolved in concert with the huge bison herds, which churned the grass to dust beneath their hooves, and with prairie dogs. Today they are among the rarest of their family, perhaps numbering no more than ten thousand individuals, and it appears that the valleys around La Soledad are one of their most important wintering areas. Even more exciting, Miguel Ángel and his colleagues have found a small, resident breeding population here, hundreds of miles south of what had been thought to be their southernmost nesting range. As we watched, the male of the pair—his colors were slightly more vivid—flared his chest feathers, lowered his head, and flew to his mate, making a sound that reminded me of a cow.

Late in the day, we got out to stretch near one of the big center-pivot irrigation rigs in the north of the valley, which sprayed water in a fine mist over a deceptively green field. The truth is that potato farming here requires great quantities of pesticides and fertilizer, and the geology of the region ensures that it isn't sustainable anyway. "This is all fossil water," Bob said. It's being pumped up from ancient aquifers hundreds of feet deep, which aren't being replenished as they're being drained. "And it's not even good water," he said. "It's highly mineralized, and eventually it just ruins the soil, so that even if someone tried to restore the prairie, I'm not sure that they could get the grass to grow in this degraded stuff."

Pronatura Noreste and the Nature Conservancy are scrambling to protect the many remaining prairie-dog towns on the llanos in this region. They've struck a deal to protect the southern third of the largest colony at La Soledad and are negotiating with the *ejido* that controls the northern third, where most of the farming is taking place. The middle is ostensibly private land, but the title seems cloudy, which makes safeguarding it trickier. For the most part, Armando and Miguel Ángel said, the local communities support saving the prairie and the dog towns, if they have an alternative to selling off the farming rights to rich outsiders. The ultimate goal of both conservation groups is to safeguard an immense U-shaped corridor encompassing the colony and surrounding land, including areas that had once been under cultivation, which will give the dogs room to expand. "This is far and away the largest, most pristine prairie-dog town left on earth, which is why we're trying our damnedest to keep it that way," Bob said.

Our last morning in Mexico, we drove out to La Soledad at daybreak, hoping for some pictures in the sweet dawn light. But overnight a chill mist had come in, and as frantic with life as the dog town had been the previous day, it was silent now—not a meadowlark sang, not an owl showed itself, not a single one of the hundreds of thousands of prairie dogs peeked aboveground as I walked off by myself, leaving Bob and Armando behind. It was like walking through an abandoned city, and the analogy—for what will happen to this last piece of the Wild West if conservationists fail—was all too easy to draw. But as I scuffed through the dusty soil, my cold hands jammed in my pockets, I heard, clear and sharp, the bark of a wary prairie dog. There was life here still—and with it, hope. It was time to turn north again.

NINE

The Sky Islands

In March 1996, a rancher and outfitter named Warner Glenn was chasing his pack of hounds in the Peloncillo Mountains, on the extreme southern Arizona–New Mexico border. The hounds, in turn, were on the trail of a mountain lion, which Glenn was trying to tree for a hunting client—except that the cat wouldn't stand its ground. Instead, it led the dogs and the mule-mounted rancher on a chase lasting for hours, before it was finally brought to bay by the pack.

But when Glenn at last caught up, rather than the cagey old mountain lion he'd expected, he was stunned to find a huge male jaguar slashing at his dogs. Recovering from his shock, Glenn grabbed a camera from his bags, took a bunch of pictures, and, physically hauling off his dogs, let the jaguar go.

Glenn's photographs—the big spotted cat perched on a lichen-covered outcropping, partially screened by a piñon pine, with the dry, rugged mountains in the background—electrified conservationists across the country. Most people, if they think of jaguars at all, consider them jungle cats, but a small population of *el tigre* remains in the Sierra Madre Occidental of northwestern Mexico, and it was from this mountain range, which reaches up into the American Southwest, that Glenn's jaguar had wandered.

While jaguars in Arizona once roamed at least as far north as the Grand Canyon, they fell afoul of hunters and ranchers early in the state's history, and when the U.S. Fish and Wildlife Service listed the cat as an endangered species in 1972, the agency only included those south of the Mexican border—on the assumption that there were none on the American side. But jaguars had, in fact, continued to show up in Arizona on a rare but periodic basis throughout the late twentieth century—and had usually been killed for their trouble. As recently as 1971, two boys hunting ducks at a stock tank in the Canelo Hills just north of the border jumped a jaguar and shot it, and in 1986 a man hunting mountain lions in the Dos Cabezas Mountains near Willcox killed another. Indeed, the most remarkable thing about Warner Glenn's jaguar was probably that it survived the encounter.*

But when, just a few months later, a second lion hunter treed, photographed, and released another jaguar, this time in the Baboquivari Mountains west of Nogales, people began to wonder: Had the largest of the Western Hemisphere's cats begun to reclaim part of its ancient home, the more than two dozen forested mountain ranges scattered across the desert of southern Arizona and New Mexico?

Unlike the Appalachians of the East, or the Rockies farther north, these mountains do not form a long, connected chain; rather, each is set alone amid a sea of lowland desert—hence their poetic name, the "sky islands." Most people know them as the scene of the fierce Apache wars, the gunfight at the O.K. Corral, and boisterous copper-mining towns like Bisbee, but naturalists revere them for their spectacular diversity of plants and animals, much of it reflecting a distinctly Mexican origin, for they are the northernmost fingers of the western Sierra Madre.

Roger Peterson had come to love the mountains of southeastern Arizona, especially the Chiricahuas, where in 1947 he spent a month with Herbert Brandt, a renowned amateur ornithologist. Of the sky islands Peterson wrote, "They are as much a true archipelago as the Azores or Hawaii, but no surf washes their talused bases; instead the desert, dry and shimmering, besieges their foothills and sweeps across the flats to the next range, twenty, thirty, or forty miles away."

It is that isolation, along with its tropical connection, that makes the sky island ecosystem such a delight and a puzzle to the naturalist. A delight because of the riches—more than 2,000 species of plants, ranging from beautiful wildflowers to unusual cacti; 265 species of birds, including 30 subtropical species; 90 species of mammals; and 75 kinds of reptiles, including some, like the Arizona ridge-nosed rattlesnake, found

*The jaguar picked the right rancher to meet. Glenn was a founder two years earlier of the Malpai Borderlands Group, a nonprofit organization of landowners in the Arizona–New Mexico border country widely praised for its science-based, conservation-driven management goals.

nowhere else on earth. A puzzle because of the oddly disjunct way in which these plants and animals are distributed. Mexican chickadees, for instance, are found no farther north than the Chiricahua Mountains (which mark the most southerly point for red-breasted nuthatches), while mountain chickadees come no farther south than the Pinalenos and Santa Ritas. The Huachucas have Arizona gray squirrels but no chipmunks, while the Chiricahuas have chipmunks but no gray squirrels; instead, they have Mexican fox squirrels, while the Animas Mountains have no tree squirrels at all, just ground and rock squirrels. At times, this biotic hodgepodge has caused big legal headaches; Mount Graham in the Pinaleno Mountains has a unique (and federally endangered) subspecies of red squirrel, which has been the cause of bitter arguments over proposals to develop some of its habitat for an internationally renowned astronomical observatory on the peak.

But sadly, the mountain islands have lost some of their presettlement luster. Ocelots are gone, as are the big flocks of thick-billed parrots that once swarmed over the mountains to feed on pine nuts; an old man Peterson once met in the Chiricahuas reminisced about seeing huge flocks in 1917, including two or three hundred swarming a single pine like Christmas ornaments—but the novelty didn't stop the man from shooting fifty-seven of them himself. The Mexican grizzly that once ruled the mountains had been killed off in the United States by the 1930s and is almost certainly extinct in Mexico as well. The Mexican wolf, a subspecies of the gray wolf, hung on in the sky islands for a remarkably long time, with credible reports into at least the early 1980s. But if the lobo survives, the deck remains stacked against it; there is an ongoing federal program to release captive-reared wolves farther north, in the remote Blue Range area of Arizona and New Mexico, but half the animals released have died under what investigators have called "suspicious circumstances." Old attitudes, especially regarding anything that might hurt a cow, die hard in these mountains.

All of this makes the continuing appearance of a few jaguars in the mountains of the Southwest so remarkable; it's hard to imagine a more potent symbol of the wild and primal past, or one more improbable in this day, and in a state experiencing such rapid growth as Arizona. But amid all the hoopla, there may be both more and less here than meets the eye. The 1996 jaguar sightings sparked epiphanies at the personal and the official level. Jack Childs, who treed the second jaguar that year, has become a vocal proponent of big-cat conservation, and the sightings finally moved the U.S. Fish and Wildlife Service, which had previously faced legal action on the issue, to list the jaguar as endangered in the Southwest (though not in Texas). They also prompted the 1997 formation of a federal, state, and private working group that drafted a jaguar conservation agreement.

But an agreement to conserve what? As the American zoologist David E. Brown and his Mexican colleague Carlos A. López González have pointed out, all this official activity overlooks the fact that there is no American jaguar population to conserve. All the jaguars killed or photographed in Arizona since 1963 have been males, often old males; this suggests the cats are not locals but strays, wandering up from the nearest breeding population in east-central Sonora, about 135 miles to the south. As with most large mammals, female jaguars rarely travel such great distances, and Brown and López believe that the chance a male and a female would migrate to the Arizona mountains at the same time, and wind up in the same area to start breeding, "is so unlikely as to be in the realm of dreams."

Recognizing that relocating jaguars from Mexico to Arizona is politically unrealistic, pretty much everyone agrees that the only hope is to preserve the relict jaguar population in Sonora. Unfortunately, while jaguars have legal protection in Mexico, this is meaningless in practice, and ranchers still kill them when they think the cats are threatening stock. The region, one of the most rugged and remote in Mexico, is also on the cusp of major new development, with proposals for large dams and open-pit mines. None of this bodes well for Sonora's jaguars—or, therefore, for Arizona's. "Should this population disappear, there won't be any more jaguars found in the American Southwest," Brown and López conclude. That loss would be more than a single species; with it would go one of the most powerful connections to the Southwest's wilderness heritage.*

Peterson and Fisher weren't thinking of jaguars when they drove up

*Jaguars have continued to be documented in southern Arizona since the Glenn and Childs sightings. The jaguar working group and private conservationists maintain automated "camera traps" along likely travel routes in the borderlands area, and in December 2001 and August 2003 they captured images of the same male jaguar, identified by his unique spot pattern, in an undisclosed location on the Arizona-Mexico border. In 2004, automatic cameras got photographs of two different jaguars in southern Arizona.

through the borderlands, having crossed from Mexico into southwestern Texas, then angling up across the boot heel of New Mexico and into southeastern Arizona. The land was desperately parched and hot, the usual spring dry season exacerbated by a long drought, and the shady canyons and cool high country made for a refreshing change—but a short one, because they only had a day in the mountains before pushing on through the Sonoran Desert and up onto the Colorado Plateau. Despite that, the avian riches of the sky islands came through; Fisher toted up eighty-four species of birds in their one-day stop, including thirty-three lifers.

I would have the luxury of a good deal more time in the Arizona mountains, and even if I didn't have a Roger Tory Peterson to show me around, I was still pretty well set for guides. I'd made arrangements to meet Tom Wood and Sheri Williamson from the Southeastern Arizona Bird Observatory in Bisbee; Sheri and Tom are exceptional naturalists, and Sheri wrote the Peterson field guide to hummingbirds. And I was especially pleased that I'd be able to spend four or five days in the field with my friend Kenn Kaufman, the author of several important field guides and a naturalist for whom the original *Wild America* has a special importance.

When Kenn was a teenage birder in Kansas, he says that *Wild America* was his bible, kindling dreams of setting out on the road, like Peterson and Fisher, to see as many birds as he could. But unlike most kids, Kenn made good on his dream. When he turned sixteen, and with his parents' guarded permission, he took to the road—first by bus and later by hitchhiking, roaming from Arizona to the Pacific Northwest, down to Texas and Florida. He hitched and birded for two years, working odd jobs in between, until in 1973 Kenn decided to do what Peterson had done twenty years earlier—set a Big Year record.

Peterson's mark of 572 species had stood for only three years; in 1956, a Brit named Stuart Keith followed the same route (with a few side trips) and managed to find 598 birds. By the early 1970s, though, birding had become much more sophisticated, with an explosion of where-to-go guides and a network of people sharing information on rare species. By the time Kenn set out on his yearlong journey, which he chronicled in his book *Kingbird Highway*, the continental record stood at 626 species, and breaking it would be a tall order for a long-haired teenager traveling mostly by thumb and surviving on about a dollar a day. In the end, he was edged out by another birder on the same quest, losing by three species, but it was the journey, not the tally, that mattered, and these days Kenn doesn't even keep a life list.

My plan was to meet Tom and Sheri in Sierra Vista, a fast-growing town near the Huachuca Mountains; Kenn would join us the following day, and then he and I would explore the Chiricahuas on our own for a few additional days, as he and Peterson had done a few years before the older man's death in 1996. Finally, I'd follow Peterson and Fisher's trail into northern Arizona, to the Four Corners region and the Grand Canyon.

Before I started chasing wildlife aboveground, however, I was anxious to see a newly revealed window on the subterranean treasures of the sky islands. In 1974, two college buddies and cave explorers named Gary Tenen and Randy Tufts squeezed through a tiny sinkhole on the Kartchner ranch in the foothills of the Whetstone Mountains, a small range half an hour west of the Huachucas. What they discovered inside astonished them: one of the most beautiful and pristine caverns ever found in North America, with two and a half miles of passages and rooms. It survives today only because of some of the most extraordinary secrecy ever to surround a land protection effort.

Tufts and Tenen—worried that if word of the find leaked out, the cave would be trashed by unscrupulous visitors—kept their mouths shut; it was four years before they even told the Kartchner family (it had not been immediately clear to them that the cave was on private land), who agreed on the need for further secrecy. Eventually, they and the family concluded that public ownership was the only option. To that end, then-Governor Bruce Babbitt was sworn to secrecy before he and his young sons were shown the cavern in 1985. Babbitt agreed that the state should buy the cave to safeguard it, and he and his two successors kept it a secret, even as clandestine arrangements were made in the state legislature to appropriate $1.6 million for its purchase. The general public did not learn of the cave's existence until 1988, when the deal finally went through.

What makes Kartchner Cavern so unique is not just its formations, which rank among the finest in the world, but the fact that it is a living cave; that is, calcite-laden water still flows over its walls, gradually enriching the already-stellar display of speleothems, the collective term for stalactites, stalagmites, and the like. Tufts and Tenen had gone to great lengths to protect the cave's delicate environment, and in the years between its purchase by the state and its opening to the public in 1999 Arizona spent nearly $30 million on an unprecedented attempt to balance access with protection.

Much of that money went into the intricate, technically challenging job of creating an artificial entry and climate-control systems to preserve the cave environment, but they did not scrimp on the visitors center, either, which is big, expensive, and beautifully landscaped with native vegetation and hummingbird plants. I was standing outside on one of the patios, watching broad-billed hummers dogfighting around a feeder as three young children climbed on and around a metal bench. One of them, a girl of about six, pointed into the gravel-covered flower bed a few feet away and said, "There's a snake."

I followed her finger and saw a three-and-a-half-foot-long western diamondback lying on the gray stone behind a clump of bear-grass yucca about six feet from us; the rattler was much the same color as the rock, motionless, its head just visible on one side of the plant. The mother came over as the kids gabbled excitedly, but she shushed them. "Oh," she said, "I'm sure it's artificial. It's probably something the park put there as a decoration." But about then, the "decoration" slowly extended its tongue and flicked it a few times, and all hell broke loose. The mother ran for the nearest ranger, a woman who waved a broom at the small crowd while barking orders to stay back. At length, another ranger arrived with a long pair of snake tongs, grasped the rattler by the head, and slowly lifted it into a wooden carrying crate, which was lidded and hauled off by two other rangers. Not until the lid closed did the snake even begin to rattle.

I couldn't help recalling that a rattlesnake was one of the few creatures on Fisher's North American wish list that he never got to see, even though he and Peterson spent more than a month in rattlesnake country, from the southern Appalachians with their timber rattlers to the palmetto flats of Florida, the mountains of Mexico, and the desert Southwest. "Tally ho, James," I thought to myself as the rangers lugged away their now angrily buzzing cargo.

With the snake gone, those of us taking the cavern tour boarded an electric tram that carried us a quarter of a mile to the entrance. About five hundred people a day may enter the cavern now, and only in carefully guided tours. We'd been read a list of rules by a blunt-spoken ranger: no backpacks or purses; no gum, cigarettes, chewing tobacco, or candy; and under no circumstances were we to touch *anything*—we were not even to rest our shoes on the curb of the trail once we were inside. Any place we touched, even inadvertently, would have to be scrubbed down with bleach to kill microorganisms and to remove an invisible sheen of skin oil, which otherwise would, for thousands of years, prevent the water from depositing calcite, the building block of cave formations.

We entered through double steel doors that looked uncomfortably like those of a meat locker into a squared-off chamber with fake-rock walls where the humidity soared to a squirmy 99 percent and the temperature was in the low seventies; unlike many caves, Kartchner is quite warm. "Like Florida but without the bugs," the ranger said, laughing, but it's no joke; given that the dry desert air outside has eight hundred times the evaporation rate of the saturated atmosphere in the cave, any intrusion of surface air could alter or destroy the cave environment by wicking away the moisture essential for speleothem growth.

Once the outer doors were closed, we exited the air lock through a fine waterfall of mist designed to damp us down even more, trapping lint, fungal spores, and the other invisible dandruff humans constantly shed, which might foul the cave. The first several dozen yards were through a man-made tunnel, which workmen had first blasted and then, as they approached the cave, hammered through the hard limestone by sheer manual labor.

Passing through still another set of air locks, we at last found ourselves in what Tufts and Tenen had dubbed the Rotunda Room, a lofty, round-domed space ending below in a mud lake, through which their first explorers' trail could still be seen. We, on the other hand, walked along a sloped path with stainless-steel handrails and drystone walls. Lighting was low, from special halogen bulbs designed to limit the growth of algae, and the lights dimmed behind us as we left each room. For the next hour, we wound our way through tight passages and into the Throne Room, past dazzling cave formations that trumped any I'd ever seen: flowstone, stalactites and stalagmites, sheets of calcite that looked for all the world like bacon, "soda straws" thinner than their namesakes and ten or fifteen feet long that hung from the ceilings in curtains—all of it with the wet, colorful sheen of living stone, still accreting at impossibly slow rates.

The state had obviously taken pains in trying to protect the cave; during construction of the walkways, for instance, workers labored inside sealed plastic tents, with the dust and welding fumes vented to the surface. But there was, and still is, great controversy over whether Kartchner Caverns should ever have been opened to the public at all.

Part of the criticism is based on the argument that any intrusion into the cave will compromise its environmental integrity, and that already seems to be the case. Comparing baseline data collected before construction with readings from a battery of sensors now in the cave, some experts contend the cavern is already changing for the worse; humidity is down while the temperature has risen, and some speleothems in the main rooms are now visibly drier than before the cave was opened. This may be part of a natural cycle—the region is in a prolonged drought—but it's hard not to worry about the implications.

But there is a larger question: Even if the cave's environmental parameters remain unchanged, is Kartchner Caverns the same place that Tufts (who died in 2002) and Tenen discovered? A state park ecologist named Matt Chew, involved in drafting a management plan for the cave's bat colony, wrote a stinging editorial in *The Boston Globe* a few months after the park opened, in which he compared it to an underground theme park, especially the painfully saccharine musical light show that ends each tour. Chew was quickly fired by the state for his comments, but, under pressure, the parks department later rehired him.

In many ways, the wrangling over Kartchner echoes the older debate about any wild, remote place: What is its value if few people can see it? There is a tradition in this country, largely unspoken but visible with every new backcountry road or picnic table, that a place that the average person can't reach with a minimum of fuss isn't worth having. In fact, though, Americans hold deeply conflicted notions on the values of wilderness for the sake of wilderness. On one hand, you have the sentiment, famously expressed by Wallace Stegner in 1969, that wilderness is not just a place but an ideal, "an intangible and spiritual resource" that is important, he argued, "even if we never do more than drive to its edge and look in." On the other is the relentless pressure to make every place of wild beauty accessible to the lowest common denominator, the person who is unwilling to step more than a few feet from his or her vehicle—or, in this case, an electric tram.

Kartchner is perhaps the most extreme example. This is a place that took 200,000 years to form and would (all things being equal) continue to thrive indefinitely if undisturbed. Should it have been left unchanged, accessible to the literal handful of people with the skills and guts to squeeze through openings the size of basketballs, or was it worth the risk to the cave to open it to hundreds of people each day, even knowing that changes to the cave environment are inevitable and, in the long run, almost certainly detrimental? To put it baldly: Is a wilderness no one can visit "worth" anything? At Lechuguilla Cave in New Mexico, perhaps the largest cavern system in the country, the National Park Service opted for preservation; all but a few researchers are barred, and the public may only visit vicariously, through photographs and video. At Kartchner, the calculus played out very differently, and only generations to come will be able to judge if it was the right decision.

It was cool and overcast at daybreak the next morning, when I met Tom and Sheri for a day of birding in the Huachucas. Tom is a beefy guy with long, curly gray hair and a thick silver beard, while Sheri is slim, her long brown hair pulled back in a bun and tucked, this morning, under a cap with a rattlesnake embroidered on the crown. Tom and Sheri are a couple of transplanted Texans who ran the Nature Conservancy's famed Ramsey Canyon Preserve in the Huachuca Mountains for years, before starting the Southeastern Arizona Bird Observatory in 1996.

Although the Huachucas are one of the smaller mountain ranges in this part of Arizona, like all the sky islands they rear with no prelude from the lowlands around them, giving them an unmistakable air of drama. Their flanks are a transitional semiarid grassland, one that has elements of both the Sonoran Desert to the west and the Chihuahuan Desert to the east and, thanks to generations of overgrazing, far more creosotebush and mesquite than there was before settlement. A bit higher are stands of live oaks, then a wide band of mixed oak and pine; above that come more stately forests of ponderosa pine, Douglas-fir, and white fir all the way to the top of Miller Peak, at 9,466 feet the highest of the Huachucas. The only incongruous part of the picture was a huge white dirigible known as an aerostat, permanently tethered several thousand feet above the Huachucas, a radar unit in its belly scanning the sky for low-flying aircraft—part of the government's effort to interdict drug traffickers and others coming in illegally from Mexico.

Seabirds swirl around the stark cliffs of Cape St. Mary's Ecological Reserve on the Avalon Peninsula of Newfoundland, the starting point for Roger Tory Peterson and James Fisher on their original *Wild America* journey in the 1950s.

Thickets of rhododendron shade Curtis Creek in Pisgah National Forest in North Carolina. The hardwood forests of the southern Appalachians are unmatched for their diversity of life, which scientists are still attempting to catalog even as emerging threats, such as alien diseases and pests, place them at greater risk.

A small herd of elk grazes near a historic barn in the Cataloochee valley of Great Smoky Mountains National Park, where the animals have been reintroduced after more than 150 years' absence.

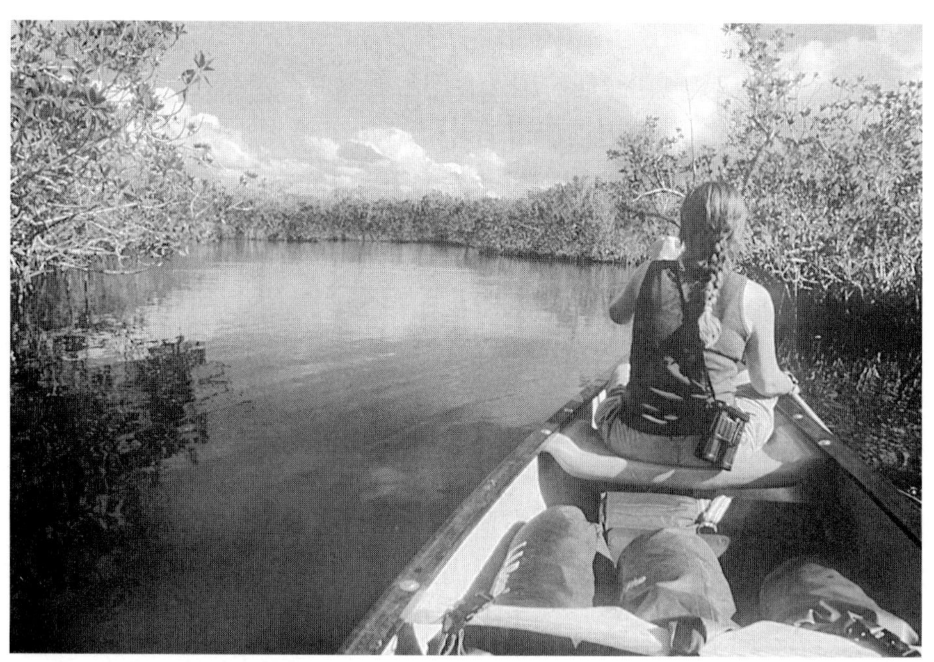

Drifting in a canoe deep in the mangrove forests of the Everglades, one easily forgets what a fundamentally artificial ecosystem this has become, its natural rhythms and pulse largely gone, its hydrology radically altered.

One of the largest masonry structures in the Western Hemisphere, Fort Jefferson sits among the coral islands of the Dry Tortugas, one of the most isolated outposts of the national park system and home to tens of thousands of nesting seabirds.

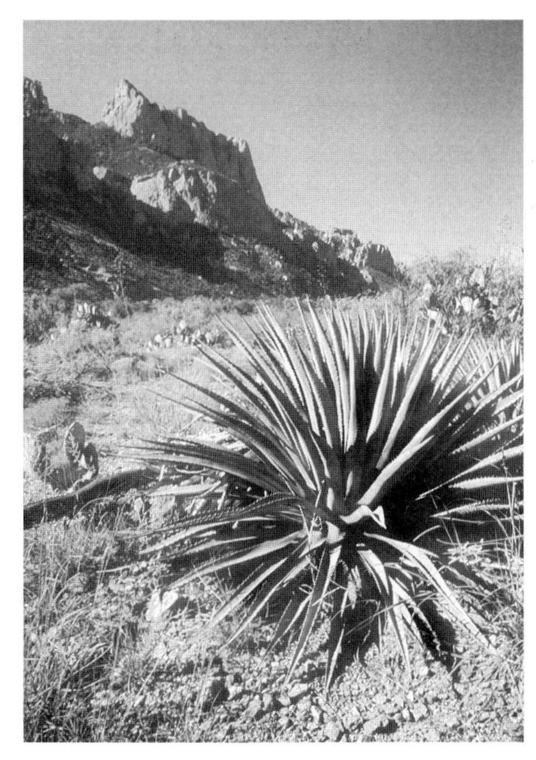

Cave Creek Canyon cuts deep into the Chiricahua Mountains of southeastern Arizona. One of dozens of "sky island" ranges that dot the borderlands and isolated by a sea of desert, the mountains are oases of unique species of plants and animals.

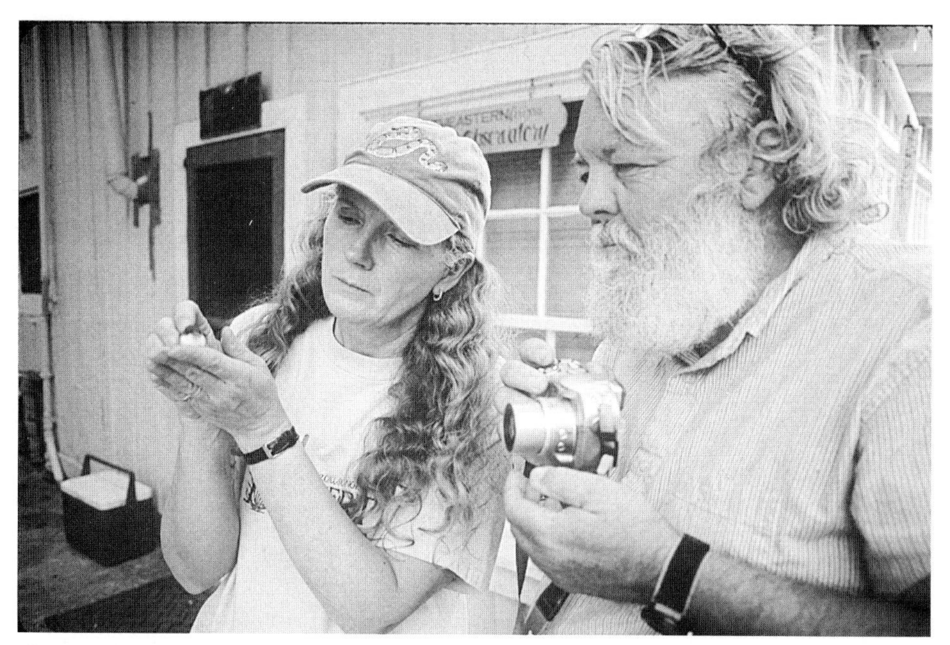

Sheri Williamson and Tom Wood, of the Southeastern Arizona Bird Observatory, prepare to release a violet-crowned hummingbird they've just banded.

Less than forty miles from downtown Los Angeles, Hopper Mountain National Wildlife Refuge and the Sespe Condor Sanctuary encompass some of the wildest land in Southern California and provide a remote stronghold for the endangered California condor.

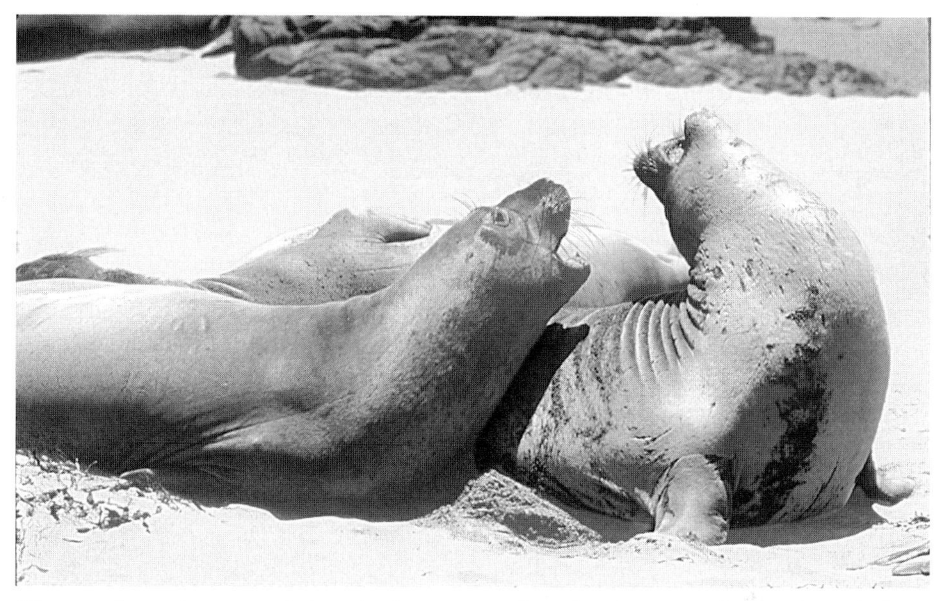

Given up for extinct several times in the nineteenth and early twentieth centuries, elephant seals have made one of the most remarkable comebacks in conservation history, with more than 160,000 of the enormous mammals now living along the Pacific coast.

Steam leaks from the cone of volcanic Mount Shasta, reflected in the Klamath River near Klamath Falls, Oregon.

Once compared with the Everglades for its wetlands and wildlife, the Klamath Basin was converted to agriculture in the early twentieth century, its scarce water largely diverted to irrigation. Today, as elsewhere in the West, the battle over that water grows increasingly strident.

Logs lie stacked along a dirt road in Rogue River–Siskiyou National Forest, where the 2002 Biscuit Fire set the stage for fierce disagreement over how best to manage wildfire and its aftermath in western forests.

Old stumps lie scattered across a fresh clear-cut in the Hoh River valley, a short distance from Olympic National Park in Washington, on land that, until the early twentieth century, held some of the greatest temperate rain forest on earth.

Hikers rest on Hurricane Hill in Olympic National Park, where alpine meadows are starting to disappear as the climate warms and trees encroach from lower elevations, and glaciers in the Pacific Northwest—as across the world—are in rapid retreat because of climate change.

A midnight sunset illuminates the Orthodox church in the village of St. George in the Pribilof Islands, three hundred miles out in the Bering Sea, as fog wreathes the distant rocky cliffs of High Bluffs. Home to hundreds of thousands of fur seals and millions of seabirds, the Pribilofs are one of the richest marine systems in the Northern Hemisphere.

Least auklets, no bigger than starlings, jostle on a rock in the middle of a huge breeding colony on St. George, where they nest in uncountable numbers in burrows among the boulders.

The rapidly growing city Sierra Vista now nudges the edge of the mountains, and we drove through mindless strip sprawl almost to the boundary of Fort Huachuca, home to the U.S. Army Intelligence Center—and to some of the best birding in the region, especially in Garden Canyon, which is where we planned to spend most of the day.

That an army base would have great birding is not as surprising as it might at first seem. Across the country, military installations, by dint of their very size and inaccessibility, have become de facto wildlife refuges; this is especially true in places like the Southeast and the West Coast, where development has destroyed most of the natural habitat. On some bases the record of protection has been uneven, but some installations have been exemplary stewards, and Fort Huachuca is one of them.

"It's no wonder the U.S. Army owns the nicest canyon in the Huachucas," Tom said as we drove out through the sixty-thousand-acre base. "They arrived in the 1870s heavily armed and could settle wherever they wanted. But it means that the fort was never heavily overgrazed like the rest of southeast Arizona, never really logged, and it has regular fires—tracer bullets will do that—so it has a better, more natural fire regimen than on the national forest. In fact, in my opinion it's much better managed than the national forest."

Few installations have been as accommodating to the general public as Fort Huachuca, either. After the September 11, 2001, terrorist attacks, birders around the country found their access to government property sharply curtailed in the name of national security. For obvious reasons, pretty much every military base clamped down on access, but although birders waited for the shoe to drop at Fort Huachuca, the public continues to be welcomed there. Tom stopped at the main entrance gate, got a vehicle permit, and the MP waved us through without so much as a peek at our photo IDs.

I'd been to Fort Huachuca once before, several years earlier, and remembered it as one of the finest mornings of birding I've ever experienced. It was my first trip to this part of the Southwest, and up in Garden Canyon the new birds were like low-hanging fruit: Montezuma quail, elegant trogon, sulphur-bellied flycatcher, Grace's warbler, red-faced warbler, and many more. We hiked up Scheelite Canyon, famous for its Mexican spotted owls, and found ourselves staring into the sleepy eyes of two of them, perched barely head-high and only a few unconcerned yards off the trail.

On this morning, however, the birds were uncharacteristically quiet; it was the fifth year of a bad drought, and so many poor breeding seasons in a row had taken a toll on populations, Sheri said. But there was still plenty to keep our interest. The sulphur-bellied flycatchers squeaked like rubber ducks, and a hepatic tanager called from the treetops; we found Arizona woodpeckers tapping gently on the branches of big Apache pines, and a few small flocks of nuthatches, plumbeous vireos, and black-throated gray warblers. As it warmed a bit, the butterflies came out: juniper hairstreaks, deep iridescent green; California sisters with bold splashes of white and orange against dark wings; and a Huachuca giant skipper, its abdomen as thick as my little finger, an endemic of the region and one of the rarest butterflies in the United States.

We paused in a grove of alligator junipers, where the ground had been churned up in patches by white-nosed coatimundis; these long-nosed, long-tailed relatives of the raccoon barely cross the American border, except in Arizona, where they have expanded their range greatly in the past century. Tom patted the junipers, their trunks twenty inches in diameter and covered in thick, checkered bark that looked a bit like its namesake's hide. "Everyone oohs and aahs over a big cottonwood, but these trees grow so slowly that these are easily three or four hundred years old," he said. "And what they've seen . . . this is the lowest pass through the mountains, and there are Indian pictographs up here—this was a major trade route, and these trees saw it all. There were Mexican grizzlies and Mexican wolves here, jaguars, probably flocks of thick-billed parrots." Later we found truly massive junipers, trees with trunks so big all three of us could have encircled them. Such monsters, Tom guessed, could easily be a thousand years old.

The next morning Tom, Sheri, Kenn (who had arrived the previous evening), and I were driving up the road that leads to Carr Canyon, high in the Huachucas in Coronado National Forest. It was shortly after daybreak, and there were mourning doves and shrikes perched on the wires, other doves and a few scaled quail sitting on the narrow road. Here's the thing about driving around with crack birders: a guy like me doesn't stand a chance. One quail up ahead looked a little different, but before I could put my finger on it, Kenn and Tom both sang out, "Montezuma quail!" while Sheri simply stabbed her arm past my face, pointing at the bird and saying, "Oh! Oh! Oh! Oh! Oh!" in a loud, excited voice right in my ear, about sixteen or seventeen times.

And no wonder, because a Montezuma quail is worth getting in a lather about. We coasted to a halt just a few yards from this one, a plump, almost globular bird, its body black and dark brown, covered with white dots, and the head painted in a bizarro harlequin pattern of black and white, with a funky brown cap that hung over the back of the crown. Occurring in the United States only in a few borderland mountain ranges, the Montezuma quail is notoriously hard to find, and Sheri pronounced us "full of good quail karma" in consequence. (She also said, somewhat sheepishly, "That's not the first time I've made that kind of noise in the backseat of a car, but it's been a while.")

We spent the morning in the high-country forests, then dropped down into Comfort Springs, where broad-tailed hummingbirds buzzed among wildflowers and band-tailed pigeons flew through the shadows of tall pines. But from the high peaks, we could see Sierra Vista, with more than forty thousand people and growing by leaps and bounds, lapping the western rim of the mountains. "Fifteen years ago, all you would have seen from here would have been a few ranch houses," Tom said. "Now you can see the Wal-Mart from up here."

That afternoon Kenn and I said goodbye to Tom and Sheri and drove northeast to Willcox, not far from where that jaguar was killed in the 1980s. The next morning we swung east on the interstate just as the sun was rising, then dropped south along the eastern rim of the Chiricahuas. "The big blue Chiricahuas," Peterson wrote about just such a morning, "the mountain fortress of the Apache renegades and the outlaw gunmen of the old West, the sanctuary of Cochise and Geronimo and Massai; of Billy the Kid and Johnny Ringo and Sam Bass. There they were, in the crystal morning light, rising like a massive blue island from the sea of the desert." It was pretty scruffy desert, truth to tell, once we got south of some irrigated ag fields near the highway—an enormity of creosotebush and mesquite, land that had been pounded by cattle for more than a century and may never recover the semiarid grasslands it once supported. As we gained the foothills, though, the desert improved; lots of agave raised twenty-foot stalks of candelabra flowers, on which Bullock's orioles fed; desert marigolds grew among prickly pear; and the long, spiny stalks of ocotillos were, thanks to recent rains, covered with thickets of green leaves.

The Chiricahuas are a much larger range than the Huachucas, covering more than six hundred square miles of rugged, mostly empty land that climbs to more than ninety-seven hundred feet in elevation and includes an eighty-seven-thousand-acre wilderness reserve. Canyons reach deep into the hills, of which the most famous, in birding circles, is Cave Creek near the hamlet of Portal, where we turned into the mountains again. The porous limestone that underlies these sky islands acts like a sponge, taking up the winter rains and summer monsoons and meting them out through the year, so that these deep canyons are well-watered oases of cottonwoods, oaks, and Arizona sycamores, alive with the sound of running streams.

The walls of the canyon rose maybe a thousand feet, pale rosy rock with an even warmer glow from the morning light, eroded into battlements and towers on which peregrine falcons nest. The white-trunked sycamores caught the rushing noise of the creek and held it in their leaves, cool and inviting, as we walked up its banks, Kenn stopping from time to time to whistle like a western screech-owl. A canyon wren popped in and out of narrow crevices in the shady rock walls, hunting for spiders, and in the sunny openings red-spotted purples, queens, and other butterflies basked. In the branches, a painted redstart was feeding—"the most beautiful warbler in the whole, beautiful American lot," Fisher proclaimed, after seeing his first near this very spot. And he was right; a painted redstart is breathtaking, a small black bird with a long tail and the hint of a crest, with white outer tail feathers, big white wing patches, and a fire-engine-red belly. What's more, the warbler is in constant, frenetic movement, fanning its tail and flashing its wings in an effort to startle bugs hiding among the foliage.

The treetops were also busy with family groups of Mexican jays, a crestless species, plain bluish gray and raucous, as well as acorn woodpeckers, which, with their red, white, and black faces, look like Bozo the Clown in feathers. These are two of the most conspicuous species in the sky island canyons, and birds with remarkably similar life histories. Both are cooperative breeders, living in extended family groups of up to two dozen that help the breeding pair (or, in the case of the jays, up to four breeding females) raise their young. Both species are dependent on the many kinds of live oaks in the encinal forests, and both hoard acorns, but in very different ways.

The jays cache the acorns they collect underground, storing them by the thousands and unerringly retrieving them, up to months later, in an amazing feat of memory. And the jays have long memories for their land, as well. Most bird territories are ephemeral, changing from year to year or disappearing entirely when the owner dies. But because of the overlapping nature of a Mexican jay clan, in which many individuals make up the group and defend the borders, their territories are, for all purposes, immortal. Two researchers in Cave Creek, Jerram and Esther Brown from the American Museum of Natural History's Southwestern Research Station, had studied a color-banded population of Mexican jays for twenty-six years. They found that at the end of their study—after generations of jays had passed—the territorial boundaries between clans were exactly the same as they'd been at the start.

Rather than building dynasties, acorn woodpeckers build food reserves of amazing size. An old tree is chosen as a "granary," and holes are drilled in it, into which acorns are hammered. Generations of woodpeckers may end up cutting fifty thousand holes in a single tree, which are filled each autumn by the foraging clan, provender against winter hardship. But the nest holes that the acorn woodpeckers cut in the old sycamores are critical, too—for the overall bird diversity of the sky islands, Kenn explained, as we watched the flock swoop away, since many cavity-nesting species depend on abandoned woodpecker nests. This is especially true of the elegant trogon, the only member of this family to routinely breed in the United States and one of the species most highly prized by birders. The previous day we'd watched a male in the Huachucas—blood-red belly, long coppery tail, and an iridescent green head and back.

"The equation here at the northern edge of the trogon's range is that there have to be lots of woodpeckers around, digging lots of holes in the sycamores, to allow enough nest cavities for the western and whiskered screech-owls, northern pygmy-owls, sulphur-bellied flycatchers, brown-crested flycatchers, even Bewick's wrens, and finally enough left over for the trogons, which despite their size are pretty low on the dominance ladder," Kenn said.

Midday found us climbing into the high country, passing from one natural community to another—first the oaks, then a mix of oaks and pines, then stands of pure Apache and Arizona pines. We found another pair of Montezuma quail and watched them from such close range that I could see the male's coal black undertail feathers as he inched away from us, still trusting to his flamboyant but effective camouflage. Finally, up around seven thousand feet and higher, we found ourselves in mixed conifer forests with a lot of Douglas-fir, where Kenn whistled up a flock of Mexican chickadees, while pygmy nuthatches called from the branches.

Near the pass called Onion Saddle, we spent several hours poking around Barfoot Park and Rustler Park—a "park," in western parlance, meaning a small meadow set among deep woods. Both were lovely little jewels just a couple of acres in size and a couple miles apart. Kenn wanted to check out reports that short-tailed hawks were hunting around the parks; these buteos, found throughout the tropics but within the United States only in Florida, have in recent years colonized southeastern Arizona, only the latest example of Mexican species moving north across the border. Many birds that were, in Peterson and Fisher's day, only rare vagrants to Arizona are now regular breeders in the sky islands, including violet-crowned hummingbirds and thick-billed kingbirds—a trend Kenn sees no sign of ending anytime soon.

The next morning I reluctantly headed north, away from the sky islands, but before we parted ways, Kenn wanted to show me around some classic Sonoran Desert habitat, the lowland sea amid which the mountain archipelago lies. Just east of Tucson, where he lived, on the lower slopes of the Rincon Mountains (themselves a small sky island), we strolled for several hours through Saguaro National Park. When Fisher and Peterson passed through here, Tucson was about a fifth its current size, and there was a distinctly wilder edge to the place; the pair even reported seeing a jaguarundi, a quail clamped in the rare wildcat's mouth, crossing Pantano Wash.* Kenn offered to show me the spot, but warned me it's a commercial neighborhood these days, not far from his own office. And while Tucson doesn't have the same out-of-control-juggernaut feel that you get

*This sighting is today considered suspect by cat experts, for several reasons. One is the two men's lack of previous experience with this odd and secretive animal: although they noted its size and shape, it took them several hours to independently conclude that the cat they saw crossing the wash was a jaguarundi. Most wildcat specialists now believe they probably misidentified a dark house cat—which may be true, but these weren't a couple of Joe Six-Packs, but two of the twentieth century's greatest naturalists. I'd be inclined to give them the benefit of the doubt, except for this fact: unlike in South Texas, where jaguarundis were well documented until the 1980s, there are no jaguarundi specimens, even from the nineteenth century, known from either Arizona or adjacent Sonora state in Mexico. Biologists David Brown and Carlos López interviewed hundreds of trappers, hunters, and ranchers in Sonora and found no one who claimed to have seen a jaguarundi, or even recognized its picture. Therefore, they believe sight records like Fisher and Peterson's were in error.

from Phoenix (while I was there, the Phoenix suburb Gilbert was declared the country's fastest-growing municipality over 100,000 people), new walled housing developments, landscaped with cacti and mesquite, have crept right to the edge of the national park.

Inside the park's boundaries, though, you get a sense of old Arizona. The hillsides were studded with a forest of mature saguaro cactus, that almost clichéd icon of the arid Southwest, but a species that is, in fact, found only in the Sonoran, the richest and most botanically diverse of North America's four desert regions.

Although it was early morning, the sun was already a physical weight on the back of my neck, and by midday the temperature would be well over a hundred degrees. We made slow, meandering progress as Kenn sorted through the many desert plants for me—velvet mesquite and blue paloverde, fishhook cactus and brittlebush, paperflower and staghorn cholla. Side-blotched lizards, not much longer than my finger, skittered away from us, while black-tailed gnatcatchers and curve-billed thrashers moved through the brush. A Gila woodpecker, zebra-backed with a red splash on its crown, called from the saguaro, into which this species drills its nest holes—like those of the acorn woodpecker, crucial to many species.

We were walking up a dry wash, in the shade of paloverde trees growing below a forty-foot-high bank, when we turned a corner and saw, maybe fifty yards ahead of us, a small herd of javelinas—four adults and three small, impossibly cute babies. The peccaries ambled toward us, one female and her baby swinging around to our right, the rest of the herd moving to our left through the undergrowth, except for one adult that stopped about thirty feet from us, its fleshy, dexterous nose twitching in confusion as we stood perfectly still. Its fur was coarse and grizzled, its proportions, like all javelinas', a bit off-kilter—all head and shoulder, as though God had gotten cheap with the back half of the critter. It sank down where it stood, settling in comfortably on the sand, until at length one of the others caught wind of us and grunted. Then, its long bristles raised along its back—but still not sure what might be amiss—it followed the rest of the herd into the thickets.

From Tucson, the two naturalists headed north, driving for hours across desert and small mountain ranges, eventually climbing the Mogollon Rim, which forms the divide between the lowland deserts and the higher, cooler conifer forests of the Colorado Plateau. The roads were narrow, winding, and challenging, and the two men took shifts at the wheel of Roger's big Ford wagon, changing off every fifty miles or so. Peterson was driving through an area of switchbacks and blind corners, staying to the inside of the curves, much to Fisher's annoyance. As they rounded a bend, another car flashed into view, and Roger slammed on the brakes as the two vehicles skidded to a halt just feet from each other. It could have been a disaster, and Fisher, his patience gone, gave his friend a tongue-lashing for driving too fast and not staying to the outside. Peterson, himself on edge, replied tartly that he hadn't been speeding and that he stayed toward the middle because he'd once lost a friend in just such a situation, forced over the edge of a cliff by an oncoming car.

"Well, have it your way," Fisher replied. "I think you're a bloody bad driver." This episode is worth mentioning only because it was, by both men's accounts, the only cross words they had with each other on the entire one-hundred-day odyssey. And there is, of course, the sad irony that it was Fisher, not Peterson, who was eventually killed in a car accident.

For the next two days, the pair explored the Four Corners region of Arizona and New Mexico, visiting the old Anasazi ruins of Canyon de Chelly, then driving west into the Hopi and Navajo reservations. To my disappointment, I found myself short of time and unable to make the long drive up to de Chelly, which holds some of the most dramatic of the old cliff dwellings. But I did stop at the Homolovi ruins near Winslow, which during their zenith in the thirteenth and fourteenth centuries were a huge complex of apartment buildings with at least fifteen hundred rooms, perched on a low bluff above the Little Colorado River. No one is yet clear how large the city really was. "It's like a fish story," a state park ranger told me. "It keeps getting bigger every time the archaeologists dig a little more."

Unfortunately, they aren't the only ones who have been digging. Collectors ravaged the site before it became a park in 1986, even bringing in backhoes in the 1960s to slash through the ruins, destroying far more than they recovered (and rendering what they took away meaningless from a scientific perspective, since it was no longer in any kind of historical context). Homolovi II, the site most accessible to visitors, looks a bit like an old mine, with rocks heaped up beside haphazard looters' pits, although researchers have uncovered and stabilized one of the town's kivas, or underground ceremonial chambers.

The inhabitants of Homolovi eventually abandoned the site, moving north about fifty miles to join the Hopi in their mesa pueblos; to this day, the Hopi consider Homolovi part of their homeland and are working with the state park to understand the site and, perhaps, mitigate what they see as the spiritual damage inflicted by the looting.

What struck me, as I wandered through the old village, was the pottery; the ground was blanketed with potsherds, most the size of a half-dollar or smaller, along with innumerable flakes of flint, left from the making of spear points, arrowheads, and knives. The ranger had told me I was welcome to leave the trail and pick up artifacts on the surface, as long as I took nothing, but I wasn't prepared for the sheer quantity of them, or their quiet, fragmentary beauty. While some of the shards were from cooking pots, made of clay coils roughly finished (sometimes with the fingerprints of the potter still visible), most were from much more refined works, the coils sanded down with gourd skins and smoothed with pebbles, then coated with wheat-colored clay slip and painted with fine geometric designs in black, brown, white, and blue-gray.

I was completely alone at the ruins, several miles of winding road from the visitors center, and the only movements besides mine were of rock wrens and collared lizards, which, if I got too close, ran away on their hind legs like miniature dinosaurs. Wherever I turned, I saw that previous visitors had collected great numbers of potsherds and arranged them on flat rocks or along the edges of the concrete pathways. I couldn't decide if this was the mindless human instinct toward pack-rat-ism, or if it was something deeper and more ritualized, maybe an unconscious acknowledgment of the age and power of the place and a groping attempt to assuage the rape of the past. Then I found a collection of fragments arranged in the shape of a smiley face, which I suppose answered my question.

I rejoined Peterson and Fisher's path just to the north of Homolovi, in the Hopi reservation, where the men stayed with a white couple who taught in one of the tribe's schools. The nearby pueblo of Shungopavi, its adobe-brick buildings perched on the edge of Second Mesa, looked to Fisher as though it had arisen from the pink and yellowish rock, as did the other pueblos they saw at Walpi and Old Oraibi. The next day, when their hosts showed them around Shungopavi and Oraibi, the men were advised to leave their cameras behind, because the Hopi preferred not to have their daily lives photographed, though Peterson was told it was acceptable to take some long-lens photos from an eighth of a mile away.

Today the Hopi are even stricter about protecting their privacy. Although tourists are welcome to visit the communities, some of which still look as though they'd arisen straight from the rock, a visitor is quickly and firmly advised that not only are cameras not allowed, but no recording of any sort is permitted—no audio tapes, no sketching, not even the taking of notes. I knew this and was reminded of it by signs posted at the entrances to villages, on the door of the Hopi Cultural Center's museum, and on the walls of several of the Hopi-owned silver and pottery shops where I stopped. Living, as I do, near the Amish country of Pennsylvania, I've seen firsthand how a culture can become a tourist attraction against its will; given that tourism is one of the few sources of income on the reservation, it seemed to me that the Hopi had found a way to assert at least partial control over the situation.

The Hopi lands lie entirely within the much vaster reservation of the Navajo, a tribe many Hopi consider Johnny-come-latelies to the region. (Anthropologists say that the Navajo, who speak an Athabaskan language, migrated from western Canada around A.D. 1000, finding Puebloan cultures like the Hopi already well established there; Navajo traditions, on the other hand, tell of an emergence into a series of worlds, with this being but the latest.) Many Hopi consider the Navajo usurpers and say that the very much larger Navajo reservation includes a great deal of traditional Hopi land; tensions remain sharp between the two nations, especially in the wake of a federal program to transfer control of almost a million acres of the Navajo reservation to the Hopi—land both tribes hold sacred.

I drove west from the mesas, dropping down to the lower plateau, where the road rose and fell in a straight, undulating ribbon across arid grassland. Here and there, tucked below mesas and along dry washes, were small dryland farm plots planted in corn, melons, and beans; summer rains would flow here, naturally irrigating the crops. Eventually I crossed into the Navajo reservation, which is roughly the size of West Virginia and covers parts of four states; unlike the highly social Hopi, the Navajo tend to live scattered across the wide landscape, where most still graze sheep and horses. Most of the homes I passed had a hogan, the traditional eight-sided summer shelter, though these days the buildings are mostly made of cinder block or covered in wood sheeting, instead of built of logs and earth as in years past.

I drove through Tuba City, then down to Cameron and west along the Little Colorado River, which has dug a canyon that would, in a just world, be famous in its own right. But few people bother to stop, because an hour farther along, below the high Kaibab Plateau, lies the greatest of all gorges, the Grand Canyon.

The canyon was a shock to Fisher; he knew it was on the itinerary for the day, of course, but his English mind was elsewhere, since that morning in Shungopavi he'd been listening to radio reports on the coronation of Queen Elizabeth II and the conquest of Mount Everest by Hillary and Norgay. Peterson pulled off the road in a forest of low pines, and James roused from his mental fog, asking if it was time for sandwiches. No, Roger said, this was Navajo Point, and they should go down a few yards through the woods first. Puzzled, Fisher did, and he later wrote in his journal:

> The world ended; began again eight miles away. Between the ends of the world was a chasm.
>
> The chasm was awful.
>
> Awe. Time brings awe to the traveler less often, no doubt, as time goes on; for time gives him, too, the accumulated, stored, recorded experience of those who have been before him. With all of these I had prepared myself—words, music, paintings, photographs, three-dimensional color movies, even. Yet all of these were, in that first moment of shock, reduced to a whisper, whispering, "Yes, this is true; this is real; this is it; this is the greatest abyss on the face of the earth; this is the Grand Canyon of the Colorado River." The loud voice (I have never heard it louder) was the overwhelming voice of awe. I heard this voice before, in many places, some unexpected . . . But never had my awful friend, awe, stood so long at my elbow, so close, as by the rim of the Grand Canyon. Never will it come so close again.
>
> Roger, who knows that I talk too much, says that I was silent for ten minutes. So was he. The first thing I said was, "I shan't want the big lens; I wish I had a wide-angle," drying my eyes under cover of my handkerchief while pretending to dry my forehead.

Like Fisher, I was visiting the Grand Canyon for the first time, and I had often wondered, in the preceding weeks, whether my reaction would be the same as his. Like him, I came to the South Rim in late afternoon, when the light across Navajo Point picks out the rugged details of the rock walls and shadows lend even greater depth to the vast space that opens below your feet. Although there were fires burning on the North Rim and the view was hazy with smoke, it was staggering, in all ways awesome—but I am forced to admit that it was not, for me, awful. Perhaps Fisher, in his brief, powerful testimony on the almost terrifying power of true awe, set a benchmark that no modern American can hope to meet. He prepared himself, yes, but we in this country are so saturated with imagery of the Grand Canyon, in every conceivable medium, that it may no longer be possible to come to the canyon with as fresh and naive an eye as James's. Nor was I alone with my thoughts; it's difficult to find solitude anywhere in the park that's easily accessible, much less at the popular overlooks. And so while I was moved, it was not to tears; and I found this troubling, as though I were letting down the canyon, and myself.

Fisher, who throughout his hundred-day tour was impressed by the American network of public lands, saved his highest praise for the national park system, of which he felt the Grand Canyon was the best example of all. It was, he said, "a perfect park, not only in its setting, but in its administration." Visitors to the canyon left "satisfied, refreshed, emotionally stirred, and with another stake in America"; he applauded the facilities and concessions "that serve the customers without bringing any indignity to the Park." And he praised the visitors—"the great American public . . . [which] becomes in the National Parks deeply respectful, orderly, extraordinarily tidy, obedient to instruction and trustful of advice. Nobody dragoons them, shouts at them. There is a minimum of regulation and warning. Yet, with few exceptions, this public assumes, on crossing the Park boundaries, a new code of behavior—almost a new tone of voice."

I spent four days at the Grand Canyon, dividing my time between the crowded South Rim, with its half-dozen hotels, jammed tourist village, and shuttle-bus system, and the North Rim, more than a thousand feet higher, cooler, with far fewer visitors. I was curious how Fisher's "perfect park" was faring these days, when booming visitation and perennial budget cuts have stretched many parks to the breaking point.

The three-quarters of a million visitors that came to the Grand Canyon in 1953 have ballooned to 4.4 million each year, making this the second-most-popular park in the country after the Great Smokies. But that many people clog roads and parking lots, overtax facilities, and create long lines just to enter the park or stand at a scenic overlook, while their vehicles generate enough smog to damage both human health and the canyon views. Visitor behavior has also changed, generally for the worse, since the more innocent days in the 1950s. Most tourists still behave themselves, but crime and misconduct have risen sharply over the years, requiring a greater focus on law enforcement—and not just the don't-pick-the-wildflowers sort. In 2003, the park investigated more than a thousand crimes, of which almost two hundred were so-called Part I offenses, a federal classification that includes murder, rape, assault, robbery, and vehicle theft.

Although the park raises about $15 million a year from its twenty-dollar-per-car entry fee, its annual budget is $18.5 million, much of it going toward repair and maintenance, and little for resource protection like endangered-species restoration or preservation of historic structures. An analysis by the National Parks Conservation Association found that in 2001 (the latest year for which the NPCA had figures), Grand Canyon National Park was underfunded by about $8.5 million, with staff jammed in aging, inadequate trailers and almost a tenth of the park's 450 permanent positions shifted to offices in Flagstaff, eighty miles to the south. Some slots simply aren't filled; as of 2003, the park, whose very essence is geology, had no geologist on staff.

This situation isn't unique to the Grand Canyon; national parks across the country are being squeezed as never before, caught between the twin pressures of increasing use and inadequate funding. Every administration in recent memory has made political hay by pledging to help the parks, but it's largely been lip service. Although the Bush administration came into office promising $5 billion in new funding to eliminate backlogs in maintenance and resource work, it has pushed through only modest increases while announcing policies that watchdog groups like the NPCA see as antithetical to the parks' well-being—relaxation of clean-air rules, for instance, and an initiative giving state and local governments broad rights over land within park boundaries, especially where the designation of new roads is concerned. One of the most disturbing proposals, however, is the plan to shift more than half of all National Park Service jobs to low-bid contractors in the private sector—not only entrance staff and maintenance jobs, but professional positions like biologists, archaeologists, and museum curators. The NPS itself says such a move would damage visitors' experiences and reduce the diversity of the park service workforce.

The Grand Canyon is also typical of the way pressure is being exerted on parks from too many visitors. After a couple of years of flat attendance, visitation rates are on the rise again, and while shuttles move people along the buses-only Hermit Road and to other points on the South Rim, plans for a light-rail system that would keep most private cars out of the park have stalled. Even the skies above the canyon are too crowded, as they are in many parks; although federal legislation was passed to reduce noisy park overflights by commercial tour operators, the Grand Canyon was specifically excluded. As many as ninety thousand people a year buzz over the Grand Canyon in planes and helicopters, making it unlikely that the park will meet its goal of "substantial restoration of natural quiet" by 2008.*

On the North Rim, I was able to escape the smaller crowds by taking almost any trail even a short distance from the parking lots, and some days I encountered only a handful of other hikers. But on the South Rim, even the steep paths into the canyon, like Bright Angel Trail, looked like conga lines in the morning, crammed with people and mule trains, so in search of some solitude I simply took to bushwhacking through the low piñon pine and juniper forest along the plateau, moving cross-country to one anonymous point or another jutting out into the canyon, far from the road. My last morning in the park I walked through dim twilight, flushing desert cottontails and jackrabbits, and found a seat on yellow Kaibab sandstone beneath a scruffy juniper by the great and yawning drop of the canyon. The smoke haze was bad, but as the morning passed, the wind picked up and lifted the pall, so that I could see the slender thread of the Colorado River far below.

Yet even here, I could still hear the rattle of aircraft and see the drag-

*If you're curious, the park considers "substantial restoration" to mean "50 percent of the park experiencing natural quiet 75 to 100 percent of the day," every day. One of the holdups has been legal wrangling between the air tour operators and the Federal Aviation Administration over just what is meant by the phrase "of the day."

onfly shapes of brightly colored tour helicopters in the distance, spilling man-made noise on the landscape; as grating as I found it, I could only imagine the reaction of a hiker deep in the canyon, or someone who'd spent days rafting down the river. I like the idea of the natural-sound recordist Gordon Hempton, who has launched a One Square Inch of Silence campaign—a simple idea with profound implications, in which he asks only that a single square inch of ground in each national park be protected from man-made sound. Of course, for this to be accomplished, vastly larger areas would have to be shielded from noise. But if we cannot escape the twenty-first century in a 1.2-million-acre national park like the Grand Canyon, where can we? This is a place that speaks to history far more ancient than ours, if only we can hear it.

I spent five hours sitting on my lonely perch, watching the turkey vultures and zone-tailed hawks soaring past, but my eyes kept combing the sky for bigger wings—and for a very direct connection with the Grand Canyon's primal past. Twenty thousand years ago, when this region was wetter and lusher, it was home to ice-age mammals—and to California condors, which scavenged the carcasses of the giant ground sloths, mammoths, and mastodons. The great mammals disappeared, and so, in time, did the condors, which were eventually restricted to a band along the Pacific coast from Baja to the Columbia River. But when the Spanish brought horses and cattle in the eighteenth century—herds that substituted for the long-vanished megafauna—condors returned to the Southwest, and at least one lingered near the Grand Canyon as late as 1924. Then the condors, for a second time, disappeared from the Arizona skies. In fact, for a time in the 1980s even the skies of California were bereft of condors, as the last twenty-seven in existence sat in zoos and breeding facilities.

But in 1996, their population bolstered by a remarkably successful breeding program, condors were reintroduced to the Grand Canyon. It was hoped that the park, with its vast size, abundant cliff caves for nesting, and lots of wildlife for food, would prove an ideal home for the massive vultures, which can soar for hundreds of miles on wings nearly ten feet wide. As I sat there in the morning sun, I knew that more than forty of the huge birds were somewhere in the Four Corners area, and that two pairs of them were tending chicks, only the second season in which breeding in the wild had taken place.

But this day, like all the days I was at the canyon, the skies remained empty of condors. To find them, I was going to have to look to the rugged mountains of Southern California, where this most famous of endangered species almost lost its battle against extinction—and where wild America continues to exist cheek by jowl with the jarringly modern world.

TEN

The Golden Coast

In 1936, Roger Tory Peterson found himself in Los Angeles with six hours to kill between trains. In typical fashion, he didn't waste the time; five high-school students from the local Junior Audubon Club picked him up and rushed him forty miles from town and up into the rugged Sespe mountains. There he saw his first wild California condors, then raced back to the train platform with but a few minutes to spare.

These days it seems to take six hours just to get out of the airport if, as I did, you happen to arrive on one of those afternoons when L.A.'s freeway system is throwing a hissy fit. The highways were jammed, so as I tried to edge out of the city and toward the coast, I had plenty of time to think.

The unsettling thing about Los Angeles, for those of us who rarely come here, is how eerily familiar it is once we do arrive. Southern California is so embedded in the collective culture that the simplest landmarks, like the roads, seem almost mythological to an outsider. When I pull onto the Santa Monica Freeway, it isn't just a highway, it's every *CHiPs* episode I watched as a kid, and I can't drive the San Diego Freeway (where I was currently stuck) without seeing a little replay in my head of the O. J. Simpson car chase.

There's such a sense of déjà vu that it's as though a character of fiction came to life and introduced himself to you. And it's not just the familiar-sounding names, of course, but the fact that from infancy, we're awash in how Southern California looks: the palms, the loose-limbed eucalyptus trees and shaggy hedges of blooming oleander; the sun-scorched hills of dry grass and chaparral below the skyline of jagged brown mountains; the white combers rolling in on the sand beaches north of Malibu, where the

Pacific Coast Highway squeezes below high bluffs on which perch garishly big houses exactly like I've seen in a lifetime of movies and TV shows. And it's not just L.A. and its environs. The first time I walked into a coastal redwood forest up in Big Sur, I had the uncomfortable sense that E.T. was going to waddle out of the ferns. Frankly, I'm always surprised by how much California looks like, well, California.

And to a birder, the whole world *sounds* like California. On film sets that are supposed to substitute for everything from Wild West towns to English villages, alien planets, and medieval villas, those with a quick ear can pick out the calls of Pacific coast birds on TV and movie soundtracks. James Fisher, poking around with Peterson in the hills above L.A., heard the small long-tailed birds known as bushtits and immediately recognized their revved-up trills from Hollywood movies, although he'd never seen the species itself.

I guess it's no wonder that since so many millions of people have come to Southern California to reinvent themselves, in the process they've reinvented nature down here, too. The region's Mediterranean climate is great for all manner of exotic plants, from Australian eucalyptuses to the ubiquitous South African ice plant, and the Asian spotted doves that Peterson and Fisher sought out are now joined by feral populations of such onetime cage birds as orange bishops and nutmeg mannikins, as well as a bewildering array of parrots from around the world—half a dozen species of Amazons and seven or eight kinds of parakeets, as well as lovebirds, budgies, cockatoos, cockatiels, and even macaws. (Ironically, some of these parrots are doing better in California than in their native habitat, where forest destruction and collecting threaten them; by one estimate, a quarter of the world's endangered red-crowned parrots now live not in eastern Mexico but in Southern California, in flocks of up to four hundred birds.)

The traffic merged, at glacial speed, onto the Santa Monica Freeway, and my mind kept wandering. You can look at this biological mishmash two ways—as a robust melting pot or as a mongrel invasion. Most conservationists, understandably, take the latter, less charitable view, worrying about the impact of invasive alien species on native flora and fauna, which has been substantial; all those picturesque eucalyptus trees, for example, maintain their dominance by producing toxic chemicals that kill other plants trying to grow around them, while the so-called killer algae *Caulerpa*, which escaped from the tanks of fish hobbyists into coastal waters, now threatens marine ecosystems.

But however fundamentally changed the greater Los Angeles landscape has been, one of the surprising things to any visitor is how close reasonably wild country remains to the heart of the city. Along the Eastern Seaboard, there are few barriers to sprawl, and the cities fan out to the horizons. Here, in the place that invented sprawl, there are checks to its progress, like rocks in a stream. The mountains that barricade the city—the Santa Monicas just ahead of me, and the Santa Susanas, the San Gabriels, the San Bernardinos, the San Jacintos, and the Santa Anas inland—run anywhere from two or three thousand feet high to more than eleven thousand. On the map, a necklace of green all but encircles the city, a string of national forests from Los Padres in the north to Cleveland in the south.

When night falls, smart pet owners bring in their poodles and house cats so they don't become coyote food, and a few people have been injured by coyotes emboldened by people who feed them, unaware of the harm it can do. But of greater concern these days is the rising number of incidents involving mountain lions, which are coming into greater conflict with people as homes push deeper into what had been empty land and as the cats (which are no longer hunted in California) grow bolder. In January 2004, a bicyclist was killed by a mountain lion in a park near Lake Forest, just southeast of Los Angeles; a day later, what authorities think was the same cat severely injured another cyclist before it was driven off.

I finally escaped the worst of the traffic and headed up the coast highway toward Oxnard, where I pulled off the road beside Mugu Lagoon, one of the last large tidal wetlands in this part of California. Flocks of great egrets and great blue herons waded in the shallows, and one of the big terns—maybe an elegant, from the droop of its bill—hunted out over the surf line beyond the narrow barrier island. I couldn't get any closer, though, because barbed wire and sharply worded signs informed me this was a military base—in case the sight of naval AWACS radar planes, endlessly practicing touch-and-go landings at the naval air station just a mile or two away, wasn't clue enough. It was a reminder that an unlikely, and largely accidental, savior of wild Southern California has been the military, for Mugu isn't the only such oasis. The Marine Corps' Camp Pendleton, between L.A. and San Diego, contains two hundred square miles of largely open land and more than seventeen miles of shoreline—the longest stretch of undeveloped coast left in Southern California—while Miramar Marine Corps Air Station at San Diego contains twenty-three thousand acres.

As development chews up habitat outside the bases, the military has found itself responsible for some of the only remaining viable populations of rare species, from threatened western snowy plovers at Pendleton to endangered California gnatcatchers at Miramar—a responsibility, the armed services contend, that is not always compatible with training. As the Department of Defense has appealed to Congress for exemption from many environmental protection laws, conservationists argue that both missions can coexist, pointing to DoD's generally good (if sometimes grudging) record at many installations and a Government Accountability Office report that found few conflicts between training and environmental protection. Despite that, the military has already been granted exemptions from some or all of the laws protecting endangered species, migratory birds, and marine mammals, and the Bush administration has proposed further relaxation of the Clean Air Act and other federal pollution laws for defense activities.*

It was early June 1953 when Peterson and Fisher arrived in Los Angeles, driving up from the south, hunting for a local race of savannah sparrow among the seeping oil derricks at the edge of the city—"pumps seesawing and clanking, breathing sweet-sour vapor all around," Fisher wrote, "pools of black water marbled by the rainbow films of oil," a scene that he said left him shocked and disgusted. He found Sunset Boulevard a different matter altogether, though. "I had expected ostentation and vulgarity; there was little of either," he said. And if I could argue that point

*In arguing that the Navy should be allowed to kill migratory seabirds on a bombing range in the Mariana Islands of the Pacific, Pentagon attorneys advanced the truly flaky justification that bombing was actually good for birders, because "bird watchers get more enjoyment spotting a rare bird than they do spotting a common one." The courts, not surprisingly, rejected that line of reasoning, but George W. Bush later nominated the man who submitted it, Department of Defense general counsel William J. Haynes II, to the federal appellate court. (The quote, by William J. Haynes II, is cited in Emmet G. Sullivan, "Memorandum Opinion and Order," *Center for Biological Diversity v. Pirie*, 191 F. Supp. 2d 161 [U.S. District Court, District of Columbia, March 2002], p. 27.)

today, I found myself agreeing with him when he said that Los Angeles was all the more remarkable when one considered that not long before, this valley was the home to bald and golden eagles—and to condors, the main reason the two men had come to Southern California.

No other animal in North America is so wrapped in potent symbolism as this giant vulture: the last relic of the Pleistocene, sacred bird of the Chumash and other Native tribes, modern emblem of endangerment and extinction; dismissed by some as an evolutionary dead end and a bureaucratic waste of money, hailed by others as a conservation success and the embodiment of wild California; and in all these ways, a continuing lightning rod for controversy.

The condor is, in fact, a very real connection to the great ice-age fauna of ancient North America. Fifteen thousand years ago, when the continent was a Serengeti-like game park of mammoths, mastodons, camels, horses, bison, pronghorn, shrub-oxen, ground sloths, and other giant mammals, it was a smorgasbord for their scavengers. These included enormous, now-extinct vultures known as teratorns, with sixteen-foot wingspans,* as well as the somewhat smaller California condor, which with its ten-foot wings was no piker, either—a glorious black bird with white patches on each wing, a ruff of long neck feathers, and a naked head of orange and purple skin. Condor remains have been found from New York to Florida, Texas, the Southwest, and Mexico, and up along the Pacific, suggesting that this huge bird probably enjoyed a nearly continental range.

Things began to go sour for the condors when the Pleistocene megafauna toppled into extinction; their range contracted to the west, encompassing the coast and mountains from southern British Columbia to northern Baja and inland to the Grand Canyon and perhaps, at least rarely, to the Rockies. Recent chemical analysis of ancient condor bones suggests those living along the coast survived the great extinction by feeding on marine mammals, a food source unavailable to doomed inland populations. That habit survived to historical times; when Lewis and Clark spent the winter of 1805–6 at the mouth of the Columbia River, they saw California condors feeding on a dead whale.

The tale of the condor's fall has been told many times, so suffice it to

*If you think that's big, consider the teratorn whose fossils were found in Argentina, which had a wingspan of twenty-three feet—the largest flying bird ever, so far as we know.

say that they fell victim to too many rifles, too many poisoned carcasses set out for grizzlies, wolves, and coyotes, too many museum collectors, too few tender deer or Spanish sheep, and too many American cattle or horses with hides too tough to penetrate—basically, too much modern humanity that felt little need to preserve an ice-age holdover. The last Baja condors disappeared around 1937, and thereafter the only remaining population lingered in the J-shaped mountains framing the southern San Joaquin valley, barely forty miles from Los Angeles. By this time, conservationists had begun pushing to save the last condors, a flock that guesstimates placed at fifty to eighty birds. A small condor refuge was designated in 1937, and ten years later the U.S. Forest Service created the forty-six-thousand-acre Sespe Condor Sanctuary, remote and rugged, in Los Padres National Forest north of Fillmore. About the same time, the National Audubon Society hired a young zoologist named Carl Koford to conduct the first real study of the birds' habits and needs. By the time Peterson and Fisher arrived, Koford's work was being published, with such details as breeding biology, which he found to be extraordinarily slow—sexual maturity at age six, and only a single egg every two years thereafter, with the chick not leaving the nest until it was six months old. Koford also made a new, worrisome population estimate: no more than sixty condors, and perhaps as few as forty.

Some now think that Koford's estimate lowballed the true number of condors by a factor of two or even three and that there may have been as many as 150 condors in the 1950s; if that's true, it makes what happened in later decades even more awful. By the 1970s and '80s, the condors were in real trouble, if you can say that about a species that had already been holding on by its toenails for more than a century. The number of condors dropped through the floor, and it now seems clear that the weak link was food; though the birds nested in the wildest parts of Los Padres, they ranged over a sprawling landscape of some six million acres, looking for carcasses, and many of those carcasses, apparently, were poisonous. Not, generally speaking, the intentional poisons of the old grizzly-killing days, though some still died from poisoned coyote bait, but something more insidious and omnipresent—lead. Researchers now believe the wild condors were doomed by spent bullets and bullet fragments, left in carcasses and gut piles by hunters. This was confirmed by radio telemetry, which led biologists to the bodies of lead-killed condors. Time was running out for the great birds.

As early as 1950, the director of the San Diego Zoo proposed catching some of the condors for a captive breeding population, an idea that drew high-profile opposition from Koford and National Audubon, among others. In fact, the history of condor conservation is a long tale of extraordinarily bitter disagreement among people of good intentions over the best course of action to take and even the basic assumptions about condor ecology, behavior, and numbers. Koford fought the idea of captive breeding until his death in 1979, and others picked up the fight thereafter, so that it wasn't until 1982, when it became clear the condors were in crisis, that the first were taken from the wild and sent to Los Angeles and San Diego zoos to form the nucleus of a breeding flock.

It was almost too late. There were twenty-one condors in 1982, nineteen in 1983, and just fifteen the following year. Biologists were trying a technique known as multiple-clutching, in which an egg is removed from the nest and hatched in captivity, spurring the adults to lay a replacement; this bolstered reproduction, but it couldn't match an annual mortality rate of nearly 25 percent. The winter of 1984–85 was disastrous, with six adult condors vanishing, leaving only a single breeding pair where the year before there had been five. The decision was made, after lengthy and acrimonious debate, to bring the surviving condors into captivity for their own safety. The last wild bird, a male designated AC9, was captured on Easter Sunday, 1987, one of just twenty-seven California condors left in the world.

For many people, the heirs of Carl Koford's view, that moment marked the extinction of the California condor as a species; never again, they said, even if at some point a few could be returned to the mountains, would the species ever be fully and completely wild, ever fully and completely real. Though I didn't share such apocalyptic opinions, I remember what a terrible blow it was to get the news of AC9's capture, to think how badly we'd failed both the condors and ourselves in letting the situation grow so dire.

But there was, even in the late 1980s, some cause for hope. Zoo specialists had proven adept at hatching and rearing condors, and in 1988 the first chick from a captive pair was produced. The success of the captive breeding program surpassed almost everyone's expectations; by 1992 there were fifty-six condors in rearing facilities, and that year the first captive-raised birds were being returned to the wild, into their ancestral range in the Sespe. In the subsequent twelve years, condors were also reintroduced to Arizona, to the mountains near Big Sur in California, and to Mexico.

At the end of October 2004, the global condor population stood at 246, with 114 in the wild—49 in the Grand Canyon/Vermilion Cliffs region, 21 in Southern California, 36 in the Big Sur/Ventana Wilderness area along the central California coast and Pinnacles National Monument in the Gabilan Mountains to the east, and 8 at the newest release site in northern Baja. Another fifteen were in conditioning pens at the various sites, awaiting release, and experts predicted this would be the last year in which there were more condors in captivity than in the wild. What's more, one of those free-flying condors in the Sespe was AC9, who had been returned to the wild in 2002.

Most important, the condors were breeding again; in November 2003, a chick fledged from a nest in the Grand Canyon, the first wild-bred condor to do so since the early 1980s. And as I arrived in California, 2004 was shaping up to be a banner year as well, with two chicks hatched in the Grand Canyon and Vermilion Cliffs and three more in Southern California, including one fathered by old AC9—a baby that had hatched, fittingly enough, on Easter Sunday, seventeen years after its father's capture. But as impressive and hopeful as all this may sound, it can mask the fact that many of the problems that laid the condor low in the first place, especially lead poisoning, still remain; the condor's future, though brighter than many of us had ever imagined possible, is by no means secure. So like Peterson and Fisher before me, I wanted to get up into the mountains and look skyward, not just for the thrill of watching one of the largest flying birds in the world, but to get a better sense of whether the condor's fledgling renaissance is sustainable.

It was June 8, the same day that James and Roger had arrived in condor country, and as I left my hotel in Ventura shortly after daybreak, I sourly noted that my weather luck was no better than theirs had been. It was chilly and cloudy, the mountaintops hung with mist, the "June gloom" that blankets the Southern California coast at this time of the year. Condors, built for soaring, are dependent on the sun and the bubbling rise of thermal air currents to lift them; a damp, foggy day would ground them, and my chances of seeing any.

"All we can do is drive up and see what we find," Denise Stockton told me when we met at the office of the Hopper Mountain National Wildlife Refuge, which oversees the California condor recovery program. Stockton is the refuge's information and education specialist, and her husband is the supervising biologist for the condor project; she had agreed to show me the refuge, which is ordinarily closed to the public to protect the condors, an hour away up in the mountains and adjacent to the Sespe sanctuary.

We drove through Ojai, then onto a succession of small lanes before passing through the first of several yellow gates on dirt roads into restricted territory on Los Padres National Forest. Denise drove us up seemingly endless switchbacks trailing a curtain of dust, easing around oil company trucks and past wells bobbing like feeding birds; at times it was hard to remember this was national forest land and not an industrial site. But soon we climbed higher, into the pale sienna hills a couple thousand feet up, right into the belly of the clouds, where the air was clammy, the breeze stiff.

"I grew up in Bakersfield, and when we went to the mountains, it was always the Sierras—if you went to Ventura, it was to go to the beach," Stockton said. "When I moved here, I couldn't believe there were mountains like this, such wild country so close to so many people." We passed the white boundary signs with the famous blue goose symbol of the national refuge system, and I could see the marks of a huge wildfire that had swept across Hopper Mountain the previous autumn, scorching most of the live oaks that huddled in the deep gullies and killing the many walnut trees that once dotted the hillsides, though they were sprouting again from the roots in green thickets amid the blackened trunks.

Denise radioed the field crew to find out who was working where. Out of the fog materialized a pickup truck parked on what I could only assume was a high overlook, a place the biologists call Silver Tanks. Inside was Dave Pedersen, one of three interns working a ten-day shift, living out of the old ranch house down in one of the deep canyons—a job that mixes radio-tracking and nest watching with less glamorous tasks, like making the rounds of local dairy farms to pick up calf carcasses for the condor feeding stations.

Dave, a Minnesotan finishing up his undergraduate work at Northern Michigan University, had a very un–Great Lakes tan and wore a heavy flannel shirt against the chill. He was standing in the blowing fog, clutching a radio receiver, an antenna, and an open notebook filled with condor identification numbers and notations of date, time, and direction of signal. He turned on the walkie-talkie-sized radio and swung the antenna to the north, deep into mist; the receiver began to give sharp, insistent little chirps. Looking over his shoulder at the readout on the receiver's display, I saw with an odd thrill that the signal belonged to AC9. It was like catching a glimpse of a celebrity.

Leaving Dave to his work, we drove over the pass and down into the valley where the ranch house lies, finally descending out the fog. Denise pulled over to the edge of the narrow road and pointed out some of the landmarks: the city of Fillmore far below in the hazy valley; the weirdly eroded sandstone ridge known as the Pinnacles, where the condors often roost; and, somewhat closer to us on a lower saddle of open grass, the walk-in trap the biologists use to catch the condors for checkups and radio refits—a big wire cage that was baited with carcasses in anticipation of a trapping effort the following week. A bunch of ravens sat on the roof of the cage, with starlings walking around below them on the ground.

Except that when I lifted my binoculars, I realized the "starlings" were in fact big ravens and that what I'd taken for a bunch of corvids were actually thirteen condors, jostling each other like teenagers on a school bus. I didn't quite know what to say; after decades of wanting to see just one condor, I found having a baker's dozen presented in such a casual way both breathtaking and a little anticlimactic. When Fisher saw his first condor, soaring high in the blue California sky, he responded with an exuberant "Tally most incredibly ho!" My reaction was equally intense, but a bit more layered, made more complicated by the condor's tangled history and uncertain present.

The condors are trapped on a regular basis, their blood checked for lead, vaccines administered for West Nile virus, the batteries replaced on their radio transmitters, which are mounted to large, numbered wing tags. They're big and strong, and handling them is dicey; biologists have had teeth knocked out by condors, and they can bite to the bone. "These days we use four people to handle a condor—one to hold the head, one the wings, another one on the legs, and one person to work the bird. We don't take any chances. Safety first, macho last," Denise said.

It isn't just the adults. When a chick hatches in a lofty, inaccessible aerie, it gets an in-the-cave physical from a vet who has to rope down cliffs, giving new meaning to the phrase "house call." Calf carcasses are placed at remote feeding stations, mixed with crushed bone and oyster shells for calcium. One could argue about how "wild" this free-flying population is; the condor is one of the most intensively managed, handled, and manipulated endangered species in the world, and watching the giant birds sitting on the baited cage, I wondered if the old critics of the captive breeding program had a point—maybe not a conclusive argument, but a valid point—when they warned of the permanent loss of something ineffable but important, a chain that broke when AC9 was captured and that even his eventual release could not make whole.

By now it was midday, and we took a short break at the ranch house, where blue grosbeaks and ash-throated flycatchers sang from the trees. While Denise and I picked a few lemons from a gnarled tree in the yard, a pair of western bluebirds watched us, the male so deep an azure he was almost black. Inside the crew quarters, there was a cardboard tube that served as an erstwhile vase for a collection of condor feathers that people had scavenged. I picked out a long primary from the end of a condor's left wing, the vane glossy black, its white quill as thick as a fountain pen; the whole thing stretched from my fingertips to the top of my biceps, almost two feet long. In the gold-rush days, it's said, miners would use a hollow condor quill to store gold dust; you could get a real treasure's worth in a feather this size.

The fog was starting to lift up in the mountains, so Denise and I drove back up to where we could see the trapping site again. There were fewer condors on the ground; several were gone, and four were aloft, taking advantage of the warming air and better soaring conditions. They flew with flat wings, the tips turned up like a sly handlebar mustache. The juveniles were all dark, but the adults had long, narrow triangles of white under their wings, with the oldest showing a touch of white on the upper surfaces of the secondary feathers as well. The sun was starting to pry its way through the clouds, and I could see the bare yellow-orange skin on the heads of the old birds as they peeled off and drifted out of sight.

Distance means little to these huge vultures; Hopper Mountain's birds may go as far as Simi Valley or Santa Monica, right on the edge of Los Angeles proper, while the Big Sur condors routinely travel the 150 miles or more to Hopper Mountain, where two of them now breed. In effect, the condors of Hopper Mountain and Sespe, Big Sur and Pinnacles National Monument represent a single population numbering about fifty birds (those in Baja are, for the moment at least, too far removed from the main group to mix). The federal recovery plan sets as its goal two populations—one in California, one in Arizona—of at least 150 condors each and including at least fifteen breeding pairs. Impressive as the gains have been, there's still a long way to go to reach that goal, and the setbacks have been significant. Natural enemies take a toll: some condors have been killed by coyotes, and golden eagles have injured many and killed several, while three condors at Hopper Mountain disappeared in the big fire of 2003.

Chick mortality has been severe; although three chicks hatched in Southern California in 2002, none survived long enough to leave the nest. One died from eating bits of trash fed to it by its parents—pop-tops, bottlecaps, pieces of plastic, a strip of cloth, a reflection of the condor's curiosity with man-made objects—and the other two dead chicks showed evidence of trash ingestion as well. Power lines have killed or injured flying condors, though now utility wires in the refuge have been fitted with wire-spiral "deflectors" to make them more visible to the soaring birds.

Lead poisoning from bullets remains the biggest concern, though; the condors pick up slug fragments while feeding on abandoned carcasses (feral pigs are often left to lie where they're shot) or from feeding on gut piles left by deer hunters, and if the lead doesn't kill them outright, it means the bird must be trapped and undergo a long, slow detox process in captivity. As we sat and watched the far-off condors, the radio in Denise's truck was crackling with conversations among the crew about Condor 213; the day before, a hang-glider pilot had reported it was flying poorly and crashing into the ground, and it had been seen dripping liquid from its beak and eyes—perhaps the symptoms of a bad case of lead poisoning, though it proved later to be a false alarm. But not always. In Arizona a few years ago, four condors died of lead poisoning within a few weeks of one another, while several others were sickened and had to be treated, probably from feeding on a single contaminated carcass.

Enough California hunters are concerned about condor welfare that a broad coalition, including hunting and gun groups, ammunition manufacturers, and conservation organizations, has made progress in moving them toward nontoxic ammo and techniques like burying gut piles. Refuge biologists argue, in fact, that hunters can fill the niche once occupied by grizzlies and wolves, in providing a steady food supply for the condors, golden eagles, and other scavengers, but only if the ammunition they use is nontoxic, like the tungsten-tin composite bullets now available.

But not everyone with a gun is so responsible. Condor AC8—one of the last wild birds pulled into captivity in the 1980s, who produced twelve chicks and more than one hundred direct descendants during her years in captivity—was released in 2000 to serve as a mentor to the younger condors. Researchers guessed she was about thirty-five years old and, like AC9, a priceless repository of condor wisdom about the Sespe, and thus a direct link between the lost past and the new captive-reared birds. So it was tragic on many levels when she was shot and killed on a private ranch near the refuge in February 2003. A suspect was arrested, convicted, and fined twenty thousand dollars, the largest amount ever levied for a condor shooting—but he got no jail time although, as Denise admitted, more than a few locals would have been happy to lynch the guy.

Throughout the day, as we watched condors soaring, Denise told me stories about many of the birds, which have strong and varied personalities—about Condor 125, for example, whom she described as the "top hoodlum" when he was younger, but who has settled down into sober maturity now that he's eleven years old, mated, and raising chicks. We agreed that there is a twinned thrill and danger in knowing them as individuals, when the real success of the reintroduction will eventually be measured in distance and anonymity—by reaching a point where the populations are wild-bred, self-sustaining, and untagged. Denise Stockton and the others on the recovery team know invasive techniques are necessary now. "But sometimes I get concerned that we're managing *too* closely," she said. "We don't have the magic hand of God—we can't reach out and make everything right every time, and what's going to happen to these birds will happen."

It was getting late, but the clouds had risen high enough that Denise thought it was worth driving to Condor Ridge, a promontory overlooking the heart of the Sespe sanctuary. We got lucky; the bottoms of the clouds licked us with tongues of mist, but we could look northeast into the almost vertical terrain that has been the condor's stronghold for the past century.

We found another intern, a young woman named Julia Lysobey, sitting cross-legged in the dirt on a dusty jacket, a hood pulled over her head in the fierce wind, squinting through a spotting scope at a cliff wall maybe a mile away. Condor 125, the former hoodlum turned good father, was sitting below the entrance to a cave where his single chick lay, but there was no sign of his mate, a female known as 111.

The baby, born in late April, was already lucky; at one point biologists thought they saw a black bear poking around the cave, but when they checked in panic, they found a healthy chick. In a few more weeks they would check again, with a veterinarian this time to do a hands-on exam, since one of the chicks that had died from trash ingestion the previous year had been this pair's. As I peered through the scope, the male flapped awkwardly from one whitewash-stained outcropping to another. Denise was looking at the sky. "I wonder where the female is. She doesn't like anyone watching the nest, and she's usually around here somewhere."

The male, big even at this distance, settled on a ledge. I was drawing my breath to say something when Denise whispered, "Oh. There's the other one."

About a hundred yards away, 111 sat on a rock beside the dirt road, staring hard at us. I had long known that condors are intelligent, intensely curious animals, but there was an almost physical slap in her glare. Unfortunately, at that moment an oil company tanker truck, spraying water on the dusty road, came around the nearest hairpin, lumbering up toward the condor. I cursed the miserable luck under my breath, but to my surprise the old girl stood her ground, though she raised the black hackles of her neck ruff, like an Elizabethan noble staring down her regal nose. Not until the truck was abreast of her, just yards away, did she lean into the wind and open her massive wings, buoyed by the updraft, then drop one shoulder and slide down in our direction.

The condor pulled up, drawing a wide circle directly above us; if I'd tossed a pebble high, I'd have hit her, and she filled my binoculars to overflowing. I thought a condor would be a hard, angular bird, but she was voluptuous, all curves and soft edges. I could see her twisting her tangerine head to watch us, the way she lowered her pale feet to use them as rudders, the slender primaries emarginated like an open hand (each of which, I now knew, was almost as long as my arm). And, too, I could see the bright red plastic tags clipped to the leading edge of each wing, and the thin wire antennas of her wing-mounted radios, the hallmarks of a jealously monitored endangered species in the age of wildlife technology.

I didn't say "Tally ho," not even to myself, under my breath. My telephoto lens was at hand, but I didn't move to pick it up. Twice more, 111 swung overhead to stare us down, and I was pinned by her gaze. And then she was gone, over the ridge, and there was really nothing more to do but drive back to the valley.

But there's one more thing I have to say. Driving down from the refuge, we crossed back into Los Padres National Forest, through the worst of the oil fields that make a mockery of the idea of protected wildland, the sides of hills bulldozed flat into drilling pads for oil rigs, dozens of them visible in every direction, the bare hills a meandering scrawl of dirt roads and pipelines. There are only one or two wells in Hopper Mountain refuge, but the Fish and Wildlife Service couldn't buy the mineral rights to the property when it bought the ranch, so the rigs may sprout there, too. It's a fact of California geology that a lot of the southern part of the state is underlain with petroleum; the stuff oozes out of the ground in parts of Hopper Mountain. And it's a fact of modern commerce that some people want it even if it comes from a national forest or wildlife refuge.

As if this weren't bad enough, now there's a suggestion to open the Sespe itself, even though the condor sanctuary has been closed to oil drilling for generations, and just as the nesting condors have finally returned. This is a heinous idea, a truly venal new level of greed if, for the sake of a few more barrels, we can't leave this one last, small, long-inviolate core of wild Southern California alone. It's the kind of nauseating, money-grubbing idea that makes you want to grab its author by the collar and shake some sense into his worthless head.

I'm not a violent guy, but this is what seeing a free-flying condor does to a person. At least this is what hearing AC9's signal, beeping across the rough-fanged mountains, has done to me; what seeing 111 sweep over my head, her pinions parted like fingers and the small white feathers beneath her wings flickering in the backdraft of her movement, has done to me. That surge of rage answered my question, earlier in the day, about how wild these condors truly are. Seeing them has made me passionately, ardently, furiously protective of these singular birds and this singular place—and isn't *anything* that sparks such fervor worthy of salvation?*

The next morning I left Ventura, slow-poking through jammed traffic at Santa Barbara, with the hulk of Santa Cruz Island and its icing of clouds far out to sea—a private cattle ranch in the 1950s, but now owned by the National Park Service and the Nature Conservancy, which are working to restore the native communities on the sixty-thousand-acre island. I drove north on Highway 101, through the still-smoldering remains of a seventy-five-hundred-acre wildfire south of Buellton, where dozens of fire crews stood on call and the air had a sharp smoky reek, then into pasture country with vineyards here and there. The dry hills were seamed with cattle trails, like topographic lines inching out the elevation. Yellow-billed magpies, big long-tailed birds in evening dress, flew from fence to fence—a species found nowhere else but central California. Some miles later, I came across one freshly killed beside the road and pulled over to look; the black tail and wing feathers were actually iridescent green and blue, like an oil slick, and the bill was the color of the remaining Hopper Mountain lemon on my car seat. The bird was warm and limp in my hands, its eyes half-closed, its feet cool and scaly.

At San Luis Obispo I turned onto Route 1 again, and just north of the Hearst Castle at San Simeon, at Point Piedras Blancas, I pulled off among a welter of vehicles on a bluff overlooking a wide, crescent beach. I walked toward a small crowd gathered by a wire barricade at the edge of the drop, maybe twenty feet above the beach, and as I did, I heard what sounded like a long, satisfying belch combined with the bass chatter of a two-by-four rattling along a slat fence. No one seemed offended, for the noise came from below, from one of a heap of otherwise sleeping elephant seals.

There is something inherently daffy about a seal the size of a minivan, so northern elephant seals never fail to charm a human audience, and

*It's still a rough world for the condors. After my visit, AC9's chick fell from its nest cave and broke its wing; rescuers found it had swallowed trash but was doing well otherwise, and were hopeful for its eventual recovery and release. The chick from Condors 111 and 125, on the other hand, died from unknown causes after also falling from its nest and being unable to return. The third chick from the Sespe, however, successfully fledged, the first condor to do so in California in twenty-two years.

they have plenty of admirers these days, because they're showing up in places where they haven't been seen since the official language of California was Chumash. Looking down the beach, I could see hundreds of the seals hauled out in the chilly sun, like the bleached trunks of enormous trees, flipping plumes of sand over their backs, lolling belly up, or arrayed in great lardish rows that reminded me of those pictures you used to see of Soviet women sunbathing on the Black Sea.

As it was early June, most of the seals I counted were subadult males, just getting the flabby, pendulous snout that (with their size) accounts for their name, and still well shy of their adult bulk of sixteen feet and two and a half tons. Most were molting, a physiologically more complex process than just shedding old fur; the upper epithelial layer of skin actually sloughs off, a process that biologists call "catastrophic molt," and leaves the seals looking as though they escaped from a leper colony. There were some females on the beach, ten or twelve feet long and weighing about a ton, and one silvery "weaner," a pup born a few months earlier, which by rights should have been out at sea, learning to hunt for itself; perhaps one of the females was its mother and hadn't yet cut the apron strings by abandoning her baby to go feed, like most of the females had. Unfortunately, there were none of the biggest, fully adult males, which wouldn't start hauling out to shed until later in the summer; when one of those old beach masters rears up, his head may be seven or eight feet above the ground.

These immense pinnipeds once dodged extinction by the thinnest of margins, and their story is one of the most extraordinary in the annals of conservation. No one is sure what the original range of the species was, but by the early nineteenth century, elephant seals were being pursued all along the California and Baja coast for their blubber, which, like that of whales, could be rendered for its oil by boiling it in huge vats known as try-pots. The seals were already rare by 1850, and twenty years later, after the sealers kept coming back empty-handed, they were considered extinct. Then, in 1880, a small herd was found hauled up on the Baja coast, and over the next few years more than three hundred seals—every one that the sealers could find—were killed. Again they were consigned to oblivion, but again a small number were found, this time on remote Guadalupe Island off Baja; all eighty-four were slaughtered over two years. Extinction was again declared, until nine appeared on Guadalupe in 1892, where they were encountered not by sealers but by a team from the Smithsonian.

If this sounds like a lucky break for the seals, think again. Reasoning that "the species was considered doomed to extinction by way of the sealer's trypot and [because] few if any specimens were to be found in the museums of North America," the Smithsonian team killed seven of the nine, and yet again the final doom of the elephant seal was pronounced. And yet again they persisted. Through 1911, small numbers of elephant seals continued to be found—and continued to be killed by museum collectors, an embarrassing black eye for science. Not until 1922 did the Mexican government finally cut the seals a critical break, declaring Guadalupe a biological reserve.

Though it was long thought that at its lowest point the population had dropped to fewer than a hundred seals, and perhaps only twenty, scientists today think there may have been more elephant seals than anyone realized, maybe as many as a few hundred; different ages and sexes haul out at different times of the year, so some missed the killers and the counters. But whatever the number, it was a vanishingly small pool on which to bank the survival of the species—but survive they did. Once protected, elephant seals began to increase in number and range, slowly at first, then with spectacular speed. In 1953, Fisher and Peterson had to go to the Coronados Islands, about twenty-five miles off San Diego (and just inside Mexican waters) to see them, but by 1960 the population was estimated at fifteen thousand, and elephant seals were being spotted as far north as the central California coast; by 1978, they were breeding at Año Nuevo, north of Monterey, and a decade after that, up to two thousand pups were being born there each winter. From its nadir, the northern elephant seal population has risen to about 160,000 in a century, a staggering rate of increase for such a large mammal. They continue to push north, with pioneers as far as Vancouver Island in British Columbia, where experts will not be surprised to eventually find them breeding.

The molting seals looked scabrous, with big patches of the old, felted skin coming off in sheets; those that had covered themselves in layers of dry sand looked uncomfortably like decomposing carcasses, until another seal would come too close and the sleeping giant would lift its head, bellow, and sometimes take a slashing bite at the intruder. Other young males, feeling their oats, would rise up, chest to chest, and grapple with each other—a prelude to the bloody, searing fights in which the mature bulls engage each winter, their bodies torn and gory from their opponents' sharp canine teeth.

Elephant seals are remarkable for more than just their size. On feeding forays they may travel as far as the Aleutians, spending more than 90 percent of their time submerged on hunting dives for squid that take them deeper and for longer periods than almost any other marine mammal. Some instrumented seals have been logged diving to depths of more than a mile, staying down for as long as two hours—a feat only possible because, as they sink, they exhale to collapse their lungs, depending on oxygen already in their muscles and bloodstream, which shuts down to all but essential organs as their heart rate drops to just a few beats per minute. Returning to the surface, they breathe for a couple of minutes, and then plunge down into the crushing deep again, a pattern they repeat endlessly for months—the only way they can build up enough fat reserves to get them through the long fasting periods while onshore breeding.

One of the reasons elephant seals are secure today is the 1972 Marine Mammal Protection Act, one of the toughest federal environmental laws on the books. With a handful of exceptions, like the traditional take of seals and walruses by Alaskan Natives, the MMPA banned the killing or harassment of seals, sea lions, manatees, sea otters, polar bears, dolphins, and whales and the importation of marine mammal products; this had a profound and almost immediate effect on many populations and marked a turning point for some previously beleaguered stocks. It has not been without its controversies and ironies; fishermen grouse about competition with growing herds of seals and sea lions, and fisheries managers trying to salvage critically endangered West Coast salmon runs have had to shoot sea lions that were scarfing down the few remaining fish as they tried to return from the ocean.

Elephant seals only started using Point Piedras Blancas in 1990, with the first pup coming two years later; by 1998, nearly seventeen hundred babies were being born there each year, another example of the explosive recovery this once-endangered species has enjoyed. But even in the presettlement days, it would have been unheard-of to find hundreds of elephant seals lounging on a mainland beach. They were originally creatures of the distant offshore islands, where they were protected from Native hunters and, perhaps most dangerous of all, from California grizzly bears that would have made a meal of all but the biggest bulls.

The humans today are friendly and the golden grizzlies became extinct in the 1920s, but the seals must still watch for danger. In one of the most dramatic encounters between predator and prey anywhere in North America, massive great white sharks now stalk the huge seals, especially in the waters off the Farallon Islands west of San Francisco. Biologists from the Point Reyes Bird Observatory conducted daily observations from a lighthouse on Southeast Farallon Island, watching for attacks on sea lions and immature elephant seals, and then approaching the scene by boat as the sharks, some up to twenty feet in length, fed on the floating dead pinnipeds. By photographing the sharks with underwater cameras on poles, they were able to identify individuals by their unique scar patterns and begin to piece together the numbers and movements of these most awe-inspiring of marine predators.

North of Piedras Blancas, the Santa Lucia Mountains crowded in, and for the next sixty or seventy miles, up through Big Sur and beyond, I followed a narrow, sinuous road that demanded constant attention—and, paradoxically, seemed designed for the sole purpose of distracting a driver with endless panoramas of achingly beautiful coastline. This part of the California shore has justifiably been called one of the loveliest on the planet, with the high chaparral-covered mountains piled up to more than five thousand feet, cut by deep canyons in which grow lush forests of tall coast redwood, and then falling sheer to the ocean with its rolling white breakers. The sea changes color by the moment as the sun and clouds and mist shuffle themselves, but this day it was mostly a deep jade green, with the dark barrier of the great kelp jungle rising to the surface in a wide brown band a few hundred yards offshore and the black rock stacks and boulders marooned by the lowering tide. Except for the thread of the road, and a few hamlets like Gorda, it is also wild and empty country, most of it part of Los Padres National Forest, including the Ventana Wilderness, where the condors have been released—and from which they sometimes fly down, as in the old days, to feed on the carcasses of seals and whales that wash up on the lonely beaches.

The central California coast is famous for another lost-and-found mammal, one that Fisher and Peterson came here to see—the sea otter. Once found right around the Pacific Rim from Hokkaido, Japan, to the southern tip of Baja, sea otters are wholly marine members of the weasel family, big (males may weigh up to sixty-five pounds), and covered in one of the most lustrous pelts in the animal kingdom. Once the Russians and Spaniards saw that fur in the early eighteenth century, the otters became the focus of ferocious hunting pressure. By 1911, perhaps as many as 800,000 sea otters had been killed, and the species had been reduced to no more than thirteen isolated remnants scattered between the Kuril Islands off Kamchatka to Islas San Benitos in Baja.

Although commercial otter hunting ended in California in 1840, when there were too few left to support the industry, the otters never recovered, and were considered extinct by the start of the twentieth century. Then, in 1914, a small group was spotted on the central coast; again they seemed to vanish and, like the elephant seals, were given up for dead. Thus, in March 1938, when Howard Granville Sharpe peered through a telescope from his ranch balcony at some odd shapes in the kelp beds far below, he wasn't at first sure what he was seeing. At length he decided the strange animals could only be sea otters—fifty of them, no less. He dutifully told the state fish and game folks, who blew off the report as misidentified seals or sea lions; the scientists at the Hopkins Marine Station in Pacific Grove, whom he approached next, were equally dismissive.

It wasn't until the following week that a state wildlife officer, his curiosity piqued, came out to look through Mr. Sharpe's telescope for himself and almost fell over in shock. Word spread to general surprise, delight—and a few red faces. The marine station staff apologized for their skepticism, telling Sharpe, "Had you reported dinosaurs or ichthyosaurs running down your canyon swimming about, we couldn't have been more utterly dumbfounded."

Although northern sea otters reclaimed much of their original range in Russia and Alaska, the southern population, which numbered about ninety animals in 1938, has had a harder time of it; almost seventy years on, they are still restricted to a 250-mile-long stretch of coast, from about Half Moon Bay below San Francisco to Point Conception north of Santa Barbara. But despite that slow recovery, this was always one of conservation's iconic feel-good stories, featuring one of the world's most appealing wild animals. Today, though, the story is woven with darker threads; the population, which had been growing steadily to about twenty-five hundred animals, dipped significantly, and mortality of breeding-age adults has spiked worrisomely. While no one is sure just how much trouble the sea otter is in, and there are more questions than answers, this much is certain: if you'd numbered the perils facing sea otters, until recently, no one would have put kitty litter on the list.

The next morning I walked down Monterey's famous Cannery Row, past the old sardine packinghouses that John Steinbeck made famous, which today hold coffeehouses and twee little shops, to the Monterey Bay Aquarium, a sprawling gray complex jutting right into the tidal zone at the edge of this deep, fecund bay. Kelp beds fringed the waters around the building, and harbor seals and sea lions poked their heads up for a look at the tourists watching back from outside decks. Sea otters floated among the bull kelp, safe from great white sharks that might be in the open water, while kayakers glided silently past.

Inside, I met Andy Johnson, the manager of the aquarium's sea otter research and conservation program, one of several organizations, agencies, and academic institutions trying to figure out exactly what's happening to California's fabled sea otters. Over its twenty-year history, the aquarium has developed a unique expertise at working with stranded or injured otters, rehabbing them, and getting them back into the wild, and Johnson took me first to a small trailer that temporarily houses part of the otter team's work. Inside the trailer, Anne Tewksbury, a volunteer, was watching a video monitor and using a joystick to train the camera on a female otter moving around a large, round pool full of rocks, with a platform made of PVC pipe and mesh in the middle for basking. Johnson explained that the female is one of two surrogate mothers the aquarium uses to raise orphaned pups, like the six-week-old youngster in the tank with her this day, a baby that had come in only a few days earlier, having been found, lost and alone, on the shore.

At one time, aquarium staffers raised such orphans, taking them for daily swims in the kelp forest, trying to teach them to dive and hunt, giving them a chance to socialize with the bay's resident otters. That approach was at best marginally successful, he said; the otters often showed too much ease and habituation with humans, and many had to be recaptured repeatedly when they didn't seem to be thriving. Now, with the use of surrogates, the pups grow up much better suited to life in the wild, though the numbers are still too small to make a definitive judgment.

The baby, who would still have been nursing if he'd been with his own mother, was busy nibbling bits of clam and crab his foster mother was passing him. "He's eating like a champ," Tewksbury said, brandishing a yellow legal pad on which she'd been noting each time the female had given him food. "Just look at all these transfers!" Full, the baby climbed up onto the adult's belly as she swam on her back in lazy circles around the tank. Like many stranded pups, this one had patches of matted, waterlogged fur, and the surrogate was busy grooming the baby back to health. The two had clearly bonded—the beginning of a six-month fostering process that would end, if all went well, with the newly weaned pup being returned to the bay.

During much of the recent past, California's sea otter population had been growing about 5 percent a year, Johnson said, but then things took a bad turn. From 1995 to 2000, the population dropped from about twenty-five hundred animals to about two thousand, and from 2000 to 2004 the number of dead or ailing adult otters found on the coast spiked dramatically. When the dead otters were necropsied, they showed a staggering rate of infection; up to 60 percent carried the protozoan *Toxoplasma gondii*, while others were infected with another protozoan called *Sarcocystis neurona*.

What stunned the biologists wasn't just the extraordinarily high infection rate but the organisms involved. *Sarcocystis* is, so far as anyone knows, carried solely by opossums, while *Toxoplasma* is the same microbe that produces the disease toxoplasmosis in pregnant women, causing them to lose their fetuses. Its only known source is house cats. In otters it can cause seizures, and even if the infection doesn't kill the animal, the loss of coordination can lead to death; for instance, those infected with the parasite are four times more likely than healthy otters to be killed by great white sharks.

Scientists know that the *Toxoplasma* cyst is almost indestructible; it can survive for months in the soil and passes through even tertiary sewage treatment unharmed. When further research showed that sea otters living near the mouths of streams and rivers were three times more likely to carry the protozoan, researchers began to wonder if the problem might not lie upstream—in the homes of California's millions of cat owners. Many people now use flushable kitty litter, and it may be that *Toxoplasma* cysts are passing down sewage pipes into waterways, and then into the ocean, where by an as-yet-unknown route it moves into the food chain and finally into the otters. There's no smoking gun, no conclusive proof that would warrant asking cat owners to change their feline hygiene, and Johnson isn't sure that would be a solution anyway. Because so many live otters have been exposed to the disease, there's even speculation that the surviving population will develop a resistance to it.

But the problem may be bigger than just one or two parasites. Another hypothesis among otter researchers is that the infections are, in a way, a symptom themselves. The otters, we know, are picking up increasing levels of contaminants from the environment, and the buildup of toxins may render them immunosuppressed, less able to fight off what would otherwise be fairly mild diseases.

Fortunately, things appear to be looking up, at least for the moment. Andy Johnson told me that the count in 2003 showed that the otters were back up to about twenty-five hundred again, and the latest tally, just completed, showed another gain to about twenty-eight hundred, which would be reassuring, except that both years also had much-higher-than-normal mortality, with infectious disease continuing to strike a huge proportion of the population, and record low pup production.

"So how do you reconcile those facts?" I asked him.

"I can't reconcile them," Johnson said. "And that worries me. Every year is a head-shaker, asking yourself, What does this count mean?"

One of the people trying to make sense of that paradox is Teri Nicholson, a research biologist with the aquarium's otter program. Her office is a block down the street from the main aquarium, in an old cannery building with a Thai restaurant next door, and when she got up from her desk to say hello, her small collie mix, Wanda, nosed out of the dark cave behind the chair to sniff my hand.

When I'd arranged with Andy Johnson to spend some time in the field with his crew, I envisioned us bobbing in a boat far down the coast, but Nicholson gathered her equipment, including a radio receiver and antenna, a tripod, and a pricey, high-magnification telescope, called to Wanda, and walked down the street, through the aquarium and across the campus of the Hopkins Marine Station on the point next door. One of the beauties of being a researcher on Monterey Bay is that your study area is barely a stone's toss away, and within a few minutes we were looking through the telescope at a dozen sea otters floating on their backs among the kelp beds. Even before she got a radio signal from the receiver, Nicholson recognized the otters—many of those in the bay have been individually color-tagged with plastic tags attached to their huge, webbed hind feet, and she was soon giving me brief life histories of several of the animals we were seeing.

As often as I've visited Monterey Bay, I'm amazed each time by the sheer fertility of the place, the riot of life on every hand. It was low tide, and the exposed rocks a hundred yards out were blanketed with harbor seals, Teri Nicholson's former research subject. A few California sea lions patrolled the edge of the kelp bed, porpoising with smooth grace, and one rose up in the water and barked like a dog. There were black oystercatchers, Heermann's gulls with soot gray bodies and pure white heads, plump little pigeon guillemots with ivory-patched wings buzzing as they flew. The huge granite boulders that guarded the point like sentinels were white with guano from the flocks of cormorants. And it was all the more striking since what's visible is only a tiny fraction of what's seething beneath the surface of this unique marine ecosystem.

One of the otters was a former rehab patient; the researchers are supposed to use a numbered code when referring to the otters, but Teri still knows this one by his old name, Pirate. She was one of the people who cared for the orphaned pups in the days before surrogate mothers were used, and while she thinks the new method produces pups with a better chance of survival, she does miss the kind of times she had with Pirate—taking him for swims a couple times a day in the kelp bed to introduce him to his future home.

"Did I love it? Are you kidding me?" She grinned a huge grin. "I had the most incredible experiences free-diving with these otters. Other otters would often approach us, females or territorial males who look at you like they know you're not supposed to have a pup, or females with their own babies that swim over—it's like a playdate. And watching the pup figure out what it's supposed to do in life—the first time they dive down and touch the bottom, it's like a light goes on, and all they want to do is dive. All of a sudden they have this three-dimensional mobility, and everything just opens up for them."

Out off the point, the rising tide was booming over the outermost rocks, torn to white foam, and not far away a female otter had hauled herself out with her pup, their fur glossy and slicked with water. The baby had what Teri called a lion's mane, a ruff of blond fur on the head that it would lose in a few weeks. The female dove into the surf, and the pup, temporarily alone, yelped a series of high-pitched cries as Pirate glided past. Then the mother surfaced and nuzzled her baby before crunching down whatever food she'd found herself.

I left Teri and Wanda not long after that; I had to keep chasing Peterson and Fisher's trail to the north, away from the coast this time and up into the snowcapped mountains on the California-Oregon border. Frankly, I was late, but it was hard to pull myself away from the Pacific, and I sat for a sweet and empty hour on the rocks beside Monterey Bay, watching sea otters roll and dive in the kelp as waves surged through. And if I did not find them as shocking as dinosaurs or ichthyosaurs, then neither has the passage of almost seventy years dulled the sense of our good fortune, however accidental and unmerited—that this world, which squandered so much that is lovely and wild, managed to hang on to this one, incredibly engaging creature. Long may it be so.

ELEVEN

Fire and Water

It was hard to leave the coast, but I am in my heart a lover of mountains, and as the day passed and I drove inland, I could feel the pull of the Klamaths and the Cascades to the north, urging me on. I turned across the head of San Francisco Bay and into the Sacramento valley, through grassland and farms, trying to make time. The Sacramento is the northern chunk of the enormous Central Valley, which covers almost seventeen thousand square miles of the state. Once quilted with native bunchgrass prairie, it was grazed by tule elk and pronghorn, which were, in turn, hunted by wolves and grizzlies, and lush with wetlands that attracted one of the largest concentrations of waterfowl in the world. Today, however, the Central Valley is one of the most altered ecosystems in California, given over largely to agriculture, with less than one-tenth of 1 percent of its original grasslands remaining, and less than a tenth of its wetlands.

I got a hint of those past glories when I took a midday break at Sacramento National Wildlife Refuge, part of a six-refuge complex in the valley near Willows. The ponds and marshes that dominate the refuge are almost entirely man-made, but they serve as an oasis for wildlife, especially the legions of geese and ducks that crowd the skies here in winter. In early June the refuge was much quieter; many of the seasonal wetlands were dry, though marsh wrens with their machine-gun calls were almost always in earshot, and the big impoundments were a smorgasbord of waterbirds—white pelicans, pintails, gadwalls, teal, shovelers, grebes, a few shorebirds. Broods of ring-necked pheasants shadowed their mothers along the dirt road, where jackrabbits hunched in the shade, and western meadowlarks sat in the tule beds and on the roadside brush, pouring out their justifiably famous songs—that single, clear clarion whistle, then a cascading rush of falling notes.

As I climbed out of the valley, the land became hillier, higher, the savannas covered with parklike forests of oak, their foliage green-gray. North of Redding, the highway rose into the mountains, and the oaks were almost immediately replaced by thick stands of pine, with Mount Shasta, its fourteen-thousand-foot peak draped in snow and clouds, playing hide-and-seek behind the lower ridges. The angle of the light, I thought, would make for an unusually good picture of the mountain, and I was looking for a place to pull over when a SCENIC VISTA sign appeared around a bend. I parked, but found the overlook was grown up with trees that all but blocked the view, and a tall chain-link fence made it impossible to get farther off the road. Stymied, I jogged up the highway a few hundred yards, my photo bag slapping against my side, and found a place where, if I stood tiptoe on the raised berm of the road, I could frame a shot of the mountain without the top of the fence intruding—but only by leaning a bit to the right, so as not to clip the edge of a blue highway sign, a tricky feat, since I was being buffeted by the backwash of eighteen-wheelers pounding by. But I managed to squeeze off a few frames.

Suddenly I stopped, feeling equal parts foolish and dishonest. The idea of photographing the mountain was, of course, to capture its beauty and wildness, virtues that would, if not evaporate, at least be tarnished if the viewer realized the pedestrian surroundings from which the photo was taken. The photograph would be a half-truth if I blocked out the crap in order to isolate the bright, wild core. So in the interest (I suppose) of full disclosure, I turned around and also took a photograph of the ratty old fence, the gravel-strewn pull-off, the line of semis and campers laboring up the grade.

At Weed I turned northeast, hooking around the top of Shasta and its two volcanic cones leaking steam into the cool air, and found myself in yet another, very different landscape, much drier, one of sagebrush, juniper, and orange-trunked ponderosa pines, their branches drooping down and then arching up again like the arms of candelabras. The road climbed to five thousand feet, then leveled off at Grass Lake, hundreds of acres so thick with sedges and bristling with nesting birds that no open water was to be seen. When a red-tailed hawk swooped low, a squadron of yellow-headed blackbirds, their white wing patches flashing, rose in squawking rage to meet it, harassing the predator away from their nests.

Nestled below the mountains lay the grasslands and fields of the Butte Valley, which I'd visited some years earlier to see the dense concentration of nesting raptors, especially Swainson's hawks and golden eagles, which feast on the millions of ground squirrels that thrive among the irrigated alfalfa fields and sagebrush flats. The Butte Valley sits at the edge of the Klamath Basin, an area the size of Massachusetts drained by the Klamath River and split by the California-Oregon border. Although gold miners were rushing into this part of northern California in the 1850s, the Modoc Indian wars of the 1870s slowed settlement, and the basin was among the last places in the Lower 48 to be opened to homesteading. At the time, the Klamath Basin ranked among the greatest wetlands complexes in the world, with more than 290 square miles of marsh and shallow lakes that attracted an estimated 6 million ducks, geese, and swans, earning comparisons with the Everglades; and like the Everglades, the Klamath wetlands were dismissed as a hurdle to overcome and a resource to exploit rather than a treasure to preserve.

With barely a foot of rainfall each year, the dry, often alkali soil of the basin was hardly ideal for farming, so to help make the region more attractive for agriculture, the federal Bureau of Reclamation in 1905 embarked on the ambitious Klamath Project to "reclaim" what was seen as wasted land and water. What eventually became a complex web of half a dozen dams, 185 miles of canals, and 500 miles of ditches emptied most of the basin's natural wetlands, leaving only a fraction of them intact. By 1920, Lower Klamath Lake had simply disappeared into the desert scrub, while Tule Lake had shrunk from its original hundred thousand acres to little more than a damp patch. What had been a western Everglades became a vast farm, growing grain, potatoes, and sugar beets, and the great migratory flood dried up.

Nor were birds the only wildlife that depended on the Klamath or suffered from its development. The Klamath River held the third-largest salmon run in the West, so rich in coho, spring chinook, and steelhead that in 1892 a federal official looking for hatchery sites recommended setting aside the river as a federal salmon preserve. The U.S. Fish Commission ignored the suggestion, and the Klamath's salmon—overfished, their way blocked by dams, the river's water diverted for irrigation—began a long slide toward oblivion. Along for the downward ride were the *qapdo* and the *c'waam*, the shortnose sucker and the Lost River sucker, which live only in the Klamath Basin and had been a major source of food to the Klamath and Modoc tribes living in the area. As badly as the salmon fared, the suckers have held on by an even slimmer margin, so that today the Klamath tribes, which revere them the way other tribes revere salmon, are permitted to kill just two of the rare and precious fish each year for ceremonial purposes.

Even after part of Lower Klamath Lake was designated as the nation's first waterfowl refuge in 1908, drainage and farming kept chewing away at what was left of the valley's wetlands; more refuges were added, particularly Tule Lake and Upper Klamath NWRs in 1928, but even within the refuges wildlife played second fiddle. Part of the reason for this is Tule Lake's unusual status—the only refuge jointly managed by the U.S. Fish and Wildlife Service and the Bureau of Reclamation, agencies with historically contradictory missions. It's been an uneasy partnership, with wildlife often the loser; as late as 1948, the federal government yanked twenty-three hundred acres out of Tule Lake NWR in order to open the land for homesteading.

The biggest problem for wildlife, however, was water—no water, no wetlands, no ducks. Peterson was especially anxious to see the Klamath refuges when he and Fisher arrived in 1953, because just a few years earlier, the USFWS had finally been able to divert enough water to reflood parts of the refuges that had long been dry. Even with restoration, the lower Klamath refuges were a shadow of their former selves, with Tule Lake just 13 percent of its original size. But the results were spectacular; Fisher said the concentration of inland-nesting waterbirds at Tule Lake eclipsed anything he'd seen outside of Iceland. "A pack of white pelicans floated on a bay like paper boats," he wrote. "Double-crested cormorants dried their black scarecrow wings on the rocky piles of rip-rap that has been dumped to act as loafing bars for the ducks . . . Herons were all over the place."

Yet even on the refuges, agriculture's imprint was unmistakable; while Tule Lake had thirteen thousand acres of wetland and impoundments, it had even more—seventeen thousand acres—of irrigated cropland, almost all of it leased to local farmers and irrigated with scarce Klamath water. This is still the arrangement and is in fact set in congressional stone. To prevent further diversion of Tule Lake's land for agriculture, Congress in 1964 passed the Kuchel Act, which finally made clear that the refuge was intended primarily for waterfowl, not reclamation. But it also mandated that "optimum agricultural use that is consistent" with waterfowl management continue and left the administration of the farming leases in the hands of the Bureau of Reclamation.

While this seemed to be another victory for birds, it actually solidified the ongoing use of Klamath Basin refuges for farming—hardly the best use for the majority of the land in a national wildlife refuge. Furthermore, Tule Lake was imprisoned behind its dikes, silting up steadily, unable to shrink and expand across tens of thousands of acres with the seasonal snowmelt in a cycle critical to keeping the marsh soils fertile. The result, over the past fifty years, has been a continual decrease in the fecundity of the refuge. Peterson and Fisher saw so many thousands of western grebes, "bodies low in the water, heads on thin necks like waterlogged boats," that the two men simply gave up trying to count them all. In contrast, I saw relatively little on the refuge's impoundments (inelegantly named the North and South sumps) except for a scattering of eared grebes, picking up the trillions of harmless midges that were emerging from the lake, and a fair number of coots. The northern migrants that once flocked here have also voted with their feet; cold-weather use by waterfowl has dropped enormously since the late 1960s, from about 3.5 million birds to barely 300,000 today. The main lake is so shallow the few remaining suckers must hide in a connecting canal.

The refuge is experimenting with creating a rotating cycle of seasonal flooding, continuous flooding, and farming that could help; and given the miserable condition of Tule Lake's wetlands, anything would be an improvement, as a visit to nearby Lower Klamath NWR makes clear. Like Tule Lake, Lower Klamath is devoted as much to farming as to wildlife, but its wetlands have been more intensively managed, its marshes allowed to periodically dry out and then reflood, keeping the soil richer in nutrients. While conservationists justifiably criticize farming practices on the refuge—not only that it and Tule Lake are the only two national wildlife refuges in the country to permit commercial agriculture* but also that the crops grown here, like onions and sugar beets, provide little wildlife food—it is clearly a healthier place for birds than Tule Lake.

*Many refuges permit farming, but under a kind of sharecropping arrangement in which the refuge managers decide what can be planted, when, and where, with the aim to create food and

Driving around the dirt roads that snake between the marshes at Lower Klamath, moving just fast enough to stay ahead of my own slowly drifting dust cloud, I put up a perpetual bow-wave of ducks, which flushed from the weeds along both sides of the road—mallards, gadwalls, cinnamon and blue-winged teal, lesser scaup, a few redheads. Killdeer crouched on the gravel shoulders, rushing away from me with piteous cries and dragging wings, trying to lure me from their perfectly hidden eggs; all I could do was stay as scrupulously as possible in the old tire tracks on the one-lane road and hope the nest scrapes were off to the side. In the open water, flocks of fifty or sixty white pelicans jammed together in masses of plunging heads and gaping orange bills, dipping up the fish they'd corralled, while American avocets and black-necked stilts whizzed by, their knitting-needle legs trailing behind. At one point, a mob of male red-winged blackbirds started screaming bloody murder, balled up in a frenzy just a few feet over the thick grass beside the road, from which emerged a long-tailed weasel, which stood on its hind legs and jumped up at the birds several times like a pogo stick.

But if conservationists thought the tough battles were in the past, bringing water back to the Klamath Basin marshes more than fifty years ago and preserving it from reclamation, they were wrong. As I drove out of Tule Lake, past green alfalfa fields and long center-pivot irrigation rigs spraying silvery mists in the morning air, I passed two signs I'd noticed on the way in, hand-painted in block letters on white wood and hammered onto stakes along the road. NEW ADDITION TO THE ENDANGERED SPECIES ACT: TULELAKE FARMERS, one said, while the other, around the next bend, read HONOR YOUR OATH: *PEOPLE* BEFORE FISH. The real battle for the

---

habitat for wildlife. The Klamath situation is a whole different animal. The land is leased out under congressional mandate, with leases overseen by the Bureau of Reclamation and with no control by the refuge's experts on what can and can't be planted, or how it can be grown. The result is a gross misuse of land that's supposed to be dedicated to wildlife, along with the heavy application of more than fifty kinds of pesticides and production of crops like onions that don't do a grebe or a duck a whit of good. It's also a huge draw on water supplies, which wind up on the croplands instead of the wetlands—up to sixty thousand acre-feet a year diverted to irrigation. (An acre-foot is the volume of water required to flood one acre to a depth of one foot, or 325,851.43 gallons; sixty thousand acre-feet is 19.5 billion gallons, which is why they calculate these things in acre-feet—it doesn't make your eyes bulge out quite so much.) Attempts to phase out farm leases on the refuges, and prohibit the growing of alfalfa and row crops with no wildlife value, have been stopped in Congress by Klamath farm supporters.

lifeblood of the Klamath Basin—its water, and whether it should flow for agriculture or wildlife—has only recently been engaged.

The Endangered Species Act is one of the few federal environmental laws with real teeth, a fact that draws plaudits from conservationists and attacks from those who consider its effects draconian. But both sides agree, when the ESA is brought to bear, it gets bureaucratic attention.

In 2001, the USFWS and the National Marine Fisheries Service issued rulings designed to protect the two species of endangered suckers in the Klamath Basin and the threatened run of coho salmon remaining in the river. (Chinook salmon in the river are somewhat more plentiful, and not federally listed.) Known as "biological opinions," the rulings required the U.S. Bureau of Reclamation, which runs the Klamath Project water system, to maintain higher water levels in Upper Klamath Lake and higher water flows in the Klamath River.

The summer of 2001 was a dry one, even by the standards of the Klamath Basin—the droughtiest in a quarter century. Ordinarily, farmers would use Klamath Project water to irrigate about 220,000 acres of crops, but with rain and snowmelt in short supply, and the new rulings in place to protect the fish, there wasn't enough to go around. Although the Bureau of Reclamation argued that lower minimum flows were sufficient, it turned off the tap to about fourteen hundred family farms, with widespread crop losses. (Farmers on another 240,000 acres outside the Klamath Project per se, however, received their full allocations.) The anger that the move generated was stark; there were rallies and protest marches, including a "bucket brigade" of fifteen thousand people in Klamath Falls. After protesters repeatedly broke the locks to open the canal headgates that control water flow, and after local law enforcement officers refused to arrest the perpetrators, federal marshals were called in. Riots and violence were widely predicted. The Klamath Basin found itself a national flash point, and an administration friendly to agriculture listened.

The next spring Secretary of the Interior Gale Norton and Secretary of Agriculture Ann Veneman stepped in, flying to the Klamath Basin and opening the headgates themselves. The farmers received their full allocation that year and celebrated a victory, but the drought continued, and by late summer the Klamath River was at a record low, with more water going into fields than down the river. The sluggish, warm water in the main channel promoted disease and carried little oxygen, and as the big chinook and coho salmon pushed up the Klamath in September to spawn, they began to die. Some thirty-five thousand fish, most of them chinooks weighing up to thirty pounds but including some threatened cohos, perished in what has been called the biggest salmon kill in history. Now the protesters were massing on the other side, including tribal members marching on the Bureau of Reclamation office in Klamath Falls carrying signs reading BUSH KILLS SALMON and the Yurok tribe, which filed a lawsuit against the Bush administration, claiming the salmon kill violated treaty obligations.*

It was a battle played out on many fronts—in the courts, on television, in the halls of the state legislatures and Congress, and on sophisticated Web sites like klamathbasincrisis.com and that of the Oregon Natural Resources Council. The Klamath became the nexus for grudges reaching far beyond the basin; property-rights activists, pro-salmon environmentalists, politicians, tribal leaders, and others from around the country piled on. That much of the brouhaha involved fish commonly known as suckers provided an easy punching bag; a writer for the conservative *National Review* dismissed the endangered fish as "unappetizing garbage recyclers" without usefulness or beauty and claimed the feds were "starving people to save suckers."

The reality, of course, was much more subtle and heartbreaking than that. Almost a hundred years ago, the government made a bad decision, to create an agricultural oasis in the middle of high, cold desert, maxing out the Klamath Basin's water in the process; today everyone and everything in the basin and downstream, from farmers to salmon to birds to

*A 2003 report by the National Research Council found no scientific support for either the higher river flows originally called for by the USFWS and National Marine Fisheries Service or the lower minimum flow levels later suggested by the Bureau of Reclamation. It also said the salmon kill could not be blamed entirely on water diversions and urged managers to look at other threats to salmon and suckers, including water quality and temperature, the loss of streamside vegetation, the impact of dams, competition from hatchery salmon and non-native sport fish, and algae blooms and low dissolved oxygen levels in Upper Klamath Lake from farm runoff. The council's report, especially an interim document released in 2002, was in turn criticized by prominent fisheries scientists as being overly simplistic and setting an unrealistically high standard of scientific proof before action should be taken, thus effectively dooming restoration efforts.

commercial fishermen to Indians, are paying for it. Good intentions don't balance bad judgment, and at some point you have to question whether it's worth it to keep sustaining the unsustainable.

Most of the families that farm the Klamath Basin were lured here generations ago by the government, offering free land and abundant irrigation in perpetuity. The homesteaders were often veterans of the two world wars; it isn't their fault that the deal is proving untenable in the long term. But this isn't a story of black and white. As sympathetic as a family farmer may be, there's been a widespread damn-the-government intransigence that has scuttled commonsense steps like buyouts of farmland and water rights from willing buyers or the phasing out of ag leases on refuge property. At the very least, if there must be farming in the Klamath, does it make sense that the crops being grown are especially thirsty ones like onions, potatoes, and alfalfa? Nor should one overlook the perks that come with farming here, especially the tens of thousands of acres of refuge land made available at fire-sale lease prices; some of those planting on the refuges were able to lease the land for a buck an acre while receiving $129 per acre in federal drought aid in 2001. The subsidies grew even richer with the 2002 federal Farm Bill, which included $50 million specifically for irrigation work in the Klamath.

But against the very visible economics of Klamath Basin farming must be weighed the other, often hidden losses of draining the basin's wetlands, damming its waterways, and diverting its flow. It's hard to quantify the loss to wildlife—how much would a robust, functioning western Everglades be worth today in tourist dollars?—but the U.S. Geological Survey took a stab at it, in a report that the Bush administration managed to squash until the details were published by *The Wall Street Journal*. The USGS concluded that while farming in the basin generates about $100 million per year, recreation, including wildlife watching and fishing, accounts for eight times that figure, and restoring a healthy river would more than treble it. The most cost-effective way to deal with the Klamath water crisis, the report's authors said, was to buy out the farmers for $5 billion, a move that would save $36 billion in the long run.

Instead, in October 2004, a few weeks before the presidential election, Norton announced the creation of the Klamath River Watershed Coordination Agreement, under which Oregon and California would take the lead in working out solutions to the water shortage in cooperation with four federal agencies. Farmers in the basin hailed it as a step in the right direction, while others, like the ONRC, dismissed it as an election-year smoke screen. Representatives of the Yuroks worried it might only maintain the status quo, and some experts saw it as ignoring the central issue, too little water for too many users. As one fisheries scientist told the Eureka, Oregon, *Times-Standard*, "The fundamental fact of life in the Klamath Basin is that it's overcommitted and oversubscribed by government programs." Meanwhile, the Yurok and two other tribes were pursuing a lawsuit against Scottish Power, whose subsidiary controls the hydroelectric dams on the lower Klamath, seeking $1 billion in damages for the historic loss of the salmon runs and the construction of fish passage facilities before the dams are relicensed. It's unlikely the war over the Klamath will end anytime soon.

And when all the shouting was over back in 2002, after water had been apportioned for farmers, tribal rights, and fish, the Bureau of Reclamation informed Lower Klamath NWR that there was no water left for the refuge. This is a common tragedy up here; wildlife still plays second fiddle. The refuge had to ask local farmers to cut back on plantings to donate some of their water allocation, as they often do, even though the refuge managers acknowledged it would be too little, too late.

This is the case across much of the West—only in the wettest years can all the appropriations called for under water law be filled. And those wet years are scarce and getting scarcer. The drought that brought the Klamath situation to a head isn't local; for half a decade or more, most of the West and the Southwest have been laboring under a drought of historic proportions. In some parts of the country it is the worst dry spell in a thousand years, based on analyses of ancient tree rings, but scientists now suspect what we're experiencing may be more the norm than the exception—that in the past century, when the West's water rights were being allocated, the region was actually wetter than normal.

However this fits into the record books, the drought has brought inevitable conflicts among water users, including agriculture, rapidly growing cities, and wildlife. It's clear that the West's water has been spread too thin, for too long, and the old approaches to partitioning it out, enshrined in the convoluted and arcane water laws in most states, are no longer sufficient. Changing it will be difficult, but the conflicts are only going to grow sharper and more strident, especially as cities push for a larger slice of the pie. Yet confrontation is not the only model for dealing with conflicting priorities in water use. Four hundred miles to the east of the Klamath, in the mountains of Idaho, regulators and landowners have tried a more collaborative approach to meeting the needs of wildlife and people.

When Meriwether Lewis crossed the Continental Divide in August 1805, he and his men camped by a small, fast river in what is now eastern Idaho, where they watched the Shoshone make nets of brush and drag them through the water. Lewis had his men do the same, and in two hours they caught "528 very good fish," including "ten or a douzen of a white species of trout . . . of a silvery colour except on the back and head, where they are of a bluish cast." These were the first steelhead and Pacific salmon the expedition had encountered, which had swum from the ocean more than 750 miles up the Snake River, and then into the tributary known today as the Lemhi River, which the Shoshone called Ag-gi-pa, or "Fish Water," because of its extraordinary salmon runs.

Salmon populations in the entire Columbia River system, including the upper Snake River basin of which the Lemhi is a part, have suffered appalling losses, first from overfishing and then from dams. There was an early power dam on the Lemhi that almost wiped out the salmon there, but when that obstruction was removed in 1957, the Lemhi's run of spring chinook recovered to about four thousand spawning adults. As more dams were added farther downstream on the bigger rivers, though, the migration was all but choked off, and the fish that did make it found the headwater streams sucked dry for farming. Often, the migrating fish followed the current into unscreened ditches, only to wind up flopping out their lives in a field. By 1994, two years after the run was listed as federally threatened, only seven salmon made it to the Lemhi to spawn. Thanks to informal efforts among landowners to keep the water flowing, the fish recovered a bit once more, up to six hundred adults in 2000.

The problem is, the Lemhi is another watershed that's over-appropriated, with more users than there is water to supply them in all but the very wettest years; when there's a shortage, parts of the river can be left bone-dry. This isn't unusual; in the neighboring Pahsimeroi River drainage, all the major tributaries have been so completely diverted for irrigation that they dry up every summer, and some have all but disappeared, so that it's hard to tell exactly where the stream once flowed. When, in 2000, part of the Lemhi was dewatered (that is the clinical term for the death of a river), killing endangered salmon, steelhead, and bull trout, the feds stepped in, and the National Marine Fisheries Service was considering the kind of action it later took in the Klamath.

But instead of a water war, NMFS and sixteen other state and federal agencies, along with the Shoshone-Bannock Tribe, sat down and hammered out a more collaborative approach. Irrigators agreed to lease water rights to the government in dry years, to keep a minimum flow of twenty-five cubic feet per second moving down the river in late spring and early summer, to help young salmon as they migrated to the ocean to mature. The farmers, in turn, had a pledge that the federal government wouldn't prosecute if small numbers of salmon died. The state legislature passed bills permitting this limited departure from normal water law, just on the Lemhi.

The agreement was hailed as a model for cooperative, pro-active conservation. Even after the two-year pact ran out in December 2003, and the drought deepened the following year, some landowners continued to flush water into the river at night, to push young salmon downstream. The pact isn't perfect; conservation groups (some of whom were frozen out of the negotiations) said that the farmers could save far more water by replacing their old, leaky irrigation ditches with pipes and sprinklers, and few people pretend this will solve the ultimate problem for salmon on the upper Snake River drainage—huge hydroelectric dams far downstream on the Columbia and lower Snake. But it shows that cooperation in the tense, divisive world of western water is at least possible and may be a small but vital step in keeping the "very good fish" in Ag-gi-pa.

I left the Klamath Basin, driving north along Upper Klamath Lake, which had a soupy green tinge—another algae bloom was under way—and into the heart of the Cascades. The clouds lowered, scraping the tops of the mountains, dragging wisps that looked like steam—which wasn't much of a stretch, here in this volcanically active range, where Mount St. Helens to the north was rumbling again after almost twenty-five years of quiet. But the most iconic image of the Cascades' violent past is Mount Mazama, which blew up ten thousand years ago with inconceivable force and formed Crater Lake, now hidden in the mountains ahead of me.

Crater Lake is where James Fisher, mild-mannered Brit and wide-eyed traveler, finally lost his patience with the American landscape. For two months, he'd been racing from one breathtaking expanse of scenic wonder to another, each more dazzling than the last, and at Crater Lake, which lay serene and impossibly blue, it finally all just seemed, well, unfair. "It's a plot, another geological plot, this place," he wrote in his journal:

> Western North America has these staggering examples of luck. Arizona's Grand Canyon, California's Yosemite Valley, and now Oregon's magic lake. All that geology can do, paraded in the pages of a living textbook . . . As I looked into the crater, I felt like a boxer coming up for the last round. "Makes you think," I said to Roger. But it hurt to think. The intolerable blue of the lake, which infected every snow slope, swamped with emotion my flickering power of analysis . . . Aloud I said, "Roger, we'd better not go to any more National Parks. I can't stand the pace."

In contrast to the dry, hot sun of the basin, it still felt like winter six thousand feet up at the lake, with a harsh, damp wind under the clouds. The snow lay deep even among the big ponderosa pines at the lower elevations, and by the time I'd driven to the rim, the drifts were several feet above my head, with the strata of the winter storms preserved in discrete layers of snow separated by thin bands of dust and dirt. At the lodge, the vertical face of the plowed drifts had been decorated with temporary graffiti—names, initials, slogans, some quite elaborate designs—cut into the white. When I mentioned it to a ranger, she sighed and said, "Well, it's better than spray paint. In a few weeks it'll be gone."

The lake, sitting below in the old volcanic caldera, was as gray as the sky, impressive but with none of the magical splendor of a sunny day. The monochrome feel even extended to the only birds in sight, the gray and black Clark's nutcrackers, a kind of large jay renowned in ornithological circles for its memory. In the fall, each nutcracker buries up to two thousand separate caches of whitebark pine seeds and is able to remember the locations of all those caches for up to nine months, apparently using landmarks and triangulation to return to precisely the right spot. And I can barely remember where I've put my car keys.

From Crater Lake, Fisher and Peterson drove north to the Columbia River Gorge, then west through Portland and into the rugged finger of land that pokes up into Washington's belly at the mouth of the great river. "Swinging around the wide curve on a hill, we suddenly left the living forest; we found ourselves in a ghost forest, where great tree skeletons, charred black, gestured in agony with their stubs of branches," Peterson wrote. "Groves of smaller trees—where the fire had rushed through without consuming the heartwood—had slipped their bark and now stood bleached and naked, like forests of weather-beaten sticks."

The two men were entering the Tillamook Burn, site of one of the most famous wildfires in American history. Actually a series of conflagrations, the first began during a hellishly hot spell on August 14, 1933, when, at a timbering operation in Gales Creek Canyon, one huge log was dragged over another, creating friction—and from it a spark, which caught in the dry slash.

Over the next ten days the fire grew to 40,000 acres—big, but nothing compared with what happened next, when the winds came. The fire exploded out across a fifteen-mile front, sending smoke forty thousand feet into the air, the suction of the rising heat generating a firestorm that uprooted trees. In the next day and a half, another 200,000 acres burned; the smoke plume was visible as far east as Yellowstone National Park and five hundred miles at sea in the Pacific, ash fell like snow in the Willamette valley, and it formed drifts two feet deep along the coast. Six years later, in 1939, fire struck again, burning another 190,000 acres; in 1945 almost the same area burned, while in 1951, 32,000 acres, mostly land that had been scorched in the earlier fires, burned—a frightening clockwork that locals dubbed "the six-year jinx."

Peterson commented on the "healthy green growth under the tall snags," much of which was the result of a massive replanting effort: 72 million seedlings of mostly Douglas-fir, with western hemlock, spruce, and redcedar, planted over the course of twenty years; for decades, it was a childhood tradition in Tillamook County for schoolkids to pitch in with tree planting up in the hills. Today the wayside signs along Route 6, not far from where the fire started, boast of "this handmade forest," which wouldn't strike any casual passerby as looking at all out of the ordinary except for its uniform age and height. There are still a few stands of old stark gray snags, especially on high, inaccessible crags, and even a few unburned areas along the Wilson River with magnificent old growth, to give a traveler a sense of what was lost.

I would follow Peterson and Fisher through the Tillamook myself, in a day or two, but first I needed to make a detour. The original Tillamook Burn was, for many years thereafter, called one of the biggest fires ever in the West. In 2002, another place in the Oregon mountains earned that dubious distinction, and I wanted to see the aftermath.

In the early afternoon of July 13, 2002, a thunderstorm moved in from the Pacific Ocean near the mouth of the Klamath River in northern California, then drifted north into the Klamath Mountains of southwestern Oregon, firing off lightning as it went.* It was one of many storms to roll across Oregon over the next several days, and many of the lightning strikes hit the Kalmiopsis Wilderness, starting fires in this 180,000-acre tract at the heart of Siskiyou National Forest. In keeping with the convention for such things, the blazes were named for landmarks where they began—the Biscuit Number 1 and 2 fires for Biscuit Creek, the Sourdough Fire a couple miles away for Sourdough Camp, the Carter Fire on a ridge near Carter Creek, and so on.

The fires grew, and within a few weeks they had coalesced into a single massive blaze that became known collectively as the Biscuit Fire. Feeding in part on trees killed by a large 1987 fire, it ballooned over the next two months. Firefighting efforts were hampered by the rough terrain, low humidity, and high winds, and by the time the fire was finally declared under control on November 8, it had covered 499,965 acres—almost 740 square miles—and involved seven thousand firefighters and support personnel from as far away as New Zealand. Forty percent of Siskiyou National Forest was hit, including virtually the entire Kalmiopsis Wilderness, along with parts of Six Rivers National Forest and neighboring Bureau of Land Management property. The cost of fighting it reached almost $155 million.

The Biscuit Fire was the largest blaze in 2002, but it was by no means the only one; nationally, that summer ranked as one of the worst fire sea-

*The Klamath Mountains, part of the Coast Ranges, are about ninety miles west of the Klamath Basin.

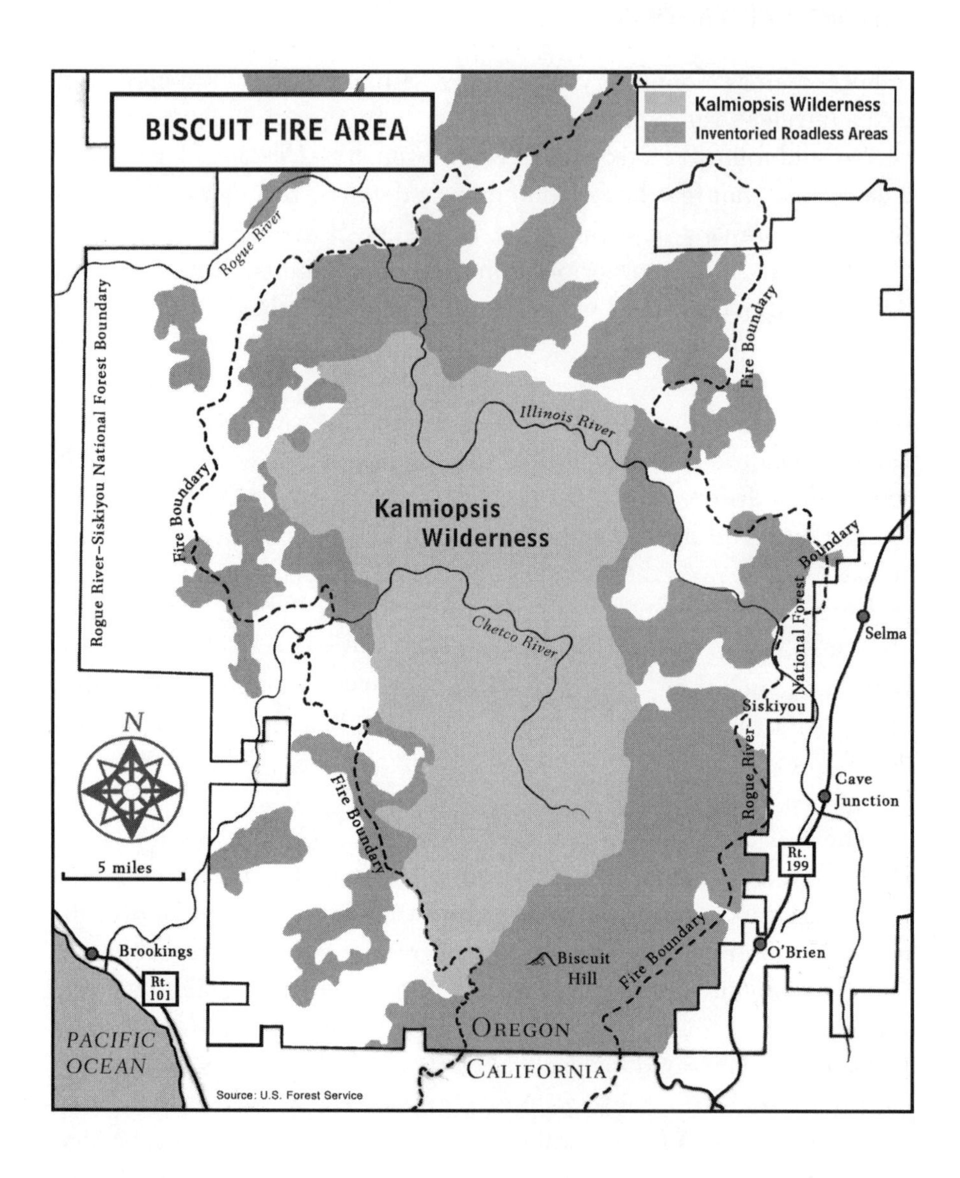

BISCUIT FIRE AREA
Kalmiopsis Wilderness
Inventoried Roadless Areas
Rogue River
Rogue River–Siskiyou National Forest Boundary
Fire Boundary
Illinois River
Kalmiopsis Wilderness
Chetco River
Rogue River–Siskiyou National Forest Boundary
Selma
Cave Junction
Rt. 199
O'Brien
Biscuit Hill
N
5 miles
Brookings
Rt. 101
PACIFIC OCEAN
OREGON
CALIFORNIA
Source: U.S. Forest Service

sons in the past half century, burning 7.2 million acres, costing $1.4 billion, and taking the lives of twenty-one firefighters. But the size of the Biscuit Fire, and how the U.S. Forest Service dealt with its aftermath, have highlighted the ferocious debate over forest and wildfire management in the West.

Fire has always been an integral part of the ecology of many western forests, which are often dominated by conifers like lodgepole pine and giant sequoia that need fire to open their cones, or by fire-tolerant (or even fire-dependent) hardwoods. The chaparral of Southern California is one such ecosystem, as are the ponderosa pine stands of the Southwest and the Rockies. In some places, the fire interval may be as little as two to five years; in other habitats, like moist coastal rain forests, thousands of years may elapse between major fires. But fire casts a shadow on the makeup and functioning of almost every western forest community.

For at least four thousand years, Native tribes in the Klamath Mountains used fire in a remarkably sophisticated way to maintain forest openings for hunting, to chase game, or to enhance certain food plants; this was especially true of fires lit to maintain the open oak woodlands, whose acorns were a major source of food for the Tututni, Chetco, and other bands in the region, or to scorch sugar pines into releasing sweet sap. White settlers had a different view. While miners set fires to clear the ground for gold prospecting, and ranchers did so to maintain meadows for their stock, the general inclination was to see fire as an enemy rather than a tool. Forestry was an infant science, and the ecological benefits of fire were little understood—the way in which low-intensity fires pruned back brush beneath the big, widely spaced ponderosa pines, for example, creating the grassy parklike settings that marked many old-growth forests.

Still, in the early years of the U.S. Forest Service there was an internal debate over whether the agency should conduct prescribed burns and allow some natural fires—the "light burning" school of thought, or zealously extinguish any fire, the "total suppression" philosophy. A series of huge fires, starting with the notorious Big Blowup of 1910 that saw 3 million acres burn in the northern Rockies, soon convinced forest managers and landowners that suppression was the way to go. After the first Tillamook Burn in 1933, it became USFS policy that any fire be extinguished or contained by 10:00 a.m. the day after it was spotted. Tillamook and other 1930s mega-fires also prodded the development of more aggressive firefighting approaches, including highly trained fire crews and airplane-deployed smoke jumpers, which began operating first in Siskiyou National Forest around 1940.

In 1944, the idea of total suppression got a powerful symbol, when a fire-prevention ad campaign for the USFS came up with a bear mascot (though at first he wore neither clothes nor the famous ranger hat); six years later, an orphaned black bear cub was rescued from a fire in New Mexico and nicknamed Smokey, and the publicity cemented Smokey the Bear's place as a national icon. Between almost universal support for fire suppression and a growing arsenal of tools forest managers could use, including aerial water bombers and chemical retardants, fire played less and less of a role in western forests. In Siskiyou NF, for example, one analysis of fire scars in the rings of trees dating back to the eighteenth century showed an average fire interval of about twelve years in some habitats. Between 1910 and 1939, 600,000 acres burned, while between 1940 (the start of the modern firefighting era) and 1986 an average of only 300 acres a year burned in Siskiyou. All the fuel that normally would have flamed out in frequent small fires simply accumulated, and the forest itself changed. Ponderosa and sugar pines were crowded out by smaller trees, and the once-open understory became choked with vegetation. Trees moved in to fill the open oak savannas that the Indians maintained for generations. Plantations of conifers seeded in the wake of logging operations created dense thickets prime for burning.

All across the West, Smokey had been too successful; the forests were piled high with fuel and waiting for a torch. In 1987, the Silver Fire caught hold in Siskiyou National Forest, burning 96,000 acres, but most Americans didn't pay attention to the question of fire management until the great Yellowstone fires of 1988, which burned some 700,000 acres in and around the beloved park.

It was an unforgettable thing to see; I found myself in Yellowstone at the height of the fires, with a group of teachers and environmental educators. The south entrance, through which we'd come, had been closed right behind us, and we'd spent the previous day with the park's chief naturalist discussing Yellowstone's ecology while watching the great billows of white smoke that dominated the horizon to the north, south, and west; at night, we stood in awe of the crawling snakes of orange light that outlined invisible, burning mountains. The next morning one of the teachers slipped and fell, cracking her head against the bus in which we were traveling; she needed immediate medical attention, so the decision was made to allow us to evacuate out the otherwise closed west entrance, along the Madison River, where some of the biggest fires were burning. We were escorted out by firefighting vehicles, but it was at times almost impossible to see or breathe; in places the flames were just a short distance from the road, the fire leaping from treetop to treetop, bombers roaring overhead to drop flame-smothering slurry. It was spectacular and frightening, yet in the midst of this inferno I was stunned to see small herds of elk and bison, some confined to just an acre or two of grass among the sea of fire, placidly grazing or chewing their cuds.

This was, of course, the lesson of the Yellowstone fires; what seemed a calamity to us was simply the natural order of things. Fewer than three hundred large mammals died, mostly elk out of a herd that numbered more than thirty thousand, and in the decades since, the ecological changes brought by the fires provided a needed jolt to species from aspen to moose. Such huge, stand-clearing fires occur every 250 to 400 years in the lodgepole pine forests of the Rockies, as essential to their long-term health as winter snows and summer rains. They bring, literally, a rebirth, though the notion was alien to many people; the governor of New Jersey, for instance, offered to send Scotch pine seedlings for emergency reforestation. But a year after the Yellowstone fires, scientists recorded native lodgepole pine seedlings sprouting up from the blackened soil at a rate of up to a million per acre.

Not all forests depend on such rare, enormous fires, however. The problem is that by short-circuiting the regular cycle of fire in many forests, we've traded frequent, relatively cool fires for rarer but hotter ones. One of the 2002 blazes, the 138,000-acre Hayman Fire in Colorado, burned the Cheesman Conservation Area, an old-growth stand in which specialists had completed a detailed history of fire cycles, based on old tree rings. Historically, fires came through the ponderosa pine and Doug-fir forest every twenty to forty years, only killing about 20 percent of the trees. But because of aggressive suppression, up to five fire cycles had been missed, and a great deal of fuel had accumulated; as a result, the Hayman blaze severely burned 55 percent of the forest, including 95 percent of the conservation area's old-growth stands.

And because so many more people now live in the forests—what demographers have dubbed the "wildland-urban interface"—the stakes are even higher, and the debate over how to manage wildfires grows, if you will forgive me, ever more heated. But nothing is quite as it seems, and even as one group used enormous fires like the Biscuit to argue for greater intervention in the western forests (including a cynical backdoor push for more logging in previously protected areas), others cautioned that the long view shows an ecological value in even the biggest blazes. "There are two kinds of forests in the West," fire ecologists like to say. "Ones that have just burned, and ones that are going to burn."

The Klamath and Siskiyou mountains overlap Oregon and California, rising from sea level to more than seven thousand feet. The rugged topography and local microclimates (as little as twenty inches of rain annually in some places, five times that amount in others) have created a botanical wonderland of unmatched diversity, such that the Klamath-Siskiyou ecoregion has been designated a site of global botanical significance by the World Conservation Union and proposed as both a World Heritage site and a UNESCO Biosphere Reserve. Many of the plants that grow here, like the threatened California pitcher plant, or cobra lily, are found few other places in the world. What's more, the Kalmiopsis Wilderness and surrounding roadless areas are the largest such tract of land along the entire Pacific coast between Mexico and Canada.

From Grants Pass I drove to the small village of Selma, on the eastern edge of the Klamath Mountains, turning up a dirt road into Siskiyou National Forest (now renamed Rogue River–Siskiyou NF). Almost immediately, I started passing a few burned patches, and then sizable areas, like the flank of one ridge covered with bare pines, their lower trunks scorched. I pulled over; below me, the hillside dropped steeply to the Illinois River several hundred feet below—rapids-strewn, its water clear green, like antique glass. Turkey vultures swooped past, below eye level, and a male western tanager sang from a nearby snag, his head glowing orange against the brown backdrop of the burned hillside. An osprey glided among the dead treetops, pulled up into a stall, and reached out to snap off a branch, with which it flew off, carrying it lengthwise as it would a fish, to add to its nest somewhere downriver.

But it wasn't until I'd driven seven or eight miles up into the national forest that I began to appreciate the scale of the Biscuit Fire. It's one thing for an easterner like me to read a figure like 499,965 acres, but quite another to see ridgeline after ridgeline rendered a bi-chromatic scene—the dull brown-gray of the burned trees and the pale orange of the exposed soil. The legendary capriciousness of fire was evident: a single pine standing unscathed among thousands of others that burned; fingers of untouched forest that seemed to wander haphazardly across the firescape; tongues of blackened trunks that reached deep into otherwise intact stands—the beneficiaries of wind that blew this way and not that, of embers that fell here but not there.

The result, as with most wildfires, was a mosaic. By the USFS's calculations, about 16 percent of the land within the Biscuit perimeter experienced high-severity fire, much of it in dense stands of mixed conifers on ridges. Another 23 percent ranked as moderate, with tree mortality between 40 and 80 percent. More than 60 percent of the almost half-million acres touched by the fire either burned at low intensity, with vegetation lightly scorched and few trees killed, or not at all.

Even before the fire was declared under control, the second Battle of the Biscuit was under way, this time over what to do with the land that had been burned. Should it be allowed to recover naturally, with minimal disturbance, or should there be an aggressive attempt to salvage dead timber and revegetate the scorched mountains? Complicating this question was the patchwork of designations of different parcels in the national forest, which included a federal wilderness and a national recreation area; inventoried roadless areas that could, in the future, receive wilderness standing and had been given special protection under a Clinton administration rule; riparian reserves and late-successional/old-growth reserves; backcountry management areas; botanical areas; and special wildlife areas—each of which carries its own set of regulations on what should and should not be done within its boundaries. All this argued, in the view of conservation advocates, for a go-lightly approach.

When the Siskiyou NF managers released their draft recovery plan for the Biscuit Fire, however, conservationists were shocked by the unprecedented scale of what the USFS was proposing: cutting roughly half a billion board feet of logs from twenty-nine thousand acres, one of the largest public timber sales ever, and a quantity of lumber that was 20 percent greater than the total annual production from all the federal land in Oregon and Washington combined. Two-thirds of the timber volume would be taken from old-growth reserves and twelve thousand acres of inventoried roadless land, potentially rendering the latter ineligible for future wilderness protection.* It also called for artificial planting on fifty thousand acres, claiming that so few trees were left in some areas that natural regeneration would take fifty years or more. Independent forest ecologists countered that fifty years is not an especially long period for natural recovery—forests sometimes take a century or more to bounce back from a large fire—and that the USFS's own mapping shows that 95 percent of the most severely burned areas are within a few hundred meters of live trees, a handy source of seeds.

The final plan, released a few days before I arrived in the forest, shrank the proposed sale to 372 million board feet on more than nineteen thousand acres, along with thirteen thousand acres of "fuel management zones"—miles of artificial firebreaks that opponents said would fragment the forest, provide entry for invasive species and off-road vehicles, and do little to curb future fires. It also cut artificial planting, but still called for thirty-one thousand acres to be seeded. While the local forest-products industry welcomed it, environmental groups vowed protests and legal action. Privately, observers on both sides admitted that with delays and the natural deterioration of dead timber, relatively little salvaged wood might ever come out of the Biscuit area.

Some logging was already under way along the road as I drove high above the Illinois River; I passed machinery (silent, as it was a Sunday) and several log piles stacked at road crossings, some of the trees four or five feet in diameter and giving off a rankly resinous smell that reminded me of stale urine. In the forest, the circles of freshly cut salmon-colored

*The Roadless Area Conservation Rule, proposed under the Clinton administration and adopted by the USFS in January 2001, extended sweeping protection from road building, mining, and logging to 58.5 million acres of federal land that lacked wilderness status but had not yet been penetrated by roads. The Bush administration froze it on its first day in office, along with a number of other environmental initiatives, then nibbled away at it for three years before abandoning the rule entirely in 2005, leaving it to state governors to decide whether they should develop or conserve federal land (and leaving open the option that the federal government could overrule them even if they tried to protect their forests). The notion that western governors—many of whom have a history of supporting extractive industry over ecological values—should decide what happens to land held in trust for all Americans did not, understandably, sit well with a wide range of conservation, environmental, and sporting organizations. Nor did it please many of the 2.2 million Americans who had written official comments regarding the original roadless rule, 90 percent of them favoring it—the largest number in the history of government rule making.

stumps shone against the carbonized trunks of smaller, commercially worthless trees and the bright green of new growth. Surveyor's tape fluttered from snags, strips of yellow, orange, or red, to complement the bright blue paint on the stumps of the trees that had already been cut.

The road was getting a lot of use from heavy equipment and wasn't in the best shape, so about eleven or twelve miles in, I found a pull-off and parked, then hiked up an old logging road that climbed above the main track, hugging the folds of the ridges. When that path petered out in an old landslide, I bushwhacked until I hit another, and then another; this particular area, outside the wilderness, was seamed with old logging roads, the legacy of decades as a working forest. I figured that as long as I kept an eye (or an ear) on the river, I wouldn't get turned around. It was bright and warm, with shade understandably scarce, but the hike was a colorful one. The dominant ground cover was woolly sunflower, a bright yellow species about knee-high, which left my jeans stained with pollen, but there were other wildflowers in great abundance, taking advantage of the unexpected sunshine—red Indian paintbrushes, pink campions, dozens of species I couldn't identify. Western tiger swallowtails floated among them, but the most common butterfly was Lorquin's admiral, black with white bands and a spot of bright orange at the leading tip of each wing.

I was surprised by the abundance of birds as well. Hairy woodpeckers—the large dusky Pacific Northwest form—worked on the dead trees, and black-headed grosbeaks and Lazuli buntings sang from many perches. I was scolded by Steller's jays, dark blue and black, as though they'd been colored by the Biscuit Fire's soot. In a small gulch where a trickle of water flowed, I heard something scurry at my feet and, expecting a ground squirrel, was surprised when two mountain quail hurried away, the male's long head plume nodding as he jogged.

One of the unusual features of the Klamath Mountains, compared with Oregon's other ranges, is the diversity of evergreen hardwoods here, species that are adapted to fire and come roaring back from the roots almost as soon as the embers cool. Even though it had been less than two years since the fire, already some of the shrubs and small trees, like Pacific madrone, had formed thickets as high as my chest. In what had been a cool gulch, now harsh in the midday sun, stood a big redcedar, its furrowed bark carbonized the first forty feet above the ground, but remarkably, its delicate foliage still clung to the lower branches, though the feathery needles were brown and dry. I scrambled up the slope, using my hands as much as my feet, stones and loose soil tumbling down, setting off little avalanches. I couldn't imagine working such terrain in a fire crew, broiling under protective clothing, lugging a Pulaski tool and a pack of emergency gear; the heat, the smoke, never quite knowing what the unpredictable fire would do, whether your retreat would be cut off by a wind-driven fireball. Surrounded by blackened snags and fallen logs reduced to charcoal, I could easily imagine the adrenaline and unease of such a situation, and I found myself thinking of the fire only as an adversary, an enemy.

There is a large and growing body of evidence, however, that suggests even huge, seemingly catastrophic fires like the Biscuit aren't the disaster they seem on the surface, nor are they necessarily out of line with the region's history. Studies of tree rings, along with those of charcoal and tree pollen deposits in soil sediments, have given scientists a historical window on fire activity in the Cascades and Oregon's coastal ranges. They've found that the sixteenth century was marked by extensive fires, while the next two hundred years were not. The nineteenth century was another period of great fires, while the twentieth saw a reprieve. While human causes obviously affected the cycle in the last century, before that it may have been driven more by pulses in the climate, such as the warm, dry trend that settled in around 1840. Within this context, the Biscuit Fire was not unusual.

Fire ecology is an emerging science, incompletely understood and not always wisely applied. One of the hardest lessons is that different forests, even different plant communities within the same forest, evolved with different fire regimes, and wise management must take that into account. It is risky, for example, to use computer models based on ponderosa pine forests in the Southwest and apply them to the very different conditions in the Klamath, yet a panel of forest ecologists and fire specialists, assembled by the World Wildlife Fund to review the draft Biscuit recovery plan, said that's exactly what the USFS was doing. The panel found that the plan overstated the severity of the fire and overestimated the benefits of salvage logging and reseeding while generally ignoring their ecological costs to a functioning forest. In fact, they said, there is plenty of evidence that salvage logging and reseeding, which establish huge even-aged stands, create better conditions for future fires rather than the reverse. They pointed out that in 2002, areas that were salvaged after the 1987 Silver Fire burned at severe levels twice as often as areas allowed to revegetate naturally.

It is clear, however, that many western forests are overloaded with fuel, sometimes to the extent that even prescribed fires can be a dangerous gamble. One solution is to carefully thin them, removing the crowded brushy trees while preserving older, more widely spaced specimens. Thinning is not an end in itself, fire experts say, but only a first step to then reintroducing regular fires, and in some ecosystems it's probably not appropriate at all; lodgepole pine forests in the northern Rockies, for instance, depend on stand-clearing conflagrations like the 1988 Yellowstone fires, not small, frequent blazes.

But thinning is labor-intensive and expensive, costing up to seven hundred dollars an acre, and because the trees that most need to be cut are small and commercially almost worthless, timber companies are unlikely to line up for the chance to do it themselves. But even as the Biscuit Fire still burned, George W. Bush came to the smoldering hillsides of southwestern Oregon to announce legislation that made forest thinning and logging a centerpiece. With the 2002 fire season fresh in its mind, Congress quickly passed what became known as the Healthy Forests Restoration Act, which Bush signed into law.

The Healthy Forests plan drew immediate praise from the timber industry, and flack from conservationists. The administration, saying that 190 million acres of federal forest and rangeland were at risk of catastrophic fire, and that administrative and legal delays had stalled thinning and restoration work on much of that land, stripped down the process for environmental assessment, speeding reviews for proposed timber harvests and relaxing regulations. Critics, on the other hand, including most major conservation organizations, called the law a cynically named disguise for increased logging, noting that it actually includes relatively few incentives to direct thinning operations at the kinds of young, dense forests that need it most, or to target the land closest to burgeoning human habitation. The law also greatly limits public involvement, they said, by reducing the scope of the review process and the ability to file administrative appeals. Coupled with the reversal of the roadless rule, and the administration's subsequent gutting of the 1976 National Forest Management Act, the Healthy Forests initiative promised more cutting in what had been protected roadless areas and old-growth stands.

At the same time, the USFS was coming under attack from some of its own employees for fighting too many wildfires, regardless of the cost in money, in lives, and to the environment. The Oregon-based Forest Service Employees for Environmental Ethics filed suit in 2003 against the agency, charging among other things that it had never assessed the impact on aquatic life and endangered species of the millions of gallons of chemical fire retardants it uses.

In terms of how the agency's resources are divvied up, firefighting is now the USFS's main job, accounting for much of its more than $4 billion annual budget, which has grown dramatically in recent years as more and more money has been devoted to firefighting. While the agency says the increase is due to the growth of private homes on neighboring land, FSEEE and others say it has become an end in itself, now that timber sales have declined by as much as 80 percent on federal land. They point to 2003, when out of more than 10,000 fires on national forest land, only 110 were allowed to burn themselves out naturally. Smokey the Bear, it seems, is alive and well in American forest policy.

I hiked a couple more miles, to the top of the ridge, where I found an overlook with a sweeping view to the west along the Illinois River. Nearby, the Biscuit Fire had burned very hot, incinerating the trees along the tops of the closest ridge, while the lower slopes were mottled with green and brown where the flames had skipped and swirled. In the distance, a timber cut had sliced a big triangle into the dull black-brown of a dense stand of fire-killed trees; the exposed hillside, green with new grass, was veined with fresh haul roads of bare dirt.

I watched my footing, because the slope was unstable and steep and it was a helluva long way to the bottom. I sat down, surrounded by sunflowers that bent in the warm wind, beside a big ponderosa pine stump, four feet wide and snapped off twice that high above the ground. Most of the outer bark, once pumpkin orange, was blackened and fragile, and I idly flaked it off with the blade of my knife to uncover the unburned layers beneath. Ponderosa pine bark is a natural jigsaw puzzle; the flakes grow in free-form, amoebic shapes that lock together, no two alike, the surface texture pebbled, the edges neatly beveled, as though they were crafted by an artisan. Brittle, many broke as I eased them out of the strata of bark, but eventually I had four or five wonderfully Gordian shapes, which I gently rubbed between my fingers as I looked out over the mountains.

It would be tempting to see, within the complex, interlocking, layered pieces of pine bark, an analogy to the complexities of history, ecology, economics, and politics that now drive the debates over both fire and water in the West, but if anything, that comparison oversimplifies the situation. Although we've come a long way in the last fifty years, we're still scratching the edges of what we need to know in order to rectify the mistakes of the past and prevent making even bigger and costlier blunders in the future. But that didn't stop me from slipping a couple of the bark keepsakes into my shirt pocket before I got up, dusted myself off, and headed down the flower-spangled hills.

TWELVE

In the Kingdom of Conifers

Wet from the waist down, I pushed through tangles of salal and rhododendron that dripped from a light overnight rain, trying not to lose my balance as my feet plunged into hidden holes. I'd already fallen once, tumbling lopsidedly between two long-dead trunks whose shapes had been blurred into a deceptively solid surface by the thick layer of moss and lichen that covered everything—rocks, logs, tree trunks, branches. At times, I felt if I stopped moving for even a few minutes, it might smother me, too, in a gentle prison of green.

The song I was chasing poured out of a dark nook beside the massive trunk of a giant Douglas-fir. It ran on and on, buzzes and trills, rolling tinkles like glass shards falling in a steady pile, repeated endlessly with barely a pause for breath, but always receding from me, luring me ever deeper into the forest; I was like the hero of a fairy tale, led off his path. But the only thing elfin about this singer, I knew, was its size. At last, moving more like a mouse than a bird, a tiny wren hopped into the open, its stubby tail vertical. I pished very quietly, and it half flew, half ran to a branch only a few feet from me, bobbing in righteous indignation but never stopping its loud broken-ice song.

The melody of the winter wren is a marvel, one that James Fisher commented on fondly several times during his American journey, as this is a species that North America and Europe share, one that reminded him of home. The tune is complex, lasting six or seven seconds and ending with a long trill—not musical in the sense of a thrush's fluting, but somehow more heartfelt, more powerful, a song punched into the mossy silence.

Trailing notes behind it, the wren ducked again out of sight, into the shadows of the trees—trees that were old when North America was still a rumor in Europe and that towered two hundred feet above my head. I was deep in the western Cascades, hiking through some of the finest old-growth forest left in the Pacific Northwest. Anyone who's spent even a few minutes walking among such a stand of truly ancient trees knows how the weight of age and mass presses down on you, slowing your footsteps as you crane back your head, muffling speech to a whisper. And that was happening to me. Even though the wet duff and moss ate my footsteps, I moved carefully; I even tried to clear my throat quietly. It wasn't only the great columns of the tree trunks and the high, dimly green space far above that made "cathedral" the first word I reached for in describing such a forest, as Fisher did when he first walked with Peterson in a temperate rain forest. It was also the recognition of being in the presence of something extraordinarily old and a little bit intimidating.

And yet at the same time, the trees brought out in me a wholly unexpected reaction. I had a fierce desire to leap into the air, to throw my arms high with some guttural, paganistic bark of exuberance, to scramble among the moss-draped nurse logs and the buttressed redcedars cackling like a happy child. The woods were thrumming with life, if lived at a slower and yet more majestic pace than mine. And the tension between those two impulses—reverential silence and wild exhilaration—had me vibrating like a plucked wire.

I stopped, listening again to the wren's ceaseless song, and suddenly I understood why he sang as long and as well as he did. Beneath such beauty and such life, how could he possibly sing anything less exultant?

The Opal Creek watershed lies about two hours southeast of Portland, in Willamette National Forest. You drive for twelve or thirteen miles on a scant two-lane up the Little North Santiam River, and when the macadam runs out at the national forest boundary, you drive gravel roads for another five or six miles, clinging to the sides of steep mountains where the morning mist rises like smoke from the dark spires of the trees.

Once there was a gold-mining camp up at the headwaters of the Little North Santiam, a place called Jawbone Flats, but today the road is gated, and the only way in is by foot, but as you hike, you pass reminders of those days, like an old mine shaft about a mile in from the gate, a hole head-high and two men wide, dripping wet, the exhalation from its open mouth cold and smelling of claustrophobia. The forest around the mine is, at first, nice but nothing exceptional—some good-sized Doug-firs, a lot of them tossed about by wind.

But within another half a mile, I was hiking through big, old trees—how old, I couldn't guess, because these Pacific Northwest giants grow at a remarkable rate, pushing for the sun through narrow light gaps that force them tall and straight. Three species dominated the forest: western hemlocks, their fanned branches tipped with fresh new growth, yellow-green like the lights of millions of fireflies; western redcedars, their trunks smooth and gray, spreading out into heavy, rounded buttresses near the ground; and a lot of Douglas-firs, that signature species of the western mountains from the Sierra Madre to British Columbia, with its thick, corrugated bark. And if that's not enough of a field mark, there is another rule of thumb up here in the Northwest: if you see a really big, truly breathtaking tree, it's probably a Doug-fir.

The trail crossed the river on a new wooden footbridge still heavy with the smell of creosote, and for the rest of the day I hiked beneath what one guidebook calls "truly exceptional old-growth"—which is as good a description as any, because words always wither when thrown against a forest like this. The wet winds come off the Pacific, the Cascades wring the rain from them, the summers are warm, the winters mild, and the richly watered mountains grow trees that are the envy of the world. The Pacific Northwest is the kingdom of the conifer, with more than thirty species of needle-leafed trees just in Oregon: more conifers in this region than anywhere on the planet, dominating almost every habitat from seaside to alpine treeline; trees of tremendous stature, with trunks that may be ten feet across the base, in forests that produce more biomass, more sheer weight of living tissue, than even the most fertile tropical jungle. And the Douglas-firs are the giants among giants, second only to coast redwoods in height, with some topping three hundred feet.

The Pacific Northwest without evergreens is impossible to imagine, but there's nothing predetermined about this particular mix of trees; like all natural ecosystems, it is the result of a happy coincidence of climate, geologic history, and the raw material of species available to colonize it. The forest makeup of this region has shifted throughout the millennia; during the Pleistocene, when the climate was considerably colder, the forests were dominated by smaller, short-lived spruces, while several thousand years ago, during a period of unusual warmth, it was a drier, more fire-prone region. The Douglas-fir, hemlock, and redcedar forest of today only became established about three thousand years ago, after massive fires swept the region, and the forests we see today are, as the trees would judge such things, only two or three generations old. It now appears that the ancient forests of the Northwest may have been the result of a unique concurrence of climate and disturbance that could not be replicated today—and that makes these forests even more precious.

How big are these trees? I wondered. It seemed almost presumptuous to ask the question—look at the damned things, I thought, they're *huge*, but I know that a reader needs something a bit more concrete to go on than that. My arms span a little more than six feet, and I wrapped them around a succession of trees like a crude tape measure, growing wet and muddy from waist to chin in the process—a tree hugger in fact as well as in metaphor. Several of the redcedars, with buttressed roots forming clefts almost deep enough to hide in, had circumferences of about fifteen to eighteen feet, while a big Doug-fir a couple yards away took four spans to reach around, twenty-five feet or so. I don't remember much junior-high math, but I know that to find the diameter of a circle, you divide the circumference by pi, so 25 divided by 3.14 . . . that's almost 8 feet across the trunk. Huge.

I slid down a wet, tricky slope to the edge of the stream, washing my stained hands in the frigid water. A dipper bobbed on a rock, then plunged into the torrent—the first of these birds I'd seen since the Sierra Madre, which gave me a little dislocating lurch. I had a clear notion, just for that moment, of the long miles and months that stretched behind me and of those yet to come in these final weeks of my journey, up through Washington and across Alaska. The stream flowed down a very narrow, very deep gorge, with the gray trunks of long, toppled trees angling down into the crevasse like jackstraws. The smallest of breezes stirred among the living trees, which were all but invisible beneath fright wigs of long, stringy moss, brownish liverworts covering the branches in heavy padding, lichens and mosses tufting the trunks, clumps of sword ferns rising from the wet sphagnum-covered ground. Perhaps, I thought, this isn't really the kingdom of conifers after all, but an empire of the lower plants, the older antecedents that knew the world in its earliest days.

There are relatively few places where you can walk in an ancient forest of any size, like Opal Creek; most of the continent's old growth is long gone, and while the Pacific Northwest has most of what's left, even here it is besieged by clear-cuts and a legacy of short-term profits and unsustainable logging. By the late 1980s, when the battle over the fate of the last old stands was reaching its fiercest level, scientists estimated that only about 17 percent of the original Douglas-fir forest, the dominant plant community in this region, remained. Even in 1953, when Peterson and Fisher passed through northern California, Oregon, and Washington, they were made uneasy by the pace at which the old growth was being logged.

"When a huge diesel-powered truck roars away bearing the body of one of these slain giants, no one with a conscience can help but wonder whether we have the right to take the life of a thousand-year-old tree for our ephemeral uses," Peterson wrote. "Already, about half of the original stand of coast redwood, estimated at 1,600,000 acres, has come to earth; another forty years will see all of the primeval trees cut except for those which the Save-the-Redwoods League has been able to buy up . . . We need wood; we must cut trees; but what bitter resentment coming generations would feel if our generation dissipated their entire inheritance and left none of the old trees standing!"

He wasn't far wrong about the coast redwoods of California; most estimates put the total remaining old growth there at about ninety thousand acres, of which eighty thousand are in public ownership, thanks in large measure to pressure over the past fifty years to buy up private holdings for parks or reserves. (Those still in timber company hands have been the focus of ferocious controversy, including tree-sitting protests in the 1990s to save the Headwaters Forest, at the time the largest privately owned stand of old-growth redwoods and now in public control.)

The Opal Creek watershed is one of the largest stands of old growth of any sort remaining in the Lower 48, but it almost met the same fate as much of the rest of the ancient timber in the Northwest. Old growth, with its high-quality lumber, was traditionally a sawyer's first choice, and traditionally, the U.S. Forest Service was eager to please by making stands on federal land available for cutting. The idea, in the decades after World War II, was to clear-cut the ancient wood, remove the snags and logs, burn off the slash, and plant a crop of a single commercially valuable species, creating vast, even-age, monoculture tree farms ready for harvest in as little as forty to eighty years.

Here in the Willamette, a lot of old growth was cut through the 1980s, and only the pressure from local activists trying to protect Opal Creek prevented the national forest from logging this grove as well. It was a battle that stretched across decades; as early as the 1950s, one family used an unusual technique—staking unpatented mining claims along Opal Creek—to thwart logging plans. But in the 1980s Opal Creek became the national symbol of the fight over old growth after George Atiyeh, a nephew of Oregon's governor and a scion of that same mine-claim family, got into the fray. At one point, Atiyeh is said to have held off timber-marking teams with a shotgun, but mostly he and his allies used lobbying, persuasion, and public outrage to make Opal Creek the poster child for the continuing rape of old growth in the Northwest. They did it so successfully that when U.S. senator Mark Hatfield, a rock-ribbed champion of Big Timber and foe of most wilderness protection, retired from the Senate in 1996, one of his last acts was to push through a bill creating the thirty-five-thousand-acre Opal Creek Scenic Recreation Area and Wilderness.

I was there on a weekday in early summer, but the trails were far from empty. There were plenty of day hikers like me and a few backpackers entering or leaving the Opal Creek or Bull of the Woods wilderness areas. Twice I encountered groups of a couple-dozen teens, muddied and intense, with instructors from the environmental education center at Jawbone Flats run by the Friends of Opal Creek; they had packs of laminated flip cards with field-guide illustrations and were discussing wildflowers and mosses. Near the trailhead, I rounded a bend and found myself wading through a multicolored clot of grade-schoolers in bright raincoats and slickers, loud with enthusiasm, and a couple of instructors who gave me apologetic grins while talking to their charges about old-growth forests.

This may seem an odd question, but what, exactly, does the phrase "old growth" mean? Everyone can form a mental picture: huge trees, shafts of sunlight, a page from an Audubon calendar. But that's simplistic, and there isn't a simple answer. Much depends on the kind of forest—its species composition, its immediate environment, the region's climatic history. Size may not be important; some old-growth trees are diminutive, like the relict pitch pine or scrub oak barrens of eastern ridgetops. Even in the Northwest, where the trees grow tall, does "old growth" refer only to forests of immense stature, undisturbed by humans (and does that mean only post-settlement pressures like commercial logging, or does it also include the kind of intentional fires set by Indians?), or can it mean any forest that is allowed to reach great age? And if the latter, what is that age? Two hundred years? Five hundred? A thousand? Should it refer to mixed-aged stands, those molded by fire or windstorms, where some of the trees are very old and others are much younger? (And if we say yes, would we include forests shaped by natural forces but exclude those rendered of mixed age by man?)

What are the characteristics, structural, ecological, or otherwise, that distinguish old-growth from merely mature forests? There's no right or wrong answer, as the variety of scientific publications on this subject over the past twenty years demonstrates. Some authors have taken a very specific approach. In 1986, a group of USFS and academic researchers tried to craft a definition of old growth specifically for Doug-fir, hemlock, and redcedar forests in California and the Pacific Northwest; by their reckoning, a true old-growth forest in that region, with those dominant species, would have at least eight trees per acre greater than thirty-two inches in diameter at breast height (dbh, a common forestry measurement) and at least twelve (mostly hemlock or redcedar) greater than sixteen inches dbh. They gave criteria for the number and size of dead snags and fallen dead trees and the composition and structure of the understory layer. Other experts eschew a scale using only size and rank old growth based as well on age: an "early" old-growth forest, in their view, has at least eight trees per acre that are thirty-two inches dbh and at least two hundred years old; a "classic" old-growth forest has trees at least forty inches wide and three hundred years old; and a "super" old-growth forest features monsters more than six feet in diameter and seven hundred years old.

The attention given to old-growth forests is a relatively recent development. For most of the last century, they were viewed in strictly economic terms, a source of high-grade wood (and lots of it), but dismissed as having limited ecological value. Many foresters, in fact, were actively derisive of old growth, considering it a waste of space—geriatric trees that were better cleared for younger, more vigorous plantings. "Storing wood on the stump," it was called. In published reports, the forests were described as "over-mature" or "senescent," but in shoptalk among forestry workers, they were "cellulose cemeteries," and even wildlife managers called them biological deserts when compared with younger forests.

By the 1970s, however, researchers had begun to actually look at the inner workings of old forests and found them vastly more interesting—and more ecologically important—than anyone had imagined. Far from being deserts, these forests have a high diversity of plant and animal life, including many specialists found only in this ecosystem. One important difference the researchers found between younger forests and old growth is structural complexity, which engenders biological richness—the deep, layered canopies and sub-canopies of ancient giants, adolescent upstarts racing to claim a light gap, young saplings, and on down through the shrubs and ferns and tangles of fallen trees to the thick layers of organic humus at ground level. As in tropical rain forests, where epiphytic plants like bromeliads and orchids growing on the branches of trees provide layers of ecological intricacy, the mosses, liverworts, and lichens that may all but smother the branches of a temperate rain forest tree likewise produce habitat for other organisms, while the lichens fix precious nitrogen for the entire system; the densely vegetated limbs condense fog and moisture, slow rain runoff, and increase water retention in the soil. Standing snags provide nest sites and perches; fallen dead logs (some taking centuries to rot) provide a seedbed for new trees and home for animals, while those that clutter streams slow the flow, creating pools and hiding places for trout and salmon.

As scientists have plotted the ecological relationships among the forest's inhabitants, they've found tangled webs within webs. The schools of salmon that once ascended rivers of the Northwest, for example, clogging headwater streams to spawn, brought enormous loads of nutrients high into the mountains, fertilizing the forests as the salmon died and were eaten by bears, ravens, and other scavengers. Scientists analyzing the growth rings of ancient trees have found the unique isotopic signature of marine nutrients within their wood, proof of the bridge between the deep Pacific and the high Cascades that is now largely severed by dams.

Another fascinating relationship involves truffles, which are really the fruiting bodies of a group of subterranean fungi known as mycorrhizae. Found almost everywhere vascular plants grow, mycorrhizae live a symbiotic existence, coating a plant's rootlets and drawing off carbohydrates while increasing the plant's ability to absorb water, phosphate, and nitrogen and producing a complex of antibiotics and growth regulators to help the plant. The fungi, in turn, depend on nitrogen-fixing bacteria within their own tissues, just as the bacteria depend on yeasts. In the nutrient-poor forests of the Northwest, mycorrhizal fungi and their symbionts are especially important to the survival of conifers, but the mycorrhizae can't spread themselves without help. So they produce the truffles, which sharp-nosed mammals like northern flying squirrels, red-backed voles, and Douglas' squirrels dig up and eat. The fungi spores, bacteria, and yeasts pass unharmed through the animals' digestive systems and are deposited in a new site to begin to grow. The fungi, in turn, are an important part of the rodents' diets—in the case of flying squirrels, almost their exclusive diet.

And the small mammals, in their turn, are the prey base for the most famous and contentious animal in the old-growth forests of the Northwest, the northern spotted owl, on whose feathered shoulders the fate of the old trees was laid. This large, rather tame forest raptor inhabits ancient conifer forests from northern California to southern British Columbia (other subspecies are found in California, the Southwest, and Mexico), and in the 1980s it became the nexus of the fight over forest management and preservation.

Quiet and retiring, spotted owls are largely—in parts of their range almost exclusively—dependent on old-growth forests, and as the old timber fell and what was left became a fragmented patchwork, the owl population overall declined as well. As early as the 1970s, some within the Department of the Interior were suggesting the owl be given protection under the new Endangered Species Act, but everyone knew that would have meant a curb on logging in old-growth stands, and action was postponed for almost two decades. Lawsuits finally forced the U.S. Fish and Wildlife Service to list the bird as threatened in 1990, when a spate of new research showed that the owls were decreasing at an accelerating rate.

Environmental groups redoubled efforts to reduce logging of old-growth forests on federal land, using the owl's new status as a powerful tool in their long fight against what they saw as unsustainable cutting. To many people in the rural Northwest, on the other hand, where logging had been an economic linchpin for generations, the spotted owl was nothing less than an attack on their way of life, and bumper stickers proclaiming SAVE A LOGGER, EAT AN OWL were common. In fact, the timber industry in the region was already in transition, from forces that had little to do with owls. Mills were increasingly automated, reducing the need for mill workers, timber prices were down, and cheaper Canadian timber dominated the market. But there is no question that the owl's threatened status brought the issue of logging in the Northwest to a head.*

Beginning in the late 1980s, various groups, from the U.S. Forest Service to a congressionally appointed panel, tried to create forest management plans that would balance owl conservation with continued timbering; few were implemented, and all drew fire from one quarter or the other. By the early 1990s, logging on federal land was in a judicial straitjacket of lawsuits and countersuits. Finally, in an attempt to settle the matter once and for all, President Bill Clinton convened a forest management summit in 1993 and, a year later, acting on one of the alternatives proposed by the participants, implemented the Northwest Forest Plan. Covering 24 million acres of federal land in Washington, Oregon, and California, the plan made 80 percent of that area off-limits to further logging while leaving about 1 million acres of old growth available to timber companies; the industry expected to be able to cut about a billion board feet of lumber a year in the remaining open land, though in practice the cut has been about half that.

The Northwest Forest Plan, for all its critics—and there have been many, on both the industry and the conservation sides—was at least an attempt to take an ecosystem-wide approach to public forest management. The kind of intensive, short-rotation tree farm model the USFS had been pursuing for years was clearly not preserving many of the species that inhabited the Northwest, from the owl to endangered salmon, steelhead, and other anadromous fish, of which some 260 runs are considered at risk of extinction. Nor was it saving one of the Northwest's most mysterious residents, the marbled murrelet.

*In the midst of all the ruckus over logging, an unexpected danger to the spotted owl was moving into the Northwest—the barred owl, the spotted's closest relative and originally an eastern bird. Barred owls spread across the Canadian plains in the late nineteenth and the twentieth centuries, possibly thanks to shelterbelt plantings and fire suppression, and then across much of the Pacific Northwest by the 1970s and '80s. More aggressive and adaptable than spotted owls, barred owls will displace their threatened cousins from territories and are now hybridizing with them. While this may seem like a natural process, ornithologists believe the barred owls are doing so well in the Northwest precisely because of the habitat fragmentation from logging that spotted owls find so hard to cope with.

Given James Fisher's keen professional interest in seabirds, Peterson wanted to show his friend this pudgy eight-inch member of the alcid, or auk, family, which is found at sea from northern California to southwestern Alaska. The men finally spotted some along the coast of Washington, "small dumpy birds that skittered like skipping stones over the wavelets" as they ran along the water to take off. Fisher was "agog" at the sighting, Peterson reported, and well he might have been. Even in the 1950s, the murrelet was famous—one might even say infamous—as the last North American bird whose nesting grounds were unknown. There was one verified egg in the Smithsonian's collection, but it had been taken from the body of a female shot in Alaska, and cash rewards were offered, and long unclaimed, for a genuine nest. Peterson, noting that some tribes called the murrelet *tichaahlukchtih*, "at night passing high over the mountains," because its *keeer, keeer* calls could be heard in the dark, speculated it might nest in forested mountains near the coast, probably under fallen logs or in burrows, like most alcids.

Finally, in 1974, almost two hundred years after the species was first described for science, a marbled murrelet nest was found—not in a burrow, as everyone had expected, nor by an ornithologist, but by a tree trimmer working 145 feet up in a huge Douglas-fir in Big Basin Redwoods State Park, California. The murrelet, it turns out, is another old-growth specialist, making its nest—a small depression on a thick bed of moss, in which it lays a single egg—on a wide tree limb far above the ground. Since then, about 160 additional nests have been found, including a few on the ground in Alaska and Siberia.

Abundant a century ago along the Pacific Northwest coast, the murrelet has fallen on bad times, disappearing altogether from some areas and becoming sharply rarer in others; in the Gulf of Alaska the pint-sized seabird appears to have declined by as much as 70 percent, and off Vancouver Island its population (based on counts while the birds are at sea and relatively easy to survey) fell 40 percent in just a decade. The most significant cause has been the loss of much of its old-growth nesting habitat, but it isn't safe at sea, either, with oil spills and gill nets that snare the deep-diving birds. (Controls on netting have reduced that particular danger somewhat in recent years.) Still, the decline continues, and the murrelet's population in the Northwest joined the spotted owl on the list of federally threatened species in 1992, a further spur toward comprehensive management of remaining ancient forests on federal land, where 90 percent of the robin-sized birds are believed to nest.

But in an illustrative, if especially odious, example of the way politics can trump science when a rare species gets in the way of a powerful industry, the murrelet suddenly found itself in danger of losing its federal protection in 2004. In the wake of a lawsuit by a timber industry group and three sawmills, the U.S. Fish and Wildlife Service's Portland, Oregon, office assembled a panel of sixteen experts to review the bird's status. The panel noted that while the Northwest constitutes almost 20 percent of the murrelet's range, it now supports only 2 percent of the surviving population, about twenty-one thousand birds—and that regional population continues to decline, with extinction in the Northwest possible in a few decades. Murrelets remain fairly common in Alaska and Canada; if the murrelets in the Northwest are genetically, behaviorally, or ecologically the same as their northern brethren, this might not be a problem, at least from an endangered-species perspective. But the panel, in reviewing all the scientific evidence, found that the birds in the Northwest are in fact distinct, and thus deserve ESA protection.

In April 2004, the Portland USFWS office agreed with its hired experts and issued a recommendation that the marbled murrelet continue to be protected as a "distinct population segment" under the ESA. But that decision was overturned by higher-ups in the Interior Department, which ordered the report rewritten and announced—absent anything except their say-so, much less credible scientific evidence—that the murrelets of California, Oregon, and Washington were not distinct from those farther north. Having redefined the southern murrelets, the administration initiated a review of the bird's status across its entire range. If it finds that the murrelet is not threatened elsewhere, those in the Northwest would likely lose their protection under the Endangered Species Act—and thus some of the protection for the old-growth forests. Conservation groups like Audubon attacked the move as policy based on junk science, but that's being too kind; it was the kind of sop to industry that has been a hallmark of the Bush administration, notable only for the especially shameless way in which it was done.

I thought about murrelets as I left the Cascades the next morning and drove west, into the Coast Ranges, through the replanted forests of the old Tillamook Burn, and up the gorgeous shoreline, wondering if there were any of the little seabirds hiding in mossy fastnesses up in the hills. The weather couldn't have been better—bright sun and a chilly wind that raised whitecaps on the mouth of the Columbia as I crossed from Astoria, with breakers dimly visible to seaward. But the drive up through western Washington, which I'd been looking forward to for several days, proved a fairly dismal experience. Once the road moved away from Willapa Bay, the clear-cuts took over—the complete domination of the land by industrial forestry. I drove toward Aberdeen, a Weyerhaeuser company town, through the corporation's lands. North of there, where the signs said the big landowner was Rayonier, the logging was even heavier, with a succession of placards giving the history of each tract—cut in 1960, planted in 1961, cut again in 2000. No matter where you looked, every hillside was in some stage of rotation, from those newly harvested with rough, freshly bulldozed roads; to older plantations where the young conifers had risen above the silvery stumps of their predecessors, sometimes with a few tall old-timers scattered around, marooned and listing a bit to one side; to tracts with thirty or forty years' growth, the trees tall but whippy. But everywhere, the forest was a commodity, a crop in the most pedestrian sense of the word.

On some of the cuts, you could read the history in the stumps—the gray giants, the old-growth stubs that were still seven or eight feet tall and two-thirds that wide, with rectangular holes in the sides where the timber crews had inserted board steps, so the cutters could get up above the wide buttressed roots and cut where the trunk was narrower and more manageable. And among the big, widely spaced stumps, the very much smaller ones of the second or even third growth, a green jungle of alder and other scrub growing among the slash, all of it purple with the high, tapered wicks of foxglove flowers, an alien species that came with the first whites to settle the region.

When Fisher and Peterson drove through here in June 1953, they described a different world. Entering the Olympic Peninsula, "we had the same feeling we have often experienced crossing the desert, or the sea. Here trees were an elemental thing like the waters of the ocean. We both fell silent as the thin ribbon of concrete took us through seas of Douglas fir, Sitka spruce, western hemlock and cedar. Trees, trees, trees—an endless green treescape seldom relieved by a house, a field, or a village." They could still speak of the Olympic Peninsula, an area the size of Connecticut, as the largest unbroken forest in the United States (this was before Alaskan statehood), but once again Peterson saw the writing on the wall; in the years since his first visit, he said, much of the old growth had been cut along the highway, and he noted that Sitka spruce here was being cut ten times faster than it was able to grow back. "How long will this big timber last, I wondered, as truck after truck rolled by."

Fifty years on, the old growth has pretty much vanished from the private timber company property on the western Olympic Peninsula, as well as from a lot of the federal and state land, replaced almost completely by managed stands on a rapid rotation. I won't say that I became inured to the sight of the interminable clear-cuts, but after hours of driving I had at least grown accustomed to being able to see the far horizon on every side, over the tops of fresh clear-cuts or brushy young trees. So it was a weird kind of reverse shock to hit the fifteen-hundred-acre Quinault Research Natural Area in Olympic National Forest, a boundary I could easily see even without the brown wooden marker, because the neighboring land had been cleared right to the line. The road slid through an avenue of huge virgin trees, a slender thread at the bottom of a narrow two-hundred-foot-deep green gorge, the sunlight flickering through in random places. It was a reminder of what the entire Olympic lowlands once looked like—and just about when I got used to being close-hemmed by trees, the road popped out the other side, and I could see across the old clear-cuts again.

I am aware that you are reading these words on a paper page. We need to cut trees, I'm not arguing that we don't. But the history of Big Timber in the Pacific Northwest is a sordid one of short-term gains and long-term losses, and no other industry save coal mining has left such a profound, all-pervasive footprint on the landscape—and even mining (I speak here as a product of the hard-coal fields of Pennsylvania, so I know a thing or two about the environmental impact of extractive industries) doesn't affect virtually every square mile across tens of thousands of square miles, the way timbering has in the mountains of the Northwest. What grows back are trees, and they form a forest of sorts, but the decimation of the ancient groves, in all their richness and diversity, is for all practical purposes a permanent loss.

Look, this isn't really complicated. For four centuries, America's economic engine has bloated itself on the continent's old growth, chewing through forests that took millennia to grow, like a layabout kid burning through his inheritance money. We cut down giant sequoias to make kitchen matches, for God's sake. Today all but a fraction of those forests are gone, and what's left is too precious—biologically, ecologically, and, as outdoor recreation and tourism grow in importance, economically precious—to cut.

We shouldn't cut another stick of old growth—for damn sure, not on public land of any sort, where at least a pretense of long-range thinking ought to be pursued, and not on private land, either. If that means compensating timber companies and assuming that portion of their holdings, fine. But the erosion of North America's old-growth birthright has gone on long enough. Don't just take my word for it; the National Academy of Sciences, in a 2000 analysis of forest management in the Northwest, concluded pretty much the same thing. So did Mike Dombeck and Jack Ward Thomas, both former chiefs of the U.S. Forest Service. Noting that old-growth harvests destroy irreplaceable resources, are mired in controversy and legal challenges, and divert attention and energy from more pressing issues, Dombeck and Thomas argued, "It's time to stop fighting over what little old-growth remains unprotected." That no doubt set timber industry teeth on edge, but Dombeck and Thomas also urged the environmental community to support forest management, including thinning of overgrown stands and more intensive use of second-growth forests, like the clear-cuts and quick-rotation tree farms I found so dispiriting. America's demand for wood continues to grow; is it ethical, they asked, to import wood from nations with weak environmental controls instead of increasing sustainable harvests at home?

By now I was heading along the coast, having passed through the Quinault Indian Reservation and into Olympic National Park—not the bulk of the park, which lies up in the mountains, but a beautiful yet oddly disjunct strip running more than sixty miles along the shore, only a mile or so wide in most places and separated from the rest of the park by fifteen or twenty miles of heavily timbered private and state forest land and small communities like Forks and Beaver. A few tendrils snake out of the heart of the park, like one that runs for twelve or fifteen miles down the Queets River, almost (but not quite) meeting the coastal strip. It's a puzzling arrangement, one that doesn't make much sense when you're looking at a map, but the explanation lies within the mix of political skulduggery and civic activism that is the history of Olympic National Park—and is a bittersweet reminder of what kind of park this might have been.

The Olympic Peninsula is enormous—six thousand square miles shaped like a blacksmith's anvil, with rivers running like spokes from the massed fist of mountains that rise almost eight thousand feet at the peninsula's heart. On a map, the region always reminds me of a bull's-eye—the dark center of Olympic National Park in forest green, almost completely surrounded by the lighter shades of Olympic National Forest and a state forest, and, around that, private land and Indian reservations running to the ocean, with the narrow band of national park land through which I was driving edging the western shore like a frame.

As early as 1890, Lt. Joseph O'Neil, among the first white men to explore the interior of the Olympic Peninsula, was recommending that the land be set aside for a national park. Instead, in 1897 President Grover Cleveland designated almost 2.2 million acres as the Olympic Forest Reserve—a huge block that included half the peninsula, including the finest stands of the oldest, most magnificent forest in the Northwest, which blanketed the lowlands. It was a visionary move, and it was, predictably, attacked almost immediately by Congress, the timber interests, and Northwest boosters, who brought their considerable political muscle to bear. Three years later they persuaded Cleveland's successor, William McKinley, to reduce the reserve by more than a quarter-million acres, and as if that weren't bad enough, the next year he lopped off another half-million. McKinley's land commissioner described the timber as "not worth preserving."

McKinley was assassinated and replaced by that ever-busy conservationist, Teddy Roosevelt, who in 1909, in his final days in office, used the Antiquities Act to declare 600,000 acres to be a national monument, in part to protect the rare race of elk on the peninsula. Six years later Woodrow Wilson excised almost half of it, ostensibly to promote manganese mining, though the tracts held little manganese but lots of big timber. Pressure to finally grant park status to what was left was growing, but it wasn't until Franklin D. Roosevelt threw his weight behind the effort that the park, back to about 600,000 acres, was finally created in 1938 (FDR added another 187,000 acres two years later). It was still a place of staggering beauty and wildness, but it was only half the park it should have been, with little of the richest temperate rain forests. To get what they could, like the ribbon of land along the Queets River, conservationists had to give up equally pristine land elsewhere, especially along the eastern slopes.

Even then, the forests within the park's boundaries were not safe. I had always been puzzled, in reading this chapter of *Wild America*, by Peterson's assertion that "it would be criminal if we allowed the saw the freedom of the Olympic park; the Park belongs to all Americans, and not to a few to make a profit from." National parks are by their very nature supposed to be safe from mining, logging, and other such industries, so why was he sounding the alarm? In fact, Olympic was still too rich a prize to rest safe, even as a park. As old growth disappeared from the private timberlands around the park, much of which had once been part of the supposedly protected reserve, the park looked more and more tempting. Aided by managers who took the broadest possible interpretation of regulations designed to allow limited salvage of dead or dying trees, loggers were able to remove 100 million board feet of timber from Olympic from 1941 to 1958, before the practice was stopped by protests.*

Today, Olympic National Park (including the coastal strip, which was added in 1953) contains more than 900,000 acres, with all but 5 percent of it federally designated wilderness. Because the park is all but roadless, most visitors see only a bit of it—some of the coast or the highway along Lake Crescent, the high country up at Hurricane Ridge, or the rain forest along the Hoh River, which you reach after a twelve-mile traverse of old clearcuts studded with giant stumps before reaching the park line. The morning I arrived, having spent the night in the nearby town of Forks, the Hoh was running hard between wide banks of rounded cobblestones, the snowmelt water a pale powder blue from the rock crushed beneath the glaciers of Mount Olympus, which feed it. High-prowed driftboats glided down the current, and a few fly fishermen stood thigh-deep and cast for steelhead. I spent an hour or two walking along the river, wishing I had my rod and waders, poking through thickets looking for birds—warbling

*Among those at the forefront of the fight was Supreme Court justice William O. Douglas, who just a few years earlier had succeeded in preserving the C&O Canal back East. Douglas, who wrote the 1965 book *A Wilderness Bill of Rights*, was an early and especially eloquent advocate for the environment, also active in his opposition to damming the Columbia and in his promotion of the development of national parks.

vireos, orange-crowned warblers, cedar waxwings, hermit thrushes, and more winter wrens.

The ground was pocked with the fist-sized hoofprints of Roosevelt elk, the largest and darkest subspecies of elk and second only to moose as the largest ungulate in North America; they are named for Teddy Roosevelt, whose actions in creating the national monument probably saved them from extinction. Once found as far south as the San Francisco Bay area, they were reduced to tiny herds, of which the park's is among the largest, at about five thousand head, and the most truly wild. Today, in fact, as is often the case with grazing animals in national parks, they may be having a deleterious effect on the forest through overgrazing. One probable reason for the overpopulation is the absence of the elk's original predator, the gray wolf, which was exterminated around 1920 and has not been reintroduced, although a feasibility study by the U.S. Fish and Wildlife Service in 1999 found it should be biologically (if not yet politically) possible.

The parking lot at the Hoh Rain Forest visitors center was crowded, not just with people but with a flock of northwestern crows, a species slightly smaller, and with a hoarser voice, than the American crow found almost everywhere else on the continent. This bunch was an efficient gang of shakedown artists, and one flew up and landed on the half-closed window of my car while I was still unbuckling my seat belt, peering in for anything edible. I flapped a hand at it and it flew, a collection of colored plastic bands rattling on its dry black legs—markers for a study, which would allow the researcher to identify each crow individually. I could imagine the eventual journal article: "Effect of Junk-Food Consumption on the Thuggish Behavior of Northwestern Crows, *Corvus caurinus*, in Olympic National Park."

The temperate rain forests of the Hoh and Queets valleys are, like Opal Creek, an extravagance of green. Drenched with up to fourteen *feet* of rain annually, the Sitka spruce, hemlock, and Douglas-firs are at times hard to make out under their growing shrouds of spike moss, ferns, and lichens. The bigleaf maples are perhaps the strangest and most delightful of all, with gnarled, long-reaching limbs from which hang sheets of moss and clumps of licorice fern, like a gardener's hallucination. Comparisons with Middle Earth are inevitable. The national and international champions of several species of trees—a 302-foot Doug-fir, a 251-foot grand fir, equally immense hemlocks, western redcedar, and spruce—are found in Olympic, mostly in the protected portions of west-side river drainages. (The grand-champion subalpine fir, on the other hand, was discovered up in the Bailey Range, around six thousand feet high at the head of the Hoh drainage.) The champ redcedar has a sixty-one-foot circumference—about ten of my arm spans, or nineteen feet in diameter, two and a half times the width of the big Doug-firs that so impressed me in Opal Creek. That's not to say these were the biggest trees on the peninsula, or that their protection was intentional; by all accounts, the really big boys were outside the park boundaries and wound up in the mills. But what's left can still bring even the most jaded visitor to silence.

Signs along the trails warned that cow elk were calving in the area and cautioned that they might be dangerous: VISITORS HAVE BEEN CHARGED, CHASED, KNOCKED DOWN AND STOMPED. I saw none in my days in the park, despite a lot of looking; Roosevelt elk tend to be much shier than their Rocky Mountain relatives. In fact, when compared with the other mountains of the Northwest, the Olympics are a world apart, known as much for what they lack as for what they have. In some ways, the region is more of an island than a peninsula, almost cut off from the Cascades by Puget Sound to the east and frequently isolated during the Pleistocene by glaciers. Consequently, there are many species common in the Cascades or Rockies that were absent here, like grizzly bears, while there are a number of species, especially plants, that are found only in the Olympics. Besides grizzlies, the famous "missing dozen" mammals originally included lynx, pika, wolverine, coyote, red fox, porcupine, bighorn sheep, mountain goats, and several rodents, while the endemics include eight plants, the high-elevation Olympic marmot, the Olympic torrent salamander, and subspecies of rainbow and cutthroat trout, short-tailed weasel, yellow-pine chipmunk, and Mazama pocket gopher.

Coyotes and porcupines appear to have come in on their own in the last century, but red foxes were intentionally introduced, as was a mammal that for years was a beloved symbol of a place where it didn't belong—the mountain goat, eleven of which were released in the Olympics in the 1920s, before the park was created. The goats did so well that by 1983 there were more than a thousand of them, trampling and grazing some of the rare (and native) alpine vegetation so badly that it was imperiled. The park trapped about four hundred goats in the 1980s and trucked them out of the park, a move that was supported by ecologists but fiercely unpopular with many visitors and with animal-rights groups. Although the trapping program ended in 1989, the goat herd hasn't returned to its past numbers, holding around two or three hundred animals; nevertheless, many ecologists recommend eliminating the goats entirely, not just in the park but in the surrounding national forest.

That evening I pitched my small tent in a secluded site in the Heart o' the Hills campground, on the park's northern side, where I'd stay for the next several nights; there were few people in the place, and most of the noise was from Townsend's chipmunks and a squad of gray jays, which surpassed even the crows back at the Hoh for brazenness. I've encountered gray jays all across the North and have always been delighted by their charming boldness around humans, but these were in another class altogether. As I ate the first evening, five of them landed on my picnic table, a little goon squad feinting at my food, so that while I swatted at one, the others made flanking attacks on my chips. One finally tried to yank the whole bag away, and when I grabbed it, the bird dropped it and bit me, hard, on the thumb. I yelped, and a snowshoe hare, nibbling wood sorrel a few yards away, looked up placidly.

The next morning before daybreak—another beautiful, cloudless day that belied the Northwest's reputation for rain—I left the campground and headed up into the high country, into forests of stunted Doug-fir and subalpine fir where male blue grouse were calling, six loud syllables like someone blowing across the mouth of a jug. Patches of snow still lay in the shadows and on the north slopes, the mushy ground around their edges bright with white avalanche lilies and marsh marigolds. On the high crags, where thin trails led across alpine meadows of lupine and Indian paintbrush, I had stellar views to the north across the Strait of Juan de Fuca, where a cruise ship was coming into Victoria on Vancouver Island, and, beyond that, the still more rugged snow-clad peaks of mainland British Columbia.

An adult golden eagle soared by, cruising the meadow, and there was a sharp trilling from the chipmunks and a high, piercing whistle from a marmot that I hadn't known was there. The bird was just below eye level, wings flat, dark tail tipping and tilting to steer in the gusting breeze; it looked at me but didn't veer away. The sun reflected off its feathers as from brass sheeting; then it crossed the pass, but the chipmunks kept on screaming for quite a while.

The high vantage also gave me a sobering perspective on the park, for I could see more clearly than ever how its skirts had been carved up by clear-cuts, the old growth gone. But occasionally, some of the wrongs of the past can be set right, or at least we can try to heal the damage. I knew that to the west of my high perch, down in the next deep valley, a hopeful end was unfolding to a long and thus far unhappy story.

The Elwha River, which flows forty-five miles north out of the heart of the park and into the strait, was famous in the late nineteenth century for its anadromous fish—ten different runs of steelhead, sea-run cutthroat trout, coho, chum, pink and sockeye salmon, and both spring and summer runs of giant chinook salmon. Scientists suspect the heavy rapids of the Elwha, roaring with springtime runoff, acted as a harsh form of selection, so that only the biggest and strongest of the spring-run chinooks managed to reach the spawning grounds—one reason why the fish of this river often exceeded one hundred pounds.

But in 1910, a privately owned 108-foot-high hydroelectric dam plugged a tall, narrow gorge in the Elwha just five miles from the sea, creating Lake Aldwell; instead of the legally required fish passage facilities, a hatchery was built, on the theory that captive-bred fry were a workable substitute for wild fish. They weren't, and the hatchery was abandoned in 1922, a few years before a second dam on the Elwha, this one 210 feet high at Glines Canyon, was built in what was then national forest land. What had been a phenomenally bountiful fishery all but disappeared, and only a handful of anadromous fish continued to spawn in the few miles of river below the first dam, including a few spring chinooks. In 1940, the park boundaries were expanded to include Glines Canyon and Lake Mills behind it.

By the mid-1970s, however, when the Glines Canyon dam's fifty-year license was due for renewal, attitudes toward dams had started to change. Instead of being rubber-stamped, the relicensing process set off a protracted, nearly thirty-year maze of legal maneuvering between the dam owners, various government agencies, conservation groups, lumber mill operators, tribal councils, and others. In 1992, federal legislation calling for the restoration of the Elwha River was signed into law, and planning for the removal of both dams began, a project expected to cost $182 million. But progress stalled again when Congress didn't find the money and when a local timber mill said it needed to find alternative sources of cheap power.

Thirty years, and the dams still aren't gone; the latest timetable calls for the walls to finally tumble in 2008, almost a century since the Elwha Dam went up. Too late for the Elwha's sockeye salmon—they may be extinct—but there are still a few precious spring chinooks left. When the river finally runs free again, seventy miles of relatively pristine spawning habitat will be reopened, and fisheries experts hope the river's fabled salmon runs may eventually reestablish themselves. It'll take a long time to fully restore the river; there are an estimated 18 million cubic yards of sediment trapped behind the dams that must be safely flushed down to the sea (where it may again replenish eroding shorelines), and the riverbed in what is now lakes Aldwell and Mills will be slow to lose its bathtub ring and regrow forests. But it is a start.

In the 1950s, when Peterson and Fisher were traveling the country, dam building was still the rage, a clear indication of mankind harnessing nature's might for the common good; Woody Guthrie sang of the Columbia's hydropower "turning our darkness to dawn." But even then, conservationists were questioning—and sometimes stopping—some of the most destructive plans. In 1949, they managed to halt plans for a dam on the north fork of the Flathead River in Glacier National Park, and though they failed to save the Columbia's once-mighty salmon runs, the fledgling environmental movement did keep a dam out of Kings Canyon National Park in California. Public outrage also successfully blocked two dams on the upper Colorado River in Dinosaur National Monument, though as a compromise conservationists dropped their opposition to a replacement project, the Glen Canyon Dam in Arizona, which flooded a scenic wonder many considered the equal of the Grand Canyon. (The Grand Canyon itself also had to be defended from a loopy proposal for a series of dams in the 1960s and '70s.)

There are still plenty of ill-conceived plans for new dams rattling around out there, and every time conservationists think they've killed one for good and all, it rises in some new and mutated form. The fight never ends, but the debate about dams has taken a new direction, as the Elwha River situation shows. After a century or more of all but unhindered dam construction, the public is asking for a reckoning of the costs, not only in dollars but in wild rivers, healthy fisheries, and local economies, which often benefit far more from good fishing and white-water recreation than they do from the sale of cheap power. And as with the Elwha, a surprising number of dams are coming down. As a nation, we've not yet taken the plunge and removed any of the biggest, iconic western dams that destroyed the great salmon runs; in fact, many of the decommissioned, demolished dams have been on eastern rivers, like the Kennebec in Maine, where runs of American shad, sturgeon, and Atlantic salmon had been choked off since the 1830s. But we're talking about knocking out the goliaths, something that was unthinkable even a generation ago. Before the Bush administration came to power, there was growing momentum to remove four dams on the lower Snake River, which have been especially devastating to Columbia Basin salmon runs. (Instead, the administration ruled in December 2004, a day after proposing a massive rollback of habitat protection for endangered salmon, that the Columbia dams were an "immutable" part of the salmon's environment, like mountains, and thus not open to removal.) Nor is it just endangered fish that are driving this undamming movement; there is renewed talk about pulling the plug on O'Shaughnessy Dam in the Sierra Nevada, which drowned the spectacular Hetch Hetchy valley and broke John Muir's heart, or even Glen Canyon Dam itself.*

For now, though, dam removal is mostly a smaller-scale operation. My home state of Pennsylvania leads the nation in dam busting, having removed ninety-one and with plans to take out many more, focusing mostly on low, antiquated, and frequently unsafe structures; this opens up spawning habitat for migratory fish like shad, blueback herring, eels, and striped bass, and better habitat for resident fish, mollusks, and other aquatic life. (It also saves lives, since low-head dams are a terrible hazard

*Ironically, what conservationists could not prevent from drowning, nature has again revealed. The prolonged drought in the Southwest has lowered Lake Powell, behind the Glen Canyon Dam, by almost 130 feet, exposing landscapes hidden for forty years. Thickets of cottonwoods have sprouted; wildlife is returning; even the unsightly bathtub ring at the old waterline is disappearing. Hydrologists say that unless the region has successive years of extraordinarily heavy rain, the lake may never fully refill, and while environmental groups are using the drought to push for the removal of Glen Canyon Dam, nature may have rendered the argument in part moot.

to boaters and swimmers.) The results can be dramatic; in the five years since the Edwards Dam was removed from the Kennebec in 1999, biologists have documented a resurgence of fish that far exceeded their expectations. This was particularly significant because the Edwards was the first dam to be ordered removed, over the strident objections of its owners, since the ecological consequences of blocking the river were found to be greater than the financial benefits of producing relatively small amounts of hydropower.

The battle over the Elwha has been predictably acrimonious, but just as Idaho has a hopeful model for cooperation in restoring a salmon river, in the form of the Lemhi, so, too, does the Olympic Peninsula. The Dungeness River, which lies one drainage to the east of the Elwha, and like it has headwaters deep in the park, had been battered not by dams but by water diversions for irrigation on fifty-six hundred acres of farmland near the coast, as well as for municipal use. So much water was being removed—82 percent of the total flow, according to one estimate—that the river's once-enormous runs of salmon had dropped to almost nothing; the pinks fell from 400,000 in the 1960s to barely 3,000 in 1981, before making something of a recovery, but the summer chum and chinooks are federally listed as threatened.

Rather than fight it out in court, all the parties with an interest in the Dungeness, including the government agencies, farmers, conservationists, and the Jamestown S'Klallam tribe, which lives at the river's mouth, worked together to find ways of decreasing agriculture's impact on the river. Farmers pledged to reduce what they took, in return for guarantees that they not lose historic water rights, while the tribe helped to find money for more efficient irrigation systems. There is cautious optimism, but the trick will be ensuring enough water during the late-summer dry spells, when many of the fish are running and farmland is at its thirstiest. Getting salmon past the lower stretches will require rehabilitation in terms of improving habitat, releasing the river channel from dikes, recreating logjams, and other measures, but if they make it through that bottleneck, there is still plenty of good habitat up in the park.

The eagle was back; I knew it, somehow, even before I turned my head and saw the wide black wings against the sun, soaring high over the mountains. The marmots, which share this alpine world, were watching, too, and I heard again their sharp alarm whistles.

Below the eagle lay the heart of the Olympic wilderness, not a scratch upon its exquisite surface—the egalitarian skyline of dozens of sharp gray summits, hundreds of glaciers and snowfields, no single mountain rising noticeably higher than the others save Mount Olympus, and it only by a little, first among equals. Below them, the dark coniferous valleys reached back into the mountains, and the urge to explore them, to climb to the high icefalls—I could taste it, I could feel it like an itch between my shoulder blades. But Alaska and the end of my trip were pulling me on, and I had no time for a long backcountry sojourn. Still, I made a vow to myself that I would come back and climb up into the high, high country, maybe up to the Queets Glacier, where the Elwha is born, and, when the dams come down, celebrate the free flow of a once-chained river.

But I had another reason for wanting to stand among the glaciers—because it may not be an option too many years hence. Across North America—across the world—glaciers are in retreat. To my east across Puget Sound, the seven hundred glaciers of the North Cascades have lost a third of their mass in the past century; all forty-seven glaciers monitored closely there since the early 1980s have been steadily eroding, and several have disappeared entirely. Glaciers in the Coast Ranges of Alaska are shrinking at a ferocious rate, as are those in the Rockies; Glacier National Park in Montana, which once had 150 glaciers, is down to 26. The same scenario is playing out in the Alps, in central Asia, in the Andes, even in Africa, where the famed snows of Mount Kilimanjaro are expected to melt away by 2020. Worldwide, only the glaciers of Scandinavia appear to be holding their own.

There is no doubt the glacial retreat is a reflection of changing climate—in the Cascades, less winter snowfall and warmer summers. Glaciers waxed and waned over the past tens of thousands of years, and so the question has long been what role humans are playing in recent retreats. The evidence is clearer with each record-warm year that the gases we've pumped into the atmosphere since the Industrial Revolution have tipped the balance; glaciers are the coal-mine canaries, and they are vanishing faster than anyone thought possible.

Climate scientists have long predicted that the clearest signs of global climate change would fall first on high elevations and high latitudes. And indeed, we have already documented changing natural communities in both such locations, from the upward shift of plants in the Swiss Alps to the northward movement of Swedish butterflies and British birds. One of the most compelling studies, published in 2003 in the journal *Nature*, compiled the results of research involving almost sixteen hundred species around the world and found that organisms were moving toward the poles at an average rate of about four miles a decade and that the arrival of spring was advancing by more than two days per decade.

For species in the mid-latitudes, this may be a pace with which they can cope (although most climatologists expect it to accelerate). But for plants and animals already living in alpine zones—like the dozens of rare species of the high Olympics—there is nowhere left to go, and the future looks bleak. Up here in the high meadows where I sat, scientists have already shown that trees are creeping in, displacing the yellow glacier lilies and the pink mats of spreading phlox that make the alpine zone so colorful and unique.

Suddenly the shrill whistles of those marmots, warning of a passing eagle, seemed to have another, more disturbing meaning altogether. It was a memory I would carry with me in the coming days as I headed north for the final leg of my long journey—to the wild and spectacular coastal tundra of Alaska, where global warming is not an abstraction but an urgent fact of daily life.

THIRTEEN

Super-tundra

Deep in the watery emptiness of western Alaska, the town of Bethel sits like an afterthought of civilization, hugging the banks of the huge, meandering Kuskokwim River. Inaccessible by land, Bethel is nevertheless the administrative and transportation hub for this part of the state, its shore lined with huge barges, its small airport the third-busiest in Alaska.

I had flown into the town the night before on a commercial jet from Anchorage, the Olympic Peninsula three days behind me, and now—on the first day of summer, a cool, drizzly morning—I was snug in the front seat of an old Beaver floatplane. The pilot, taxiing to the open channel of the river, turned us into the wind and pushed the throttle; the growling engine roared, scattering a bunch of loafing glaucous gulls and lifting us off the gray water.

Bethel lies within the boundaries of the Yukon Delta National Wildlife Refuge, an immense reserve covering more than 19.5 million acres, an area about the size of Maine or South Carolina, that is second (by a whisker) only to the Arctic NWR as the largest such unit in the federal system. We flew northwest into the heart of the refuge, below a sky of high overcast against which pockets of mist rose like smoke signals, and over a landscape of modern-art doodles and swirls—absentminded rivers that doubled back upon themselves, old oxbow lakes, pewter-colored ponds that merged, amoeba-like, with one another. Pairs of swans, big even from two thousand feet up, flew with synchronous wing beats over vivid green sedge marshes, amid which were set oddly shaped patches of brown upland tundra.

The Yukon and the Kuskokwim are Alaska's two biggest rivers, both flowing west through the empty interior. The Kuskokwim rises in a mountain chain of the same name about five hundred miles from the sea, but the twenty-three-hundred-mile-long Yukon transverses the entire state, with its headwaters encompassing much of Canada's Yukon Territory. Spreading out on the flat, marshy land of western Alaska, the rivers come within about twenty-five miles of each other, then veer off to the north and the south, forming this roughly triangular delta of extraordinary scale and fertility.

We flew for an hour and a half across a world at least as much water as it was land; the sun lightened the hazy sky, and the ground below lit up with thousands of reflective shards. We crossed wide Aropuk Lake, crossed the Talik River and the Manokinak, where white canvas wall tents marked Native fishing camps, their drying racks filled with red king salmon meat; finally Ken, the pilot, turned up along the big, muddy Kashunuk River, with the Bering Sea just barely visible far off to the west. He pulled the plane into a tight curve over a small white building with wide orange stripes painted on the roof and looked hard at the choppy currents on the brown river below. He didn't like what he saw. "See there?" he said, his voice tinny in my headphones. "That's a sandbar just under the surface, gotta be a big one. Damn. And it looks like the tide's way out. I don't know if there'll be room to turn around down there, but it's the only place"—he already had us coming in low on the final upwind turn, dropping fast—"so here goes." With that he slid the plane down, the rough water smacking the metal floats, and nosed us up to the wide mudflats by a crude dock.

This was Old Chevak, the site of an abandoned Cup'ik Eskimo village, and a U.S. Fish and Wildlife Service field camp for more than five decades. The small, simple building had once been, of all things, a Catholic church, a fact that Fisher and Peterson noted when they came here with several friends in 1953, the next-to-last stop on their continental tour.* It would be my base as well for the next week as I joined a team studying nesting shorebirds—and spent one of the most remarkable periods of birding I've ever enjoyed.

By coincidence, a bunch of goose researchers departing after a month of work were waiting for their boat, so there was quite a crowd gathered on the dock, which was twenty yards of glistening mud from the plane. I

*They incorrectly referred to the church as Russian Orthodox, however.

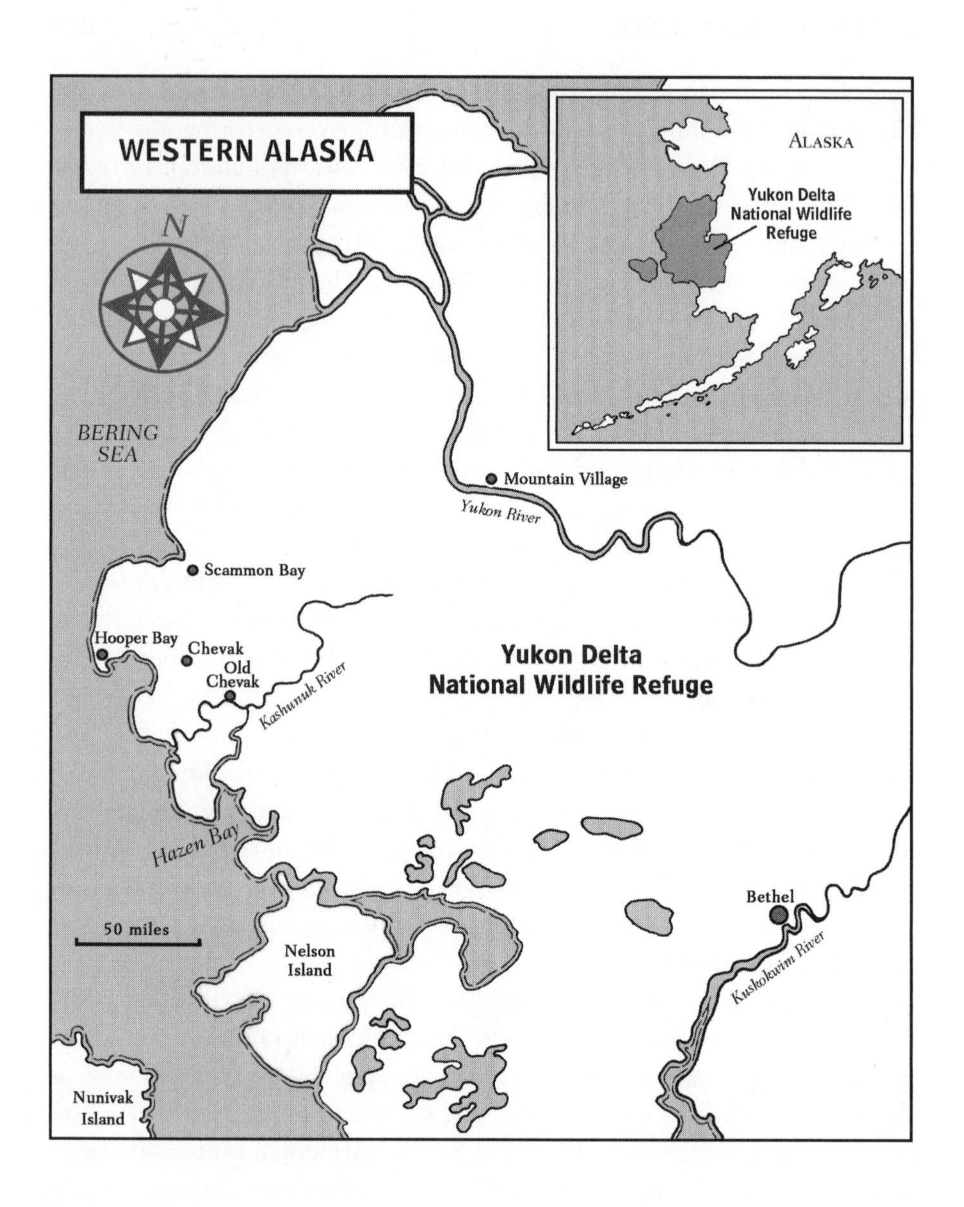
WESTERN ALASKA
N
ALASKA
Yukon Delta
National Wildlife
Refuge
BERING
SEA
Mountain Village
Yukon River
Scammon Bay
Hooper Bay
Chevak
Old
Chevak
Kashunuk River
Yukon Delta
National Wildlife Refuge
Hazen Bay
50 miles
Nelson
Island
Bethel
Kuskokwim River
Nunivak
Island

shouldered a heavy duffel bag of my camping gear and stepped off one float, instantly sinking knee-deep in the mire. I was wearing hip boots, but my leg was trapped, the boot pulling off my foot, the other leg working deeper and deeper as I struggled for balance and purchase. A couple of the goose guys jumped to help, moving with an odd quick-time shuffle to keep from sinking. "Learn to love the mud," one of them said with a smile, lifting the bag from my shoulder and offering a steadying hand as I worked my feet free.

It was chaotic for a while. The plane roared off just as the boat came in for the goose crew, and while they got things loaded and made their goodbyes, I set up my small backpacking tent on one of half a dozen wooden platforms a short distance behind the building, next to three big dome tents the shorebird researchers were using. Little western sandpipers perched like weather vanes on each rounded roof; the tents were, save for the building, the highest perches for miles and irresistible to the birds. As I unrolled a sleeping bag inside my tent, I saw, faint through the greenish nylon, the silhouette of a sandpiper a few inches above my head.

Finally, as it began to rain, I sat down inside the now-empty camp with Brian McCaffery, the refuge's nongame bird biologist, a quiet, introspective man in his forties with a gray goatee and a long dark ponytail below his brown uniform cap. With us was James McCallum, a British bird artist in his early thirties who was spending the summer volunteering at the refuge. While we looked over James's watercolors of nesting turnstones and jaegers, the other members of Brian's small crew came in, Alice Nunes and Grace Leacock, who had been out since early morning searching for the nests of bar-tailed godwits, a rare shorebird that was the focus of their summer's work. Alice, from Long Island, was an AmeriCorps volunteer based in Bethel; she wore a delicate silver nose ring and had a wonderfully convoluted Egyptian-Armenian-Portuguese heritage. Grace, whose father was the new refuge big-game biologist, was also at Old Chevak as a volunteer, having arrived there just two days after moving to Bethel from the Anchorage area. Still in high school, she was nevertheless a field camp pro, having been practically raised in the bush, and because her mother is Thai, she added to the camp's multicultural mix.

The old church was a small wooden frame with a shell of corrugated metal, an unheated entryway at one end that served as a mudroom and pantry, and a small bunk room at the other end, now vacant except for

Brian; outside was a one-holer outhouse with a plastic window in the door that gave a view across the bird-rich marshes. The bulk of the building, though, was occupied by a fifteen-by-twenty-foot kitchen, warmed by an oil stove in one corner. The walls were hung with faded topo maps pasted together to form a collage of the delta, along with old satellite photos, curling posters about waterfowl, and pictures of past field crews; nails held hot pads, graters, and dish towels over the sink, and there were shelves with half-used bottles of sunblock, rolls of duct tape, bouquets of goose feathers sticking out of jars, and old, bleached fox skulls and bird bones. James, his shaggy red hair a tousle, was cooking pasta and vegetables, trying to use up a couple pounds of ground meat (the propane freezer had died a few days earlier); everyone's wet clothes hung from cords strung above the stove, and the air was warm, fuzzy with the smell of curry and damp wool. It felt immediately like home.

The rain and wind strengthened, and so instead of heading out, we all settled down with mugs of tea while Brian talked about the refuge and its incredible wildlife, especially its birds. The Y-K Delta, one of the world's largest wetlands complexes, is the feeding and nesting ground for three-quarters of a million swans and geese, 2 million ducks, and more than 5 million shorebirds. "The central coastal section of the refuge is certainly one of the greatest concentrations of large-bodied waterbirds in the world—ducks, geese, cranes, swans, and loons," he said. Many are birds that appear to be in trouble, like spectacled and Steller's eiders or emperor geese, or that, like black turnstones, breed almost nowhere else on the planet.

"In many respects, we feed the Pacific flyway, with cackling geese and white-fronted geese in particular, but we send birds out into every flyway leaving North America but one, the flyway from northern Greenland down the Atlantic coast," he said. "We contribute birds to all four of the main North American flyways, while all of the bristle-thighed curlews and some of the plovers and tattlers follow the central Pacific flyway. Bar-tailed godwits, most of which stage here, as well as the North Slope population of dunlins, use the east Asian–Australia flyway. We even have wheatears that go to Africa, crossing all the Asian flyways to get there. And then we have sharp-tailed sandpipers, which are just weird. They're coming here from central Siberia to stage, but we don't know if they're then going straight across the Pacific to Australia, like the godwits, or if they're going around the edge.

"Western sandpipers may be the most abundant shorebird in this hemisphere, and almost all of them breed in western Alaska, mostly here in the refuge," he continued. By some estimates, the population exceeds 6.5 million, and no matter where you go in the delta, the little "westies" seem always to be underfoot. When biologists surveyed parts of the delta, they found as many as three hundred pairs per square kilometer. "That is stunningly high," Brian said. "Compared with the rest of the North American Arctic and sub-Arctic, that's just off the charts."

The superlatives go on and on. "Over eighty percent of the world's black turnstones are thought to breed on the delta, and two-thirds of them nest within two kilometers of the coast. The entire population of bristle-thighed curlews stages here. The majority of western black scoters nest on the delta as well. Where there's habitat, like in the mud volcanoes [a line of ancient hills to the east], we have some of the highest densities of nesting raptors in North America, including gyrfalcons and golden eagles. As for songbirds, there's not real high diversity, but some of the long-distance neotropical migrants breed here, including some of the classic eastern flyway migrants like gray-cheeked thrush, blackpoll warbler, northern waterthrush, alder flycatcher, and red fox sparrow." Once, while conducting surveys for other species, the Yukon Delta biologists recorded whether or not they saw any short-eared owls in their study sites; at the end of the season, when they computed the sightings, they concluded that there were as many as eighty *thousand* of the raptors on the refuge—and that was without making a particular effort to find them.

Although the flat, marshy delta is its heart, the refuge also includes uplands like the Askinuk Mountains and huge Nunivak Island off the coast, with its herds of musk-oxen. But this is not all empty wilderness; Bethel sits deep within it, and there are dozens of villages of Cup'ik and Yup'ik (the closely related Eskimo cultures of the delta) clustered along the waterways and coast, with rapidly growing populations. The land within the refuge border is a hodgepodge of federal, state, private, and village corporation holdings, and the refuge has a legal mandate to provide opportunities for subsistence hunting and fishing by both Native and white residents—all of which makes the job of managing the place unusually challenging.

Early conservationists quickly recognized the delta's importance to wildlife. In 1909, Teddy Roosevelt declared part of it the first federal bird refuge on the Alaskan mainland, but when Peterson, Fisher, and their friends Bill Cottrell and Finnur Gudhmundsson came in 1953, the area around Old Chevak still lacked any federal protection, despite its reputation as the greatest goose nursery in the world. The naturalists argued in *Wild America* that the region deserved formal recognition, which finally came in 1960, with the creation of what became the Clarence Rhode National Wildlife Range. In 1980 this was folded, along with other federal lands, into the current refuge.

The names and boundaries have changed, but the birds remain timeless. After we'd cleaned up dinner, and the rain had eased to a fine, windblown drizzle, I stood in the lee of the old church, listening to the wails of Pacific loons in the distance and watching Arctic terns plunging into the river chop. Sandhill cranes bugled somewhere across the tundra, and small groups of shorebirds and ducks—whimbrels, ruddy turnstones, Pacific golden-plovers, green-winged teal, long-tailed ducks—flew past. Sabine's gulls, their white wings marked with black chevrons, flapped languidly by. All along the muddy edge of the Kashunuk, dozens of western sandpipers probed for food; when I walked down to the dock, they scurried just a few yards away as a flock of red-necked phalaropes flew toward me, parting around my head like a stream around a rock, so close their wings hissed in my ears.

Every few moments brought new birds, many of them species that a birder waits a lifetime to see: flocks of emperor geese, their bodies dark blue-gray and their heads white, stained orange by the iron oxide of tundra mud; hoary redpolls, fluffed and pale; yellow wagtails, an Asian species with a beachhead in western Alaska, which performed looping display flights overhead; long-tailed jaegers, slim, highly predatory relatives of the gulls, their wings slender and built for speed, their outsized central tail feathers twisting like streamers behind them. Flocks of geese passed downriver, mostly greater whitefronts and cackling geese. It was an avian show whose equal I've rarely seen, and I recalled Fisher's words, that Old Chevak was "stiff with northern birds, so that never for a moment were we out of sight or hearing of crane, goose, duck, or wader." So filled to the brim with life was the delta that the admiring Fisher, who had extensive Arctic experience, dubbed it a "super-tundra."

And in the days ahead, I also found myself falling into the same schedule as my predecessors, who Fisher said were "out fourteen hours a day and splashed through bogs and ate like hogs and slept like logs." That summed up my experience, too, as I tagged along with Brian and his crew. Their focus was the bar-tailed godwit, perhaps the most impressive migratory bird in the world. There are others that go farther (the Arctic tern, for example), but no bird makes as extreme a migration as this leggy shorebird.

A bar-tailed godwit is a bit bigger than a pigeon; the breeding males are a bright cinnamon color, and both sexes have a very long, slightly upcurved bill. Bartails breed across northern Eurasia, and like many Old World birds a small population, probably numbering about a hundred thousand, has crossed the Bering Strait to colonize western and northern Alaska. Come autumn, the Alaskan godwits make the longest nonstop migration known of any bird—almost seven thousand miles across the widest part of the Pacific Ocean to Australia and New Zealand, with no evidence that they stop along the way for rest or food. In fact, research in the 1990s showed that the godwits accomplish this feat by gorging so heavily before they leave that they effectively double their weight, taking on the greatest fat load of any wild bird. To compensate, they then undergo a rapid atrophying of their digestive organs; the gizzard, liver, and kidney all shrink dramatically, shedding weight for the flight, while the heart and muscles expand.

Yet even with these physical changes, and aided by prevailing tailwinds, they must fly continuously for at least five or six days to reach Australasia, where they regrow their guts and start to eat again, spending the winter loafing and regaining their strength. Young godwits remain in the Southern Hemisphere until, at age four, they are old enough to breed, but in March or April the adults clear out for the north, taking an elliptical route up the western rim of the Pacific, stopping in places like China and Korea's Yellow Sea coast. They return to the Y-K Delta in early May, when the land is just emerging from snow and ice.

But while ornithologists are dazzled by the godwit's migratory feats, they are also concerned for its future. Given that it breeds in one of the most remote corners of the continent, this may seem odd, but McCaffery, one of the few scientists to study the species in North America, ticked off the many hurdles facing it—not just the long, naturally hazardous migration route, or the fact that godwits are so late to mature, but the loss of critical wetlands on the wintering grounds and particularly on the northbound route through Asia. One especially worrisome development is the

Saemangeum seawall in Korea, the world's biggest coastal reclamation project, which will destroy a seventy-four-thousand-acre estuary on the Yellow Sea where large numbers of godwits, dunlins, and other shorebirds now stop.

Nor is everything necessarily going well in Alaska. In early autumn, biologists have a chance to survey the godwits when the birds gather in concentrated flocks along the coast, preparing to migrate. Normally, you'd expect to find large numbers of young godwits at that time of year, but the *highest* proportion of immature birds Brian and his colleagues have ever seen is 3 percent, and some years it's a fraction of that. It's almost as though the godwits aren't breeding, or if they are, they aren't raising many chicks to flight age. By contrast, among other shorebirds the percentage of juveniles in autumn may exceed 50 percent.

That's why McCaffery and his volunteers were in Old Chevak, spending their days combing the soggy tundra and endless marshes for godwit nests, a task that calls for immeasurable patience and (given the soggy terrain and cold, often rainy weather) a fair helping of physical stamina. One possibility is that the birds are picking up a toxic stew of chemical contaminants; this may be interfering with their ability to reproduce or perhaps even altering their behavior, making it less likely that they'll successfully rear chicks. It's most likely they're being exposed to toxins in Asia, where a third of humanity lives along the godwit's springtime migration route. But it may be a local problem, too; when health specialists tested the blood of Native infants in the delta, they found the highest levels of DDE, the breakdown product of the pesticide DDT, recorded anywhere in the circumpolar region. Wind and ocean currents, we now know, concentrate a lot of the world's nastiest substances in the high latitudes, where they work their way up the food chain, building up in long-lived species like humans—and godwits, which may live for more than thirty years.

So as he and his volunteers find nests, Brian has been filching one egg from the clutch of four, for future testing. Given his worries about godwit survival, this isn't something he does lightly, but it's an essential step if we're to learn where the problem lies. By the time I arrived, they'd found eleven nests and taken nine eggs, and every one of the nests had subsequently been destroyed by predators, usually in a matter of days. In the case of one nest, visible from camp, the godwits drove off first an Arctic fox, then a sandhill crane while Brian looked on.

It may be that the researchers are somehow revealing to predators the location of a nest, though they try not to approach any more closely, or more often, than necessary. This might just be the normal way of things; between jaegers, cranes, foxes, ravens, weasels, mink, and otters, there is no shortage of carnivores, and the team also found the remains of nests that were attacked before the humans discovered them. Or perhaps the godwits are not hiding the nests as well as they should, or defending them as vigorously as they might. James, who has watched this species in Siberia and Scandinavia, found many of the Alaskan birds curiously passive when a predator came near, compared with the loud complaints and blistering attacks he's used to seeing in Europe. Perhaps this is simply a regional difference in behavior, or perhaps this is a result of chemically altered behavior.

The next several days were on-and-off rain, with temperatures in the forties, and we lived in our waterproof parkas and hip boots. On a typical day, James would head off in one direction to observe godwit pairs in hopes of finding chicks, a scope over his shoulder and his blue and white hat a receding dot of color, while Alice and Grace would head in another, scouring the study area for new nests. I helped Brian analyze the sites of the failed nests they'd already found, marking out a meter-square area around each one and tabulating the various plants that made up the ground cover, collecting samples of mosses and lichens for later analysis. The godwits make a small depression in the spongy mat of tundra vegetation, lining it with long strands of pale greenish lichen that looked like bean sprouts; the finished cup is almost invisible, and the four eggs, which are greenish brown and heavily mottled, are very well camouflaged. From time to time, we'd hear the loud *tivo-o, tivo-o* calls of the godwits and see them standing guard near a couple of small gray-brown chicks—proof that a few eggs, at least, had managed to evade the predators. Once, when a jaeger flew too close, the male sallied up in a classically ferocious attack, screaming imprecations at the other bird as he dogged it far over the horizon.

At midday we'd meet again at the camp for something quick and hot, then head back out (often James, who seemed immune to miles, cold, and rain, would stay out all day), then come back for a late meal before falling into our sleeping bags. Since I had arrived on the first day of summer, there was no real night to speak of, only a couple of hours of twilight around 3:00 a.m., when we could see the lights of Chevak, the Cup'ik village of about 850 people twelve miles to the northwest. When Fisher and Peterson came in 1953, the Cup'ik had only recently abandoned Old Chevak and its little church for the new, drier site, which sits on a high bluff above the river. In those days, many of the Cup'ik in Chevak still inhabited barabaras, or partially buried turf houses, and used skin kayaks, though there were already a handful of wooden homes and motorboats. That was then; even from my distant vantage point, through binoculars I could see that satellite TV dishes dominate the skyline in Chevak today.

Near the end of my time at Old Chevak, the rain and clouds finally cleared, and the sun came out. For a short while it looked like a bad day for bugs; the wind dropped and the sun warmed the air into the upper fifties, and mosquitoes by the hundreds rose from the sedges, forming gray clouds around me, reminding me that Fisher, with his wide experience in the Old World Arctic, considered Old Chevak's mosquitoes the worst he'd ever seen. But then a cool breeze woke, and what followed was a glorious romp. I gulped a quick bowl of oatmeal and set out alone, a tripod with a telephoto lens over my shoulder, heading out along an unnamed tidal slough that entered the Keoklevik near camp.

I moved out of the relatively higher uplands near the old church—"higher" meaning six or eight feet above the tide line—and into a miles-wide bowl of shallow ponds, marshes, and sedge meadows, through which ran a couple of deep but narrow sloughs. On some of the smaller lakes I flushed red-throated loons, while the larger bodies of water held Pacific loons, which are heavier and need a longer stretch of open water along which to run as they struggle to take off. There are common loons in the delta as well, Brian told me, restricted to the very biggest lakes, but I saw none. In a wide meadow beside the river, I lay on my stomach while male dunlin, red-backed shorebirds with soot black bellies, made display flights all around me, flying low on quivering wings, then hanging motionless in the air a foot from the ground while they poured out long, burring calls. With no high perches, many birds use the sky as their signpost; no matter which direction I looked, I could see Lapland longspurs, yellow wagtails, and others performing similar displays.

The sheer density of nesting birds around Old Chevak was both a delight and a constant worry to me; I feared that despite my best efforts, I might crush a small longspur or sandpiper nest, so cunningly hidden as to be almost invisible. Even the nests of the bigger birds were hard to see. Many times in the previous days I was startled by a pintail hen or a teal bursting up from almost at my feet, squawking and dragging its wings as though in distress, trying to lead me away from eight or nine eggs in a bowl of soft, delicate down.

With one eye on a pair of sandhill cranes watching over their lanky rust-colored chick, I almost stepped on a family of whitefronts in the tall sedges; the gander hissed and bluff-charged, while the female herded the stumbling goslings away through the marsh. I let her pass, then tried to skirt the bunch of them only to find myself mixed up with an angry pack of dunlin and black turnstones, fluttering around me with shrill cries. I moved away from them, just to be dived on by an Arctic tern that gave a rasping growl, and then by a bunch of westies and more dunlin. I wandered in aimless zigzags for a further twenty minutes, a human pinball ricocheting from one complaining bird to the next, until at last I found an uncontested spot where I could finally sit and eat my lunch in peace. No sooner had I plunked down gratefully in the soft, damp tundra, however, and taken a granola bar and water bottle from my pack than a savannah sparrow flew in, its bill full of insects for some hidden chicks, and chipped a blue streak at me from a weed stalk a few yards away. "Gimme a break," I muttered to the bird. "You can just yell at me for five minutes while I eat something."

As I chewed, I admired the Lilliputian world growing around me—greenish reindeer lichen, salmonberries with deeply cleft leaves and white blossoms, Labrador tea, and a clump of spirea, six or seven inches tall and holding a dozen flat, nickel-sized racemes of white flowers. The only trees were dwarf birches, a couple of inches high, with leaves the size of a newborn's fingernails, scalloped and edged in yellow, their surfaces pebbled and matte—like everything on the tundra, intricate at a scale we humans are prone to overlook. White dominated the upland flowers, but down in the marshy sedge meadows there was a riot of color—yellow or magenta louseworts, blue Jacob's ladder, pink brambles. Two weeks ago, Brian told me, everything here had still been dead and brown; now the grasses and sedges were waist-high. Two weeks. That's what twenty-four hours of sunlight a day can do.

A long-tailed jaeger floated by, reminding me of the frigatebirds in the Dry Tortugas, the same combination of long-winged grace and quiet menace, which, of course, the other birds recognize, so that every jaeger I saw flew with a tethered clot of little western sandpipers or dunlin mobbing behind it, calling in alarm. But it's when the nesting birds see one of the bigger, heavier parasitic jaegers that they really push the panic button. Parasitics get their name from their habit of stealing food from seabirds, but in most of their circumpolar range they are superb hunters, almost as agile on the wing as a falcon, and so the group name, *jaeger*, German for "hunter," is also well bestowed.

Earlier in the week, I'd seen a parasitic lazily flying across the tundra, when it flushed a small shorebird—probably one of the young westerns that had recently fledged. Instantly, the jaeger gave chase, dogging the small bird through a spectacular minute-long series of snap turns, rolls, cracking drops, and swift climbs. With every maneuver, the jaeger got a bit closer to the westie, forcing it lower and lower as I watched, riveted, through binoculars. For long moments they would both disappear below the horizon, lost in some small draw, and I would hold my breath, assuming it was over; then they would reappear, both still flying with such explosive energy that I could barely stand to watch. Once an Arctic tern took a swipe at the jaeger, breaking its attack for a second, another time a black-bellied plover did the same, but it wasn't until a long-tailed jaeger got into the act, harassing its bigger cousin, that the parasitic finally gave up. By then, they were so far away I couldn't be sure whether it had caught the sandpiper—nor, to tell the truth, was I really sure which outcome I'd been hoping for.

While the refuge is almost unimaginably fertile, and far enough removed from civilization that you'd expect the passage of time to have little effect, even here there have been losses and changes. McCaffery pointed out that Peterson and Fisher visited Old Chevak just a few years before standardized continental waterfowl surveys began, so we don't know exactly what the populations of many species were at the time of their visit, but some birds have been on a roller coaster since then. Cackling geese and Pacific white-fronted geese have had an especially wild ride. In the mid-1960s, the cackler population was estimated to be 400,000, and the whitefronts about half a million. By the early 1980s, both had plummeted; there were about 100,000 whitefronts, Brian said, and barely 25,000 cacklers. The cause was overhunting both in Alaska and on their wintering grounds on the West Coast, and by all but closing the season for those geese, bird managers have effected a dramatic recovery, with a fourfold increase in whitefronts and a sixfold increase in cacklers.

Most of the cacklers, which used to winter in central California, now shortstop in Oregon, "because that's where people are growing grass for golf courses," Brian said. I thought he was making a joke, but no—a lot of Oregon farmland is devoted to raising grass, and the farmers there, annoyed by the geese nibbling their crop, prevailed on the local U.S. Fish and Wildlife office to develop a depredation plan, which would have allowed the protected birds to be killed. This did not sit well with Yukon Delta Natives and USFWS staff, including the refuge manager, Mike Rearden, who had expended so much time and sweat to save the geese. When they found out, they marshaled a group of Y-K Natives to travel to Oregon and meet with the farmers. "So this farmer drives up in a Cadillac, wearing a belt buckle this big"—Brian's hands form a circle the size of a bread plate—"and he's complaining about economic hardship to a bunch of guys who are trying to feed their families on these birds." The depredation plan was quickly scuttled, and the farmers, while free to scare the geese off their fields, can't shoot.

If things are looking good for some breeds of waterfowl, the last half century has been much harder on others. Populations of emperor geese, the only species of waterfowl to spend its whole life in Alaska, are now estimated at fifty-five to seventy thousand on the refuge, only about half what was there in the 1960s. Things are even bleaker for two of the big, heavy-bodied sea ducks known as eiders. Steller's eiders, whose males have a white head and an orangish body, were never common on the refuge but have essentially vanished from the Y-K Delta as a breeding species; no one really knows why, and they are listed as threatened under the Endangered Species Act.

Things are only marginally better for the spectacled eider, which has fallen from about a hundred thousand birds on the delta in the 1970s to only about seven thousand today. "Spec" eiders are gloriously oddball ducks, the male white and black with a bright orange bill, sea green head, and huge white goggles around the eyes; even the brown females have buffy spectacles. Until recently, every spec eider on the planet seemed to vanish after the breeding season, their wintering grounds a mystery. Then, in 1995, scientists tracking satellite-linked transmitters found the birds, 155,000 of them, gathered in leads of open water among the pack ice of the central Bering Sea, where they dive for mollusks in the frigid water.

While the demise of the Steller's eider is a mystery, biologists have a prime suspect for the spectacled eider's troubles—lead poisoning from the shotgun pellets that Native (and some non-Native) hunters have been using on the delta for generations. Studies near Old Chevak have shown that one in ten female specs caught in early spring have ingested pellets of spent shot, which they pick up from bottom sediments while grubbing for food; by August, half of the hens still on the delta have lead in their bloodstreams, as have more than a tenth of their young. Even a small amount of lead, ingested by accident while the birds are feeding in bottom sediments, can cause weakness, gut paralysis, and a highly unpleasant death. Even though lead shot for waterfowl hunting was banned years ago in the United States, there's still a lethal inheritance buried in the refuge's ponds and rivers from generations of past hunting; what's more, until recently in the poverty-stricken villages of the delta, it was more common for Native hunters to use cheaper lead shells than those with expensive steel or other nontoxic shot. "It was a big psychological hurdle for people to accept that they could be putting out so much lead to affect an entire population," McCaffery said.

But the Native residents of the delta are coming to grips, however painfully, with the fact that with twenty-first-century technology, they are having an unprecedented impact on the land that has supported them for millennia. More than almost any other Native group in the country, the Cup'ik and Yup'ik have until now straddled the worlds of tradition and modernity, following a seasonal round of spring sealing, whaling, and waterfowling; summer fishing, egg collecting, and berry picking; and winter trapping. But the advent of firearms, snow machines, outboards, and ATVs has altered the old balances, as has the changing social structure of the villages—the breakdown in the authority of the elders and the growing chasm between young people and the land. Many kids would rather watch satellite TV and play video games than learn to hunt and fish. Jobs are scarce, alcohol remains an entrenched problem, and illicit drugs are a potent and growing menace.

Some of the best emperor goose habitat in the world lies within the refuge borders, but on land under the control of Hooper Bay, a Yup'ik village on the coast west of Chevak. It is, Mike Rearden told me, the best of the best. "If someone told me I had to boil down the whole nineteen-and-a-half-million-acre refuge to just one area, the best sixty thousand acres to protect, that would be it," he said. Yet people from Hooper Bay and another Yup'ik village, Scammon Bay, have been ripping the area to shreds with ATVs, much of it joyriding by kids with nothing better to do. "The damage is appalling—it literally keeps me up at night," Rearden said. Even if the four-wheeler use ended tomorrow, it would take at least a decade for wetlands to recover, while the more delicate upland tundra—where the paths used by Arctic foxes last for years—might take seventy to one hundred years to heal.

When the elders of Hooper Bay were first approached some years ago about a federal conservation easement on the land, they rejected it angrily; the Yup'ik have not always been well treated by the government, and many bristle at what they see as the sterile, disrespectful approach Western science takes toward wild animals. Nor has it always gone down easily that their homeland is now a vast wildlife refuge, most of it no longer under their direct control.

But with time, as the damage has worsened and spread, Yup'ik views about this critical parcel have changed, Rearden said. The refuge will be paying the village about $5 million for the easement, and where the village had once contested the appraisal price as too low, they no longer do so. "We're past that," Rearden said. "The last meeting we had with the village, I brought it up and they cut me off cold. They said, 'You need to understand, this isn't about money.' They can see the need to do this now; they can see that the younger generation coming up just doesn't have the connection to the land that they have."

Returning to camp late in the day, I always had two landmarks to guide me—the bright orange stripes on the roof of the old church, which were visible for miles out on this straight-edge landscape, and three tilting crosses that mark the site of the old Cup'ik cemetery. Even in the 1950s, Fisher described it as a neglected place, with skulls and femurs lying exposed, and Brian had warned me to steer well clear of it, because the old grave houses had rotted, leaving gaping holes under a skin of vegetation. There was, however, one tended grave near the edge, a low, waist-high steepled roof of tin with bunches of plastic flowers and a cross at one end, where the savannah sparrows would sit to sing. All the graves here had once been shallow and roofed like this, because just below the surface lies the permafrost, the ever-frozen soil through which no grave digger can chop, which forms a liner to the whole Y-K Delta, ponding water near the surface to create the immense wetlands.

Permafrost is, as the name suggests, permanently frozen soil, and it underlies all but the southernmost 15 percent of Alaska. The upper layer of soil, known as talik, melts each summer, but just a few feet below the ground a shovel *pings* against a rock-hard bed that in some parts of the Arctic may extend down more than a thousand feet. The ground has been frozen here since the Pleistocene, occasionally eroded by a river or a hydraulic gold-mining operation to reveal the frozen carcass of an extinct long-horned bison or some other ice-age behemoth. (Contrary to folklore, the meat is usually pretty rank, and definitely not something even the most adventuresome diner would care to sample.)

But the permafrost has been warming for a century in Alaska, and evidence suggests that south of the Yukon River, where I was, it is actually melting away; by the end of this century, government scientists project, the top thirty feet could disappear. Up here on America's last frontier, the weather's been acting squirrelier and squirrelier, and it has a lot of the locals worried. Winters have been much less severe; in villages where people's homes used to be buried under so much insulating snow that they had to tunnel out, these days the drifts don't even reach the windows. River- and lake-ice thickness is way down, and spring is coming earlier, Rearden said. In fact, the whole summer season seems weirdly accelerated these days. Each year for decades, one of the central chores at Yukon Delta NWR has been the roundup of geese for banding at midsummer, when the birds molt their flight feathers and become temporarily flightless. Traditionally, the molt occurred no earlier than the third week of July, but this year, Rearden told me, geese were already molting the last week of June.

The warm winters that Mike Rearden told me about aren't just his imagination; Alaska is on the front lines of global warming. Since 1900, the planet as a whole has warmed by roughly one degree Fahrenheit, a seemingly small rise but one that, because it is an average, actually represents a dramatically warmer climate. But in parts of Alaska, the average annual temperature has jumped five degrees, with an eight-degree increase in winter temperatures in the last thirty years. That's in keeping with increasingly sophisticated computer models, which largely agree that the impact of climate change will fall first, and most heavily, on the highest latitudes, like Alaska.

As bad as lead poisoning and ATV abuse may be, they pale when compared with the impact climate change will have on the vast wetlands of the Yukon-Kuskokwim Delta. No one knows, for instance, whether long-distance migrants like godwits and other shorebirds, which time their global movements to correspond to local bonanzas of food all along the migration route and depend on a specific suite of habitat, food, and climate on the breeding grounds, will be able to adapt. (Nor do we know what changing climate will do to global weather systems and prevailing winds, which are critical to birds hurtling halfway around the world.) Cold-adapted species, from tundra plants to fish, are already losing out to aggressive invaders from the south better able to tolerate milder conditions.

But the real worry is what the ocean will do to the vast tidal system in the refuge coastal plain, where much of the land is only a few feet above the water as is. "If the majority of climatic projections are even relatively accurate, especially in terms of sea level, then that's going to radically alter what's going on out here," Brian McCaffery said. "The gradient is so low that any significant increase will probably mean the loss of tens of thousands of acres, if not more." Storm tides, which already sweep across low-lying coastal sections of the refuge, sometimes leaving endless windrows of drowned brant eggs, will worsen. Plant communities, which are used to responding to climate shifts over the course of millennia, not decades, are unlikely to make the transition, McCaffery said, leaving the wildlife that depends on them with nowhere to go. This would be calamitous for species like the black turnstone, the majority of whose world population nests in the delta on salt-tolerant sedge meadows within a mile or two of the sea.

In the Lower 48, global warming can seem like an abstraction,* but up here in Alaska the signs are shockingly visible. Where permafrost has

*When a team of almost two dozen scientists from a variety of disciplines reported on the projected impact of climate change on California—from heat-wave mortality in Los Angeles to profound disruption of water supplies caused by diminishing snowpack and scant rainfall—almost every news story focused primarily on a single aspect of the report: the threat to California's wine industry.

been melting, the land subsides unevenly, creating "drunken forests" of slumped, leaning trees and damaging roads, homes, pipelines, and runways to the tune of $35 million a year. Shore ice in places like Point Barrow, the most northerly spot in Alaska, used to go out in July; now it may melt as early as March. The snowpack on land at Barrow is also melting far earlier than usual, from late June in the 1940s to as early as mid-May now. Temperatures on the North Slope have gotten so high that Natives can no longer hunt in midsummer, because the meat, which they used to be able to cache in snowbanks, now rots. Comparisons of aerial photographs taken there in the 1950s with conditions today show striking differences in vegetation; stream valleys that were once treeless are now crowded with bushy willows. There are robins nesting in the coastal villages of Inupiaq Eskimos, who lack a name for this new bird from the south.

NASA satellite images show that more than 20 percent of the Arctic summer ice cap has melted since 1979, and the U.S. Navy, which likes to hide submarines beneath the ice cap and keeps a fretful eye on its condition, reports that the depth and volume of the ice have declined by as much as 42 percent in the last thirty-five years. At this rate, some estimates posit that summer sea ice will be gone entirely by 2070. Leaving aside the probable impact of this change on global ocean currents like the Gulf Stream, which hinge on the flow of dense, cold water out of the Arctic, this is bad news for all manner of species, including polar bears, which depend on the ice sheet for their seal hunting.

Already, at the southern edge of the polar bear's range in Hudson Bay, the sea ice is melting earlier and freezing later than it did just twenty-five years ago. The bears, which cannot hunt seals while stranded on land, essentially fast for four or five months, living off the fat they built up all winter. Because of the lengthening warmth, the bears now spend as much as four extra weeks onshore—a small but crucial difference, for they start the fast twenty pounds lighter for every week of hunting they miss. Consequently, the western Hudson Bay population (the most intensively studied polar bears in the world) is showing signs of trouble; the situation is most critical for pregnant females, which remain on land over the winter, giving birth to one to three cubs and supplying all their food, through milk, until they leave their den in March. Birthrates have dropped 15 percent in recent years, and the cubs that are born are smaller, lighter, and dying at a faster-than-normal rate.

In the last decade, as much as 4 million acres of spruce forest on the Kenai Peninsula in southcentral Alaska have been killed by bark beetles. These small pests thrive during hot, dry summers, and whenever there are a few warm years in a row—as in the 1880s, the 1910s, and the 1970s—the bugs cause brief, intense infestations in Alaskan forests. The return of cooler, wetter conditions always reined them in, but in the late 1980s the state entered what, judging from tree rings, is now the longest streak of hot summers in at least the last 250 years. The bark beetles on the Kenai Peninsula went from sporadic pest to genuine plague—the biggest forest insect outbreak ever recorded in North America, with dead trees covering an area bigger than Connecticut and swarms of beetles moving through the air like billowing smoke. Scientists blame the outbreak directly on the unprecedented warmth, and say there is more to come as the beetles push farther and farther north. As they do, they are encountering trees already in poor shape to fight them—scientists studying spruce forests in the interior, near Fairbanks, report that outwardly healthy-looking spruces are showing signs of stress from heat and drought, including greatly reduced growth.

Fire has become much more of a menace, too—not because of decades of Smokey-the-Bearism, stamping out every blaze, as they did in the Lower 48, but because it's simply hotter and drier and there is more dead and dying timber than ever before. My days in Anchorage were dim and smoky from huge fires in the interior, and U.S. Forest Service specialists warned that the bark beetle outbreaks create "the potential for large, intense wildfires unlike any that have occurred in recorded history."

Climate experts warn that all of what's happening in the Far North is likely to create a positive-feedback loop, a self-reinforcing cycle by which warmth begets warmth in a quickening dance. Arctic snow and ice reflect back all but 20 percent of the solar energy that hits them; as they melt, the darker land they reveal absorbs the heat, warming the atmosphere. As the boreal spruce forest that covers most of northern Canada, Alaska, and Eurasia dies off, besieged by insects and stressed by heat and drought, the decomposing trees will release still more carbon dioxide into the atmosphere, worsening the greenhouse effect. Wildfires will do the same, and the loss of forest cover by whatever means will hasten thawing of the deep permafrost, and that, in turn, will free up several ice ages' worth of previously frozen peat and other organic matter, releasing hundreds of billions of tons of additional warmth-trapping carbon into the atmosphere. Melting permafrost would also release immense quantities of methane now trapped in a substance known as gas hydrate, a crystalline form of methane caged within water ice. As the permafrost melts, the methane escapes, and because methane is a greenhouse gas ten to twenty times more powerful than carbon dioxide, this could tip the climate imbalance even further.

The response of the Bush administration to all this has been a yawn of denial and inaction, often with either a dismissal of the underlying science or a cynical manipulation of results contrary to the administration's stated views. Climatologists, watching the unmistakable signs of change emerge across the North—and frustrated by the American head-in-the-sand attitude—sigh with bitter resignation and say, You ain't seen nothin' yet.

The next morning, reluctant to leave, I wedged myself into the backseat of the refuge's red and silver Husky floatplane, and Mike Rearden taxied out into the main channel of the Kashunuk, gave it gas, and nosed up with a roar. Instantly, my perspective changed, and the land over which I'd been tromping for the better part of a week became strange, until I found my bearings—there, the brown peninsula of tundra where a longspur landed almost at my feet; there, the pancake marshes where black turnstones screamed at me, and the long sedge meadows, scarves of pale green along the river, where the dunlin nest.

Rearden made a wide circle, and the orange and white camp came into view under the wing, bright in the morning light. Then we were crossing the delta, flying over a landscape that has changed not at all since Roger Peterson and James Fisher were here—and one little changed, except for the white tents of a few Yup'ik fishing camps, since the end of the last ice age. But, I wondered, what will this place look like in another fifty years? Will it even be possible to stand beside the Kashunuk and watch the arrowed flight of a tern, or will this seemingly timeless land finally have run out of time and been drowned by the rising sea?

I leaned my forehead against the window, eyes closed, the vibration of the engine making my skull buzz. When I opened my eyes a moment later, the camp was gone, replaced by an immensity of water and marsh and tundra, over which, as always, flickered the shapes of birds.

And so this is what I want to remember, and what I hold in trust for the future: the low, eternal light, and the cottongrass bowing in agreement with the wind; a longspur hanging in the air like arrested time while its song spills out; two loons keening like banshees, their heads glowing silver against the dark water; a pair of godwits, drawn back here from far New Zealand, standing guard over three chicks—he the color of a robin's breast, his bill opening with a rhythmic warning as the brown female hurries the babies to cover. Then the male leaping into flight, driving into the air with muscular wing beats to chase away a jaeger. The female, whistling an all clear, and the gray chicks rising from their hiding places among the ripening salmonberries to resume eating and growing, so that they, too, may one day ride the Pacific winds to the antipodes and back—back to a land ancient and, I pray, safe.

FOURTEEN

The Wildest Shore

On the Fourth of July, 1953, a big Reeve Aleutian Airlines DC-3 clattered to a stop on a gravel airstrip on St. Paul, the largest of the Pribilof Islands, three hundred miles west of the Alaskan mainland in the Bering Sea. On board were Peterson, Fisher, and their friends Finnur Gudhmundsson and Bill Cottrell, fresh from the Y-K Delta and making the final stop on the *Wild America* journey that had begun, for Peterson and Fisher, almost three months earlier in Newfoundland.

The four men loaded their gear onto a bus, which took them to the forty-four-square-mile island's only village, also called St. Paul. The local Aleuts were squaring off in a holiday baseball game against a team made up of St. Louis fur company workers, there for the summer fur seal harvest. The ball game was, for most of the company men, a rare diversion on a lonely, boring outpost, and when Fisher told two of them that visiting the Pribilofs was a thirty years' dream come true for him, "they looked at him as though he must be mad," Peterson wrote.

"'I'll be damned!' said one. 'There is no accounting for tastes!'"

No accounting, indeed. But among naturalists the Pribilofs loom large, and Fisher's dream is a common if rarely pursued one, luring a handful of birders and photographers each year to St. Paul and (to an even lesser extent) its smaller but still biologically richer neighbor, St. George. Reading *Wild America* as a boy whetted my appetite to see the islands, which I'd first heard of through Rudyard Kipling's classic story "The White Seal," a tale full of mysterious fogs and long, rolling Russian names, of fur seals dancing through the phosphorescent waters and Kotick, the white seal, who leads his people to a safe haven beyond the reach of humans.

And so, two days after departing Old Chevak by floatplane, I was flying back across western Alaska, looking forward to fulfilling my own thirty-year-old Pribilof dreams. Sadly, fabled old Reeve Aleutian Airlines has gone out of business, taking with it a link to Alaska's territorial past, but a little regional carrier named Peninsula Airways has picked up the Pribilofs run, using what I must admit looked, given the long ocean crossing, like a pretty flimsy nineteen-passenger prop plane. I'd almost have preferred an old DC-3, which would at least have been roomier. I had to bend almost double to squeeze into the narrow tube of the plane, and once seated, I kept apologizing to the patient Aleut woman in front of me, for my knees fit only if I spraddled them to either side of her seat, where they kept bumping her elbows.

It was a three-hour flight, first angling west past Lake Clark and Iliamna Lake, where the Alaska and Aleutian mountain ranges merge—a string of slumbering volcanoes oozing steam along the horizon, set among sinuous glaciers striped with long, thin bands of crushed rock. The winter snow still lay thick on the high country, but the glaciers ended in deep green valleys, down which were braided long, silty gray rivers. Then we came to the flat and watery land beside Bristol Bay and out into the Bering Sea, where the clouds closed in. For the next couple of hours there was nothing to do but read, daydream, and, in the absence of a flight attendant, shuttle a wire basket of bottled water and snacks among ourselves, pushing it up and down the aisle along the floor.

Given the famously awful weather of the Pribilofs, Peterson couldn't help gloating a bit that they landed at St. Paul in bright sunshine, which lasted most of their visit. The same surprise awaited me. As we banked into St. Paul, where we would exchange some passengers before continuing to St. George, the clouds gave way, and I saw the island—low and open, with a few small hills, long sandy beaches, and modest cliffs. Forty-five miles to the south, St. George floated on the horizon, higher, more jagged, and much more imposing, rising vertically from the sea. The plane bounced down the gravel airstrip, whipping up a great blast of dust, and taxied up to a big corrugated-metal hangar. The first officer squeezed out of the cockpit and, crouching, opened the hatch and lowered the ladder. Warm sunlight flooded in. "Welcome to St. Paul," he said, his voice a little bemused. "This is the nicest day I've *ever* seen here."

We were on the ground just long enough to refuel and pick up a few inter-island passengers, and for me to watch gray-crowned rosy-finches—

big, chunky songbirds that looked black unless the sun hit them right, when I could see the pink cast to their chocolate bodies—scuttling in and out of holes in the corroded roof of one of the airstrip buildings. Then back in the plane for the twenty-minute hop to St. George, a sideslipping approach that avoided the thousand-foot hills at one end of the strip, the same blast of dust from the red volcanic gravel, and a few people with pickups and ATVs sitting beside a lonely metal shed, on which was painted "St. George Airport—Dedicated 1992."

Among the waiting group was Art Sowls, the seabird biologist for the Alaska Maritime National Wildlife Refuge, the most expansive of all federal refuges, stretching from the edge of the Chukchi Sea north of the Arctic Circle, out to the end of the Aleutians, and down into the southeastern Panhandle of Alaska—a triangle that, if moved to the Lower 48, would have Georgia, California, and northern Minnesota as its corners. Within that immense area, the refuge protects about twenty-five hundred widely scattered islands and pieces of coastline totaling 4.5 million acres, including the ocean cliffs of the Pribilofs (the Aleut communities retain control of the inland areas).

We threw my bags in the back of Art's white pickup, which, in addition to the usual U.S. Fish and Wildlife Service logo on the door, carried an odd decal—the silhouette of a rat with the red circle-and-slash symbol for "no" superimposed on it. I climbed in, along with two of my fellow airplane passengers we'd picked up in St. Paul: Karin Holser, a St. George resident who coordinates the Pribilof Islands Stewardship Program, an environmental and cultural outreach project for kids and young adults on both islands; and Kris Hulvey, a grad student from UC–Santa Cruz who was going to spend the summer working with Art's crew, comparing the vegetation of the two islands and assessing how much damage introduced reindeer were inflicting on the tundra.

Art Sowls was a huge man, six foot six or better, thickly built with a dark beard, wearing old field clothes—blue jeans and a flannel shirt, whose collar, where it stuck out above a sweatshirt, was worn to nothing but frayed ends. "As far as Pribilofs weather goes, this is on the gorgeous side of gorgeous," Art said as we rattled away from the airstrip, which lies on the south side of the island, several miles from the village. The sky was unmarred blue, and a cool breeze set the yellow Arctic poppies dancing. "I'm going to drop you off, give you about fifteen minutes to get settled, and then come

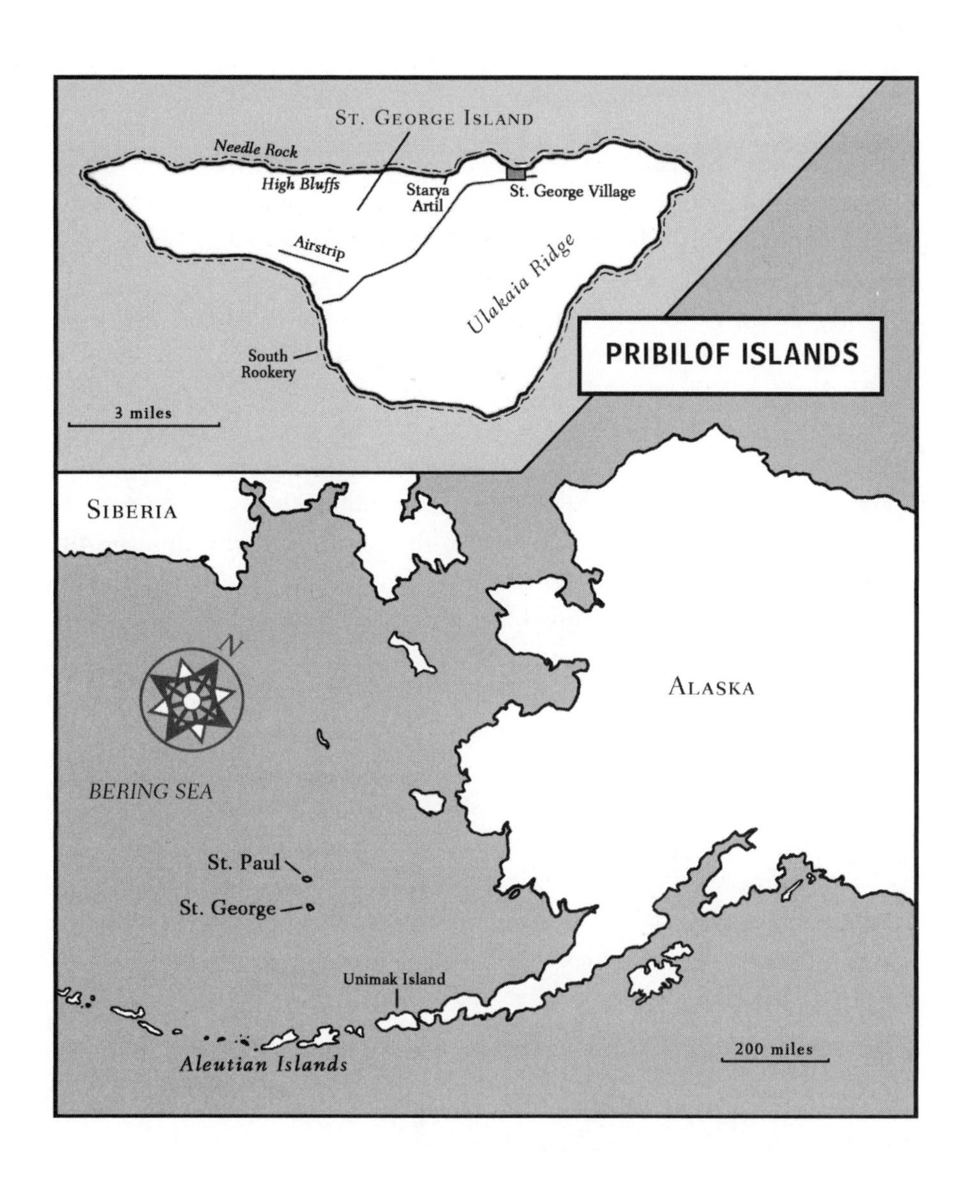
St. George Island
Needle Rock
High Bluffs
Starya Artil
St. George Village
Airstrip
Ulakaia Ridge
South Rookery
3 miles
PRIBILOF ISLANDS
Siberia
N
Alaska
Bering Sea
St. Paul
St. George
Unimak Island
Aleutian Islands
200 miles

back for you all—you and Kris need to climb to High Bluffs while it's out of the fog, because that might not happen again for weeks. If I give you a ride out to the trailhead, that'll save you a few miles of walking."

We crested the low saddle that runs across the middle of the treeless island, high ridges to the east and west of us, and the village came into view in the distance, a few dozen wooden homes crowded around a tidy old Russian Orthodox church with its onion dome. Not far away, fur seals lay scattered like brown logs on the grassy slope of Staraya Artil, the site of the original Russian settlement, founded when the fur hunter Gerassim Pribylof blundered upon the island in a summer fog in 1786 and named it after his sloop, the *St. George*. Beside the church sat the small hotel where I was to stay, a historic building that once housed federal officials overseeing the island's lucrative fur seal harvest, nicely fixed up some years ago in hopes of attracting more tourists, with satellite TV in all the rooms and a big kitchen/breakfast room downstairs. Margie, the manager, had been expecting a couple of PenAir mechanics in addition to me, but they weren't on the flight, and I was the only guest throughout my stay.

Art was true to his word, and I had just enough time to change into boots, fill a water bottle, and stuff my pack full of camera gear before he returned. It was dinnertime, but the sun was still fairly high, and while it would eventually set, I knew that wouldn't be until after midnight, and then only for a few hours. Art dropped the three of us off at a trailhead, which eased its way up the long tundra slope toward High Bluffs, another couple of miles away. St. George's volcanic past was everywhere to see; we picked our way among reddish rocks that looked like pumice, filled with airholes but dense and heavy. Karin pointed out displaying longspurs and snow buntings as Kris fell repeatedly to her knees, examining unfamiliar plants and, on occasion, picking specimens for later identification. She didn't have a plant press with her, so she secreted the leaves and flowers in the pages of her checkbook for the time being.

For an hour we climbed, easily at first, then up a steep grade to the top of the bluffs, which rose in jagged steps like saw teeth from the low ground near town. Thousands of kittiwakes rode up the incline with far less effort than we, filching the lift of deflected winds off the face of the cliff as though they were riding an escalator. At the peak of High Bluffs, several wide-fingered crags known as Needle Rock were combing fog out of the damp wind, so that white mist rolled back between them in thick rivers like steam out of a volcano, but it was sunny just down the cliff where we were. With a word of warning about unstable edges and the off-balancing winds, Karin and Kris dropped their packs and crawled to the edge of the bluff, lying on their bellies to peer over. I did the same, and what I saw will be with me, I think, for the rest of my life.

The cliff fell away for about a thousand feet, to a black rocky beach rimmed with white breakers. To the west, toward Needle Rock, the land curved slightly, creating a shallow bowl, and in this mile-long arc, backlit by diffused light, were thousands, perhaps tens of thousands, of the ethereal white gulls known as kittiwakes, soaring and gliding. Most were so far away or so far below that they were motes, yet they gave a powerful, almost terrifying depth and dimension to the great space framed by this curving edge of the island. The sea was dark blue-black and calm, its surface ruffled by the wind into dendritic patterns, though off to the far northwest, where Russia lay, the ocean was a blaze of bright yellow-white light, impossible to look at.

The cliff face was not flat, but rather ribbed with narrow buttresses of volcanic rock that had cooled into tightly packed basaltic columns. Some were cloaked in vegetation so thick it looked quilted, but most of the ribs were exposed rock, and on them nested many hundreds of thousands of birds—the white kittiwakes, black-and-white murres, smaller auklets of several species. So staggered was I by the sweep of the scene that it took me a few moments to realize that there were birds literally at my nose—dozens of thick-billed murres crammed onto the rocks just below the lip of the bluff, only eight or ten feet from me. They showed little alarm, as if serene in their inviolability.

If they were safe, it was only from us. An arctic fox, blue-gray and not much larger than a cat, barked from a rise thirty or forty yards away, then scampered down the cliff on a slope that would have given a mountain goat the vapors, leaping from one skimpy foothold to the next as it stirred up clouds of complaining kittiwakes. Not far away was its den, littered with the blue-green shells of murre eggs and the white feathers of the gulls; belowground, we could hear the female growling and yipping in agitation at us. I picked up a clump of shed fur that trembled in the wind, thick and dusky, and held it to my nose, smelling the sharp scent of fox.

Karin slipped away, telling us she'd meet us at the trailhead later, and for the next four hours I lay by the edge of the cliff, moving from time to time for a fresh vantage—here a wall of murres hundreds of feet high, there a cluster of fulmars or a city of red-legged kittiwakes, settled in their neat nests of grass and white guano. Among them were a few of the slightly larger black-legged kittiwakes, the same species I had last seen at Cape St. Mary's in Newfoundland. Unlike the cosmopolitan black-leggeds, the red-leggeds are a Bering Sea specialty, found only on four small island groups in Alaska and Kamchatka, with three-quarters of the world's population nesting on St. George. They were small and tern-like, with carmine legs and, as I soon learned, the same startling color lining the mouth, as they hung in the updrafts a few yards from us and squealed like rusty hinges. Their eyes were huge and black, for they are partially nocturnal, feeding on bioluminescent lanternfish.

Farther down, on the lower cliffs, sat the large red-faced cormorants, black with an iridescent sheen of purple and green on their plumage, their facial skin bright crimson—another Bering Sea endemic with largely the same nesting distribution as the red-legged kittiwakes. Why, in the vastness of the Bering Sea and its countless islands, should the kittiwakes be found only in the Pribilofs, on Buldir and Bogoslof islands in the Aleutians, and the Commander Islands off Russian Kamchatka? If this seems a mystery to us, it's because we do not yet know how to look at the world through the eyes of a bird.

At last, Kris and I reluctantly pulled ourselves away from the High Bluffs, just as the fog began to thicken, making the first half of the trail down spectral. Karin met us with the news that Art was off reindeer hunting, handed us each a thick pastrami sandwich that tasted like heaven, and looked at the sky, where the low sun was dropping out of a bank of clouds as we emerged from the fog. "We might get a sunset tonight," she said. "That doesn't happen too often out here."

I had Karin drop me off outside of town, the copper-clad dome of the green and white church glinting dully. I was exhausted, but the sun, which hung a hand's span above the horizon, seemed not to move, and so in the oddly suspended time of a far northern sunset I kept wandering the edge of the island, as though in a dream. Fur seals swam beyond the kelp line, sleek bodies muscling through the swells, and one young bull rose close to shore, a flatfish clamped in its jaws, shaking it like a terrier worrying a rat. Least auklets flew in from the sea, hundreds of chittering little birds in globular, ever-shifting clouds, headed back to their nests in the rocky slopes of Ulakaia Ridge, a half mile behind the village. Soon there were so many flocks pouring in from the ocean that they looked like smoke billowing off the sea. Skeins of murres flew close to the water, footballs in formal wear, and in a small cove I found nearly a hundred harlequin ducks. The drakes were simply breathtaking, dark blue-gray with white bars and blotches on their heads and sides, their chestnut flanks a perfect counterpoint to the cool palette of the wet rocks and the warmth of the evening light. Rosy-finches chased each other across the low bluffs, their shadows doubling the players in this game of tag. It was midnight, and I looked westward toward town, where Karin and a friend were putting out in two small *iqyaks*, the Aleut-style kayaks, paddling through the golden shimmer as seals played around them. For another hour I sat among the blue and white lupines and watched a sky ever more crowded with birds, until at last the sun eased almost imperceptibly into the sea. I hoped this moment, too, would stay with me a lifetime.

My journey was, in one respect, an oddity in the Pribilofs. Almost all of the visitors who venture out this far into the Bering Sea go to St. Paul, which has an aggressively marketed birding-based tourism operation run by TDX, the village corporation. In this, they are unconsciously echoing Fisher, Peterson, and their companions, who spent virtually all their time on St. Paul, save for a ride on a supply ship that allowed them to pass just one night on St. George. My plan, on the other hand, was to start on St. George and after several days—if the weather gods smiled—fly to St. Paul. Even today, very few outsiders bother to visit this smaller island, which rates only a couple of PenAir flights a week; add to this the foggier weather on St. George, and it's entirely possible to be stuck there for days or even weeks beyond your planned departure. (Bad weather can just as easily ground you in St. Paul, but the daily flights make it easier to eventually get off.) The previous year, St. George residents told me, they'd had only a single tourist.

The next morning was windy and cool, with blue sky out over the ocean but a cap of fog hanging like a breaker over the center of the island, concealing the top of Ulakaia Ridge; flocks of auklets appeared and disappeared into the cloud bank. There were no trails to the ridge, so I simply made off cross-country over what looked, from even a short distance, like a slightly unkempt meadow but that proved, once I got into it, to be a pernicious knee-high mix of grass, cow parsnip, and ferns, concealing a multitude of unseen rocks and troughs that were both tiring and time-consuming to hike across. Most of the time, I could not see where my feet were going, so I found myself probing with each step, forever thrown off kilter by a loose rock, an unexpected hole, or a tussock that turned under my foot.

But what it exacted in delay and discomfort, the land more than re-paid in beauty. The island was in full bloom, and these lusher meadows looked blue from a long way off, thanks to dense stands of big Arctic lupine and Jacob's ladder. Where the meadows gave way to moss, lichen, and crowberry, there were white cloudberries and saxifrage, purple whorled louseworts, the yellow of mountain buttercup and Alaskan poppy, and mats of deep, deep blue forget-me-not, the state flower. But my favorite of St. George's wildflowers was the chocolate lily, only a couple of feet tall, with blossoms of a glossy maroon framing yellow anthers. The first time I saw one, I stopped and stared; it seemed like a plant fashioned for a tropical jungle, not a subarctic island, but it grew by the thousands.

After a mile, the land rose more steeply on the flanks of the ridge, and the air was filled with the chirring calls of thousands of auklets, which wheeled overhead in a continuous gyre, new birds flowing in from the ocean, others dropping to the ground, still others rising from among the rocks and forming little clots to buzz back to sea. The ground was rough with big, angular boulders, smothered in moss and lichen, and half-concealed by lanky clumps of grass and cow parsnip. On almost every exposed rock sat a dozen or more of the little seabirds, each the size of a starling. Least auklets are the smallest of the world's alcids, scarcely more than six inches long—stubby birds with small conical bills and white shirt-button eyes, which feed on planktonic crustaceans well out at sea but nest in deep crevices in the rocks. From far below my feet, I could hear an ongoing buzz, almost electronic in nature, that rose and fell in pulses as the nesting auklets griped about my intrusion.

I was close to another fox den, and the male yapped monotonously from a few yards away as I picked my way among the rocks. When I stumbled upon its den, I found it a mess with auklet wings, scraps of feathered skin, and a few whole, freshly dead birds that the male had dropped for his mate. I lifted one, which filled my palm, feeling the fleshy webbed feet and looking closely at the dark gray plumage, speckled white on the lower belly, and the tiny red tip to the small bill. I replaced the dead bird, but the fox yowled even louder.

I settled myself next to a rock in the midst of the colony, and within moments auklets were trying to land beside and even on my head, aborting at the last second with a dry *thddddtttt* of wings as they flared off, little feet splayed. I tried to get a sense of how many birds might be present, but I gave up without even knowing how to make a guess. There were obviously many, many thousands in flight at any given moment, but as they eddied around and around, some constantly joining the mass or falling out, I couldn't begin to estimate their numbers. And I knew that counting the dark entrances to the nest cavities was pointless, because Art had already told me the auklets nest as deep as thirty feet in the jumbled boulders, layers and layers of nesting birds using common entrances. In fact, no one has come up with a good way to census least auklets, and the estimate of 9 million in Alaskan colonies is the baldest of suppositions, acknowledged by the experts to probably be a fraction of the true number. What's more, the Ulakaia colony on St. George is immense, running for miles along the ridge, and though I knew that compared with some least auklet colonies in the northern Bering Sea and the Aleutian Islands, this was a small one, the number of birds it contained was staggering.

But "staggering" is perhaps the best adjective for St. George's wildlife riches. This island, maybe twelve miles wide and five and a half deep, hosts one of the most extravagant concentrations of marine life in the world, including an estimated 3.5 million seabirds. (By contrast, St. Paul, which is justly famed for its birding, has a scant half-million seabirds.) There are other islands with larger bird colonies, and some, like St. Paul, with more seals, but the combination of the two, Art Sowls believes, gives the Pribilof Islands the greatest concentrated biomass of any island group in the Northern Hemisphere. Little wonder, then, that the Pribilofs are often called "the Galapagos of the North." Often, in the handful of precious days I had to hike its lonely cliffs and tundra, the sight of so many birds and seals, teeming in every direction and at every moment, gave me the overwhelming sense of glimpsing something primeval, something I had feared was lost from the world.

And yet, as awe inspiring as St. George may be, it is not what it was just a few decades ago, and experts like Art are deeply worried about the future of the Pribilofs' wildlife. The number of fur seals has dropped by more than half from the 1950s, while Steller sea lions have declined by more than 80 percent, finding a place on the federal endangered-species list. Seabird populations have roller-coastered: thick-billed murres and red-faced cormorants are down and both species of kittiwakes have fallen by about half since the 1970s, though they have recently shown improvement; for others, like the least auklets, there's simply no way to tell if the population is going up or down. "Most things that we have data on, like thick-billed murres and red-legged kittiwakes, have gone down," Art said one evening, after a meal of salmon and fresh reindeer tenderloins. "We don't have data on auklets and puffins, but there are indications that the Ulakaia colony is reduced; the elders in St. George say it used to be much bigger." As to why the trends are so worrisome, well, that's a puzzle, as we'll see.

One thing about St. George I found unchanged from Fisher and Peterson's visit was its foxes. Their only night on the island, an Arctic fox tried to steal one of Bill Cottrell's shoes right out of his room in the old government residence, where I was staying. While I had no nocturnal visitors, the foxes were like cats around the village, trotting unconcernedly among the cottages and homes, curling up beneath their tails in the shadow of the store/post office/village corporation building. Karin and others warned me not to leave anything lying on the ground unattended, even if I was close by; what the foxes don't tear open to eat, they will defecate or urinate on in order to scent-mark.

Arctic foxes are circumpolar, white in winter but brownish gray in summer. Some of those native to the Pribilofs, however, are the rare "blue" phase that remains a smoky color year-round. Pribilof foxes were taken by the Russians starting in the eighteenth century and scattered on islands throughout the Aleutians, there to reproduce and provide more furs; American fur traders did the same thing, putting red foxes on some of the islands as well, and by the early twentieth century foxes had been established on roughly 190 Aleutian islands. While this makes economic sense, it was an ecological catastrophe; on islands where no mammalian predators had ever existed, the foxes wreaked havoc with many birds, all but eliminating colonies on some islands.

For almost a quarter century it was thought the foxes had eaten one bird, the dark and diminutive Aleutian Canada goose, right into extinction, until three hundred geese were discovered on tiny, storm-tossed Buldir Island in 1962 by Robert "Sea Otter" Jones, the intrepid manager of the Aleutian Islands National Wildlife Refuge, a forerunner of today's Alaska Maritime NWR. Jones had never given up on the hope that Aleutian geese might have survived somewhere in his vast, inaccessible refuge, and even prior to his electrifying discovery he'd begun to clear foxes from one island so he'd have a haven for any geese he might find.

With safe islands on which to breed, and protection on their wintering grounds in Oregon and California, the Aleutian Canada goose began a remarkable recovery, which led to its complete removal from the federal endangered-species list in 2001, when it numbered roughly thirty-seven thousand. (The latest estimate puts the population at more than seventy thousand, and the geese have become so numerous that farmers on their wintering grounds now grumble about crop damage.) But fox eradication, though an expensive, labor-intensive effort, is a boon for more than just geese, and some 1.5 million acres have now been freed from the non-native predators. As a result, some of the lost seabird colonies are reestablishing themselves, especially whiskered auklets, which roost on land in winter, making them particularly vulnerable to predators. With the introduced foxes gone, biologists report, auklet numbers are exploding.

One spot with an unusual fox history is Walrus Island, a small islet in the Pribilofs just off St. Paul. When Peterson and his companions visited it at the end of their stay—a day or two after putting James Fisher on a plane to reluctantly start his long trip back to the U.K.—they found the island awash in sea life. So many thousands of Steller sea lions crowded the edge of the island that their boatman joked they might have to shoot their way in; once ashore, they found the island jammed with up to 2 million murres, nesting not on cliffs (for Walrus, a low, flat slab of rock barely above the sea, has none) but on the ground, like penguins. Among the multitudes of murres also nested cormorants, puffins, gulls, and several species of auklets. If the Pribilofs were the climax of the whole *Wild America* odyssey, Peterson wrote, the visit to Walrus Island was the very highest point of all, and he was sorry James had missed it.

I had hoped to visit Walrus Island myself, but months earlier, in making plans with Art Sowls, he told me not to bother. "There have been some real changes in the Pribs since the fifties, and Walrus is one of the biggest," he said. "The foxes got there, and the birds are gone." In this part of the Bering Sea, extensive sea ice doesn't usually form, but the winter of 1970–71 was brutal, and the waters between St. Paul and Walrus froze. A few Arctic foxes crossed, and after that the ground-nesting murres, and pretty much everything else, either decamped or were eaten. Today a handful of foxes remain, and only a few hundred seabirds use Walrus. But the refuge has no plans to remove the predators, which got there naturally and have probably come and gone on Walrus regularly over time. "Walrus was fox-free for eighty years, and it will be fox-free again. I figure that one of these days, a big hundred-year storm will come and wash right over the island, and that will be the end of the foxes for a while," Art said during my visit. "I'm sure it's happened before."

As isolated as the islands of the Bering Sea and Aleutian chain may be, they are not beyond the reach of modern problems. This is a busy maritime corridor, with cargo ships, tankers, and a huge international fishing fleet moving through year-round; thousands of ships thread the narrow Unimak Pass at the eastern end of the Aleutians, and many thousands more pass north of Buldir Island at the western end every year. That's a lot of traffic in bad seas and worse weather, and the potential for disaster is high.* Roughly once a year, a boat runs aground in the Pribilofs, and oil spills, like one in 1996 that fouled thousands of king eiders at St. Paul, are an ever-present menace.

"A little oil spill here, at the wrong time of the year, could kill as many birds as the *Exxon Valdez*," Art said, which is sobering, because that infamous 1989 spill killed an estimated quarter-million birds. But if there's one thing that makes Art Sowls lie awake at night, it's the specter of something far worse than oil. "It's been fifteen years since the *Exxon Valdez*, and we're still seeing effects from it, but even with a bad oil spill there's a chance at recovery. A rat spill, though? Rats are forever."

All it would take is for one infested ship to dock in St. Paul or St. George, or to run aground on the rocks—all it would take, literally, is for

*Disaster is exactly what happened about six months later, when the 738-foot freighter *Selendang Ayu* split in two on Unalaska Island in Unimak Pass, spilling some 350,000 gallons of oil into the Alaska Maritime NWR.

just one pregnant female Norway rat to swim ashore, shivering but alive, and this wildlife paradise would be a ruin. If that sounds melodramatic, consider Kiska Island, far out in the Aleutians. Kiska is one of the crown jewels of the Alaska Maritime NWR, home to an estimated 3 million to 6 million least and crested auklets. But rats got there in the 1940s, probably on troopships during World War II, when the Allies fought a bloody and largely forgotten war with the Japanese for control of the Aleutians.

The details are gruesome; Art and other scientists who have been there describe rats eating the brains and eyes of incubating auklets, consuming their chicks, destroying their eggs. Because rats are even more skilled burrowers than auklets, there's almost no place a bird can nest that the rats can't reach. Biologists figure that at this rate, Kiska's birds will be gone in thirty years; and then, as on most of the other thirty Alaskan islands that have become infested, the rats will turn to intertidal organisms, to tundra seeds, and to cannibalizing each other.

At about seventy thousand acres, Kiska is more than twice as big as the largest island ever cleared of rats, twenty-nine-thousand-acre Campbell Island near New Zealand. Eradication will require helicopters carpet bombing the place with thousands of pounds of rodenticides, with a price tag of millions of dollars and no guarantee of success. But to do nothing all but ensures the loss of what may be Alaska's largest auklet colony.

Obviously, when it comes to rats on seabird islands, the key is prevention. Until artificial harbors were built in the 1990s, ships coming to the Pribilofs anchored well offshore, and the islanders rowed out in huge, open sealskin-covered boats called *baidars* to off-load cargo or passengers—a cumbersome process, but one that made it far less likely that rats could infiltrate the islands. But now that ships can dock here, every vessel is a potential carrier. That's why Art has an almost evangelical zeal when it comes to rat prevention, including that symbolic "no rats" decal on the USFWS truck. "Building a harbor is like opening the door for them," he said, but he and his colleagues, with backing from the communities, have tried to set up safeguards.

Rat-infested ships are barred from the Pribilof harbors, and incoming ships may be inspected and briefed on their responsibilities, with rat-prevention kits—traps, videos, and printed materials—handed out to captains. Rat-control stations, which are basically wooden boxes containing snap traps and poison bait, are scattered around the waterside buildings and docks and checked regularly. Since the trapping program started eleven years ago, Sowls said, they've caught six rats out of three-quarters of a million "trap nights" (a trap night is one trap set for one night). That's not a guarantee that the Pribilofs remain rat-free, of course; it could take years for a small, unnoticed population to make its presence felt. But in the meantime, USFWS has created a rat-spill response team, which will swing into action when there's a shipwreck, just as does the agency's oil-spill response team.

It's said, in the tradition of the Aleuts (who call themselves Unangan), that the first person to find these islands was the father of a chief from Unimak, who, while hunting in his *iqyak*, was swept off by a storm and carried for days through the gale to a place where the air was full of the roaring of seals. After spending the winter alone there, he returned to Unimak with many sea otter pelts and stories of the place the Aleuts came to call Amiq.

The Aleuts did not colonize the Pribilofs, however, and the islands were uninhabited when the Russians came in 1786. Gerassim Pribylof (or Gerasim Pribilov, or Gavrill Pribilof, or Ocrassim Pribylov, depending on which source you pick and how it transliterates his name from the Cyrillic alphabet) was led by the bellowing of seals through the fog to make landfall at St. George. Fur drew the Russians; Pribylof had been dispatched from Unalaska, in the Aleutians, to search the Bering Sea for the breeding grounds of the northern fur seals, which with sea otters were the living gold of the czar's new American empire. And for the next two hundred years, until just two decades ago, seal fur drove the islands' economy. It's a sad and shameful tale—but not necessarily for the reasons you might expect.

Pribylof's discovery was spectacular, especially when, the following summer, the fog cleared long enough for the seal-hunting crew that had overwintered on St. George to realize that there was an even larger island on the horizon, which soon proved to have even more seals. The first years were a slaughter of epic magnitude; the foxes were so tame they could be caught by hand, and sea otters crowded the water in such numbers, one Russian source claimed, that they made it difficult for boats to land. When the ships left the Pribilofs after the second hunting season, they were laden with a reported forty thousand sealskins, two thousand sea otter pelts, seven tons of walrus ivory, and all the whale baleen they could carry.

Naturally, it did not last. Within six years, hunters couldn't find a single sea otter near the islands, and the species remained extirpated until a few were reintroduced in the 1960s. Attention swiftly shifted to the fur seals, with fleets of sealers descending on the islands in a free-for-all that only ended in 1799, when the czar, trying to impose some order, gave a monopoly on the fur trade to the Russian-American Company.

To create a resident workforce, the Russian fur traders brought Natives from the Aleutian Islands, first on a seasonal basis and then, starting in the 1820s, in permanent settlements. The Aleuts were initially little more than slaves; wives were held hostage to coerce the men into working, for example. But with time, the Aleut communities on the Pribilofs grew to a sort of bicultural stability—self-governing along traditional lines, fluent in Russian and Aleut, grounded in the Orthodox Church, with rights as full Russian citizens, and paid competitive wages for their work, which revolved almost entirely around sealing.

Fur seals and sea lions are members of the eared seal group, those with small, pointy external ears (the other major division, the earless or true seals, includes such species as harbor and gray seals). Among the most truly pelagic of all the Northern Hemisphere's pinnipeds, northern fur seals spend most of the year in the deep, cold waters of the northern Pacific Rim as far south as Japan and Southern California, diving almost seven hundred feet deep for fish and squid. At one time, they bred along both coasts throughout this vast region, though fur hunting by Russians and Americans wiped out many of the southerly colonies in the nineteenth century; in just two years, 1808–9, sealers killed 130,000 on the Farallon Islands off San Francisco.* Eventually, the seals could be found

*At least we think these were northern fur seals. Despite the tremendous numbers killed, not a single specimen wound up in scientific collections, and experts can't be sure, looking at the bones unearthed along the California coast, whether they came from northern fur seals or the similar, and still highly endangered, Guadalupe fur seal. In any event, fur seals finally reclaimed the old Farallon rookeries in the 1990s, when a northern pup was born there, the first of either species in at least 150 years. In the years since, the small breeding colony has continued to grow.

in small numbers only on the Commander Islands in the western Bering Sea, the Kuril Islands north of Japan, and Robben Island in the Sea of Okhotsk. The only large colonies, constituting 80 percent of the world's northern fur seal population, bred on the Pribilofs.

Now as then, the older, dominant bulls arrive in May or June and stake out territories on the rookeries (as the breeding sites are known), with the adult females starting to arrive in mid-June. There is a stark difference between the sexes: the males weigh up to five or six hundred pounds, while the silvery gray females are barely a fifth that size. Both sexes have long, pale Fu Manchu whiskers that whiten with age, and like other eared seals, they can turn their front flippers out ninety degrees from the body and rotate the hind flippers forward, giving them far greater speed and agility on land than the earless seals, which must crawl on their bellies. This speed, combined with the bull's sharp canines and irascible temper, can make them a real menace to anyone getting too close.

To protect the breeding seals, the refuge closes the rookeries to all but a few scientists, but I had a chance to tag along with Eli Reynolds, a junior from Principia College in Illinois, and Mark Merculief, a St. George high-school student, who were two weeks into a season-long project under Karin Holser's supervision to monitor the South Rookery, one of four places on the island where the seals breed. We walked a mile or so back through meadow to the edge of a low cliff, at places only fifty feet high, below which were beaches of sofa-sized rocks, black and wave-smoothed, on which hundreds of seals were hauled out. It was another blue-sky day, and although the temperature was a breezy fifty-five degrees, with the bright sun this counted as a scorcher on St. George, and the seals were flapping their leather-strap hind flippers in the air to cool themselves. The big bulls were defending their territories, roaring and bellowing at each other, and the females and immature males were making plenty of noise, too, as they scuffled among themselves—a fur seal rookery is not a live-and-let-live kind of place.

Some of the older females were easily distinguishable by their heavy pregnant bodies, and within a day or two of arrival they would give birth to a ten- or twelve-pound pup, wide-eyed and so dark brown it looks almost black. A few days after that, the female will have mated again, and having filled up her baby with milk, she will head off to sea for a fishing trip that may last more than a week. For the next three or four months, until the pup is weaned, its mother will alternate a couple of days on the shore, pumping fat-rich milk into the rapidly growing baby, with foraging trips that may take her more than 275 miles from St. George, along the edge of the continental shelf to the Aleutians.

More and more seals were arriving every day, mostly females, quickly swelling the rookery toward the thousand-seal mark the researchers expected. A lot of the seals were staying off the beach, out in the deep but surpassingly clear water a few hundred feet offshore, snaking among the long, dark ribbons of kelp, coming up from time to time with fish, or just resting on the surface in what the old sealers called the "jug handle" pose—head up, hind flippers curled forward above the water to touch one of the outstretched front flippers. It didn't look all that comfortable to me, but then, I'm not a seal.

Eli, Mark, and I whispered as we edged into position along the bluff, hunching down to keep our profiles low and unthreatening. The two young men—Eli lean and bearded, wearing a pair of nylon rain pants, Mark chunkier, in yellow fishing overalls with a red cap turned backward—began to methodically census the bawling, skirmishing, sleeping clusters of seals below. They weren't simply counting bodies; they had to determine sex (which, with the nearly identical females and younger males, often meant waiting patiently until the seal rolled over so they could get a peek at its genitals) and try to age the animals by looking at their faces—the whiter the whiskers, the older the seal. This took about fifteen minutes for the small group below us, and then Eli and Mark shifted to the next spot along the cliff a hundred yards away, leaving me ensconced in a little nest among the grass and cow parsnip.

I had a lordly view of the seal beach—high enough for a good vantage, close enough to get the full effect of the smells and sounds. The latter, in particular, were arresting and difficult to describe, without referencing socially awkward bodily functions. Let's just say the calls had elements of a roaring lion and a heavily amplified goat bleat, and that I am not kidding when I say it was wonderful to listen to. The roars of the seals blended with the name-calling of the kittiwakes, and the constant high chirring and cheeping of least auklets that swarmed the lower cliffs, to provide a soundtrack appropriate for this wildest shore.

The seals were mostly lounging in groups, the females bunched up near big, beefy males with thickly maned necks and absurdly small triangular heads with little upturned noses. But if another male came too close, or a female tried to leave, the bull bellowed a war cry and gave chase, canine teeth bared—and be damned to anything that gets in the way, including newborn pups. Twice during the afternoon Mark and Eli watched big fights unfold in which pups were ground beneath rampaging males; we assumed the youngsters were killed, but Karin and Art later said probably not. "Pups are made of rubber," Karin said. "Unless there's direct trauma from a bite, especially head trauma, they're usually okay." In fact, the females pick these bouldery beaches especially so the pups have a chance of surviving a steamrollering male; on a smooth, sandy beach they would be squashed flat.

The original fur seal population at the time of discovery has been estimated at about 2.5 million, but the Russians killed so many that a few decades later, they were forced to curb the sealing—restricting the cull to immature males starting in 1835, a move that established a crude system of sustainable harvest yielding about thirty or forty thousand seals a year. Unfortunately, when the United States bought Alaska from Russia in 1867 (with fur seal pelts a major incentive for the $7.2 million purchase), the Americans ignored the earlier example. For the first two years there was no control on sealing at all, and upwards of 300,000 seals may have been killed. Eventually the government leased the islands to the Alaska Commercial Company and set a quota for immature males that was higher than the Russians had permitted, and probably too generous.

The Americans also ignored the Russian precedent toward the Pribilovians. Although the purchase treaty specified that Russian subjects were automatically entitled to U.S. citizenship, the United States categorized the Aleuts as "uncivilized" and failed to extend citizenship to them. Initially, the Alaska Commercial Company (which had obtained a twenty-year lease on sealing rights) paid the Pribilovians a fair wage that placed them on par with workers elsewhere in the United States. But at the same time, the company and the U.S. Treasury took a disgracefully paternalistic attitude toward the Aleuts, forcing them out of their traditional homes into wooden houses, paying them only in store credit, threatening them with fines or imprisonment for such infractions as "sauciness," restricting their right to travel, even controlling when and whom they could marry. With time, government control grew tighter and tighter; the Pribilovian Aleuts were eventually classified as wards of the federal government and excluded from the 1924 act that finally gave U.S. citizenship to American Indians. What's more, it became government policy to treat the Pribilofs as a secret reservation, a situation that continued until after World War II.

Nor did their relatively high standard of living continue. By the 1890s, seal populations were falling rapidly, and as the hunting quotas were cut, poverty became entrenched on the Pribilofs. The trouble was not the land-based sealing but a new and much greater danger—pelagic sealing, in which fur seals were harpooned on the open ocean, with no limits and no quotas. It was an appallingly wasteful hunt, in which half the harpooned seals were lost and up to 80 percent of those taken were females with a baby waiting back on the beach. (Because the females mate within days of giving birth, conservationists protested, each harpooned seal actually represented three deaths.) The number of fur seals on the Pribilofs plummeted at the end of the nineteenth century, and President Grover Cleveland even had revenue cutters in the Bering Sea arrest foreign sealers, mostly Canadians, touching off an international incident with Great Britain.

For decades, the major powers in the North Pacific—the United States, Great Britain, imperial Russia, and Japan—shilly-shallied over pelagic sealing as the stocks declined and public ire over the slaughter grew. Some of the smaller breeding colonies, like those in the Kuril Islands, vanished completely. Finally, in 1911, the four nations came to an unusual accommodation, known as the North Pacific Sealing Convention. High-seas hunting was banned; instead, the United States agreed (after a period of complete protection to allow the seals to recover) to cull the Pribilof herd on a regulated basis, in keeping with the size of the population and taking primarily young males between the ages of two and five. Similar controlled culls would be conducted on Robben Island by the Japanese, and on the Commanders by the Russians, with the furs being divvied up among the four signatory countries on a proportional basis, Japan and Canada each receiving 15 percent in compensation for giving up pelagic sealing.

Unlike many international accords, this one worked remarkably well. The herd grew quickly, and by 1940 the Pribilof population was back to about 2–2.5 million seals, roughly its original size, with about sixty thousand young males killed each year. Periodically through the summer,

Aleut workers would separate out bachelor bulls, driving them overland to the killing areas. James Fisher joined a killing crew on St. Paul during his visit (Peterson, "not liking scenes of carnage," declined the opportunity) and described it thus:

> Undersize and oversize animals were weeded out. Then deftly the rest were stunned with single blows of hickory clubs that looked like six-foot baseball bats . . . Immediately other men dragged the stunned animals to a strip of lawn and laid them on their backs in rows of ten. The gauge man came round, checked the measurements of each seal and the tally-man marked it off. Seconds later the sticker followed, quickly making a belly incision from chin to tail and plunging his knife into the animal's heart [a method chosen for its presumed humaneness].

The harvests of the mid-1950s, which reached ninety-six thousand bulls a year, were the last big culls. Disease was spreading through the rookeries, which were thought to be too crowded, so a major kill of females—more than 300,000 over twelve years—was authorized. This is now considered to have been a major mistake; the herd size continued to fall even after the females were again protected. Public attitudes were also changing, with pressure from animal-rights activists reducing the market for fur coats, and then taking direct aim at sealing. Organizations like Friends of Animals did a masterful job of linking, in the public mind, the Pribilof kill with the widely condemned clubbing of baby harp seals in Newfoundland.

If you believe that killing wild animals is always fundamentally wrong, then that might be a fair comparison. But from a wildlife management perspective, the fur seal harvest was a spectacular success, with the population maintained near its historic level for decades and powerful economic incentives for the nations of the North Pacific to ensure a healthy herd. It also provided the almost sole support to the Aleuts, who by now were finally freeing themselves from much of their historic government oppression. Although the Pribilovians fought hard to save the cull, the activists won. Commercial sealing had ended on St. George in 1973, and after the Reagan administration decided to abandon the long-standing program as a way to trim the federal budget, the final hunt was held on

St. Paul in 1984. The international agreements governing sealing were allowed to lapse, and the only sealing now done on the islands is subsistence hunting by the Aleuts, who take fewer than a thousand seals a year for meat and furs.

The loss of the seal harvest was a body blow to the communities on both islands; St. Paul saw its wage base halved overnight, along with the loss of substantial federal subsidies. A trust fund was set up to help the islanders shift to private enterprise, largely through crab fishing, and the artificial harbors at both towns were built to facilitate the transition. But the crab stocks collapsed, and while things are tough on larger, busier St. Paul, which has a population of about 550, they are downright desperate on St. George. When I was there, the village of about one hundred people only had a two-week supply of fuel for its electrical generator, and the small village government staff hadn't been paid in weeks, though they kept on working.

In the meantime, St. George tries to hang on to its past while hoping for any kind of a future. Tourism is a possibility, though the vagaries of weather make it hard to attract visitors; the village also hopes to draw big-game hunters for the island's growing reindeer herd. One evening Karin and Art showed me around the old seal processing plant, a one-story U-shaped building just uphill from the water. Defunct for more than thirty years, it remains much as it was, thanks to preservation efforts by the village, which hopes it may serve as a tourist attraction. Outside leaned the wooden frame of an old *baidar*, one of the huge skin boats manned by a dozen men standing at great oars, with a pilot working a big steering sweep. Today the Pribilovians have motorboats, and this link to the past bleaches in the cold air, its ribs of worked driftwood slowly collapsing under their own weight.

"This was the washhouse section," Karin said, her footsteps echoing dully in the big room, which was filled with waist-high wooden tanks, each the size of a small bedroom. "The trucks would pull up outside, and they'd unload the seal pelts—see how there's a window above each tank? Two hundred and fifty pelts per tank, and there'd be a government guy counting each one as it went in. Then they'd flood the tank with saltwater, just to wash off the blood and dirt."

After initial cleaning, the pelts—counted carefully again by an overseer—would go to the blubbering room, a long, whitewashed, and brightly lit space that formed the base of the building's U. Here, workers with curved, double-handled knives would slip a pelt, fur side down, on a slanted metal form and quickly but carefully scrape off any bits of blubber or flesh; a skilled Aleut could clean thirty pelts an hour. Finally the skins (counted once more by the ever-watchful overseers) went into the brine tanks in the third room for salt curing. The tanks, built of redwood and probably dating to the 1920s, had machine-driven wooden paddles to keep the pelts churning while they steeped in the brine. At last, the skins were wrung out, then packed in salt, rolled into neat little bundles, and packed in wooden barrels for shipment.

Karin had a couple of tanned seal pelts—one with the silvery guard hairs still in place, another one that had been shorn, exposing the layer of impossibly luxuriant deep brown underfur, packed with more than 370,000 hairs per square inch. On the wall was taped an old poster from the Fouke Fur Company, showing a young woman in a 1960s hairdo pouting seductively as she modeled a sealskin coat. "Be one of 7,000 outrageously pampered women in the entire world," it said, a bald appeal to snobbery.

The irony is that today, two decades after the last harvest, the northern fur seals are in far worse shape than anytime since the era of pelagic sealing. The Pribilof herd is down to about three-quarters of a million, only a third of its 1950s size, and continuing to fall about 5 percent a year. Nor is it only fur seals. The Steller sea lions that once crowded Walrus Island so thickly are in even worse shape; their Alaskan population has dropped by 80 percent in the last twenty years, and they are now a federally endangered species. In all my time on the Pribs, I saw just one, a big male, tawny and powerful, hunting flatfish off a rainy point on St. Paul.

No one knows why the seals and sea lions have declined so drastically, but when you factor in similar problems for harbor seals, the steep, long-term slide in seabirds like kittiwakes and murres, the collapse of once-strong shrimp and crab fisheries, and the disappearance of half the sea otters in the Aleutians since the mid-1980s, it's tempting to look at pervasive, ecosystem-level changes in the Bering Sea as the cause—and there are a surprising number of potential candidates.

In 1977, the waters of the Bering Sea underwent sudden and significant warming by several degrees, and winter sea-ice cover has shrunk. No one knows if this was the result of a natural cycle, human-caused climate change, or a combination of the two, but it was followed by a shift in the makeup of the sea's fish community, away from frigid-water species like herring and capelin to walleye pollock and cod, which can tolerate somewhat warmer waters. Since the 1970s, the once-dominant fisheries for shrimp, king crab, and snow crab have collapsed, even as commercial groundfish landings skyrocketed, with as much as 6 million metric tons of pollock coming from the Bering Sea at the height of the fishery in the 1980s, much of it from the so-called Golden Triangle between the Pribilofs and the Aleutians.

To residents, scientists, and fishermen alike, the Bering Sea seems to be in a state of profound and alarming flux. Jellyfish appeared in plague-like numbers for a time, then crashed. Greenland turbot, once a commercially important groundfish, are rapidly disappearing, while a species of planktonic crustacean that once formed a major food for Pribilof seabirds has vanished from that part of the sea; both are cold-water species that may not be able to handle today's warmer water. In 1997, short-tailed shearwaters that migrate by the millions to the sea from their breeding grounds in Australia suffered a massive die-off; I was near the Aleutians that summer and found the carcasses washed up by the hundreds on shores near Cold Bay. Starting that same summer, and lasting the next five years, the eastern Bering Sea almost literally turned white from an unprecedented bloom of phytoplankton, which scientists think may have been linked to changes in the planktonic fauna further up the food chain.

In the view of some people, like Karin, blame for the decline in seals and sea lions lies squarely at the feet of the commercial fishing fleet, which still takes almost 2 million metric tons of fish from the Bering Sea each year. "If you overlap where the seals are feeding with the area of heaviest pollock fishing, they match perfectly," she said. "Of course the fishing is having an effect. How much evidence do you need? There are many factors that are having an effect on the fur seals and marine mammals throughout the Bering Sea, but the one thing that we can do something about is the amount of biomass being removed by the fishing industry." Others, however, think that it's not the quantity of fish available to the seals but the quality—the "junk-food hypothesis," which suggests that the replacement of fatty fish like capelin and herring by lean fish like pollock means the pinnipeds, especially juvenile seals and sea lions, aren't able to consume enough calories to survive. Recent research has poked holes in that idea, but still others argue that the problem is environmental toxins, like heavy metals or organic chemicals, which have worked their way up the food chain to accumulate in the tissues of top predators (including Alaskan Natives).

One of the most intriguing explanations for the changes in the Bering Sea was proposed in 2003 by a team of marine biologists, who suggested the root cause was the slaughter of half a million great whales between 1949 and 1969 in the waters around the Aleutian Islands. This, they believe, set off an ecological cascade effect still rippling through the ecosystem. According to the theory, the orcas (killer whales) that once preyed on baleen and sperm whales were forced to find other prey; they turned first to harbor seals, which are slow and docile—and whose numbers began to crash in the late 1970s. With harbor seals depleted, the orcas turned to fur seals and Steller sea lions; as they declined, the whales began taking sea otters; as the sea otters disappeared, the number of sea urchins on which they once preyed exploded, endangering once-thick kelp forests that provide habitat for a host of other species.

Much of this is speculative, the paper's authors admit, but it does neatly tie up a lot of loose ends left dangling by other hypotheses. The evidence that orcas are eating their way through the previously common sea otters, for example, is circumstantial but persuasive; few people have seen this happening, but while otters have all but disappeared from some islands, they remain common in places like Clam Lagoon on Adak, where a bridge blocks orcas from entering. Nor, the scientists argue, would it take a lot of hungry killer whales to force such sweeping collapse of pinniped and otter populations; if only two or three dozen orcas switched to a diet of sea lions, that would theoretically be enough to explain the sharp fall in sea lion numbers.

Of the many competing explanations for the Bering Sea's changes, which one you favor may depend on who or what you are. Small-scale fishermen often blame the huge industrial trawlers and longline operations for taking too many fish, while the big corporations say there's plenty of fish to go around, but perhaps for sea lions it's the wrong kind, brought on by warmer waters—the junk-food hypothesis. Scientists warn that solid answers are probably years away and, given the complexity of this wonderfully fertile ecosystem, unlikely to be simple. Few would be surprised to learn that all the possible pressures, from climate change to toxins to overfishing, are playing a role in the ill health of the Bering Sea, making solutions that much more challenging.

Karin, Art, and I stood in the echoey dimness of the old seal plant and talked until it was twilight, chewing over the mystery and fragility of this northern world. We finally bade each other good night out on the dirt street, where a fox loped around a corner, paused to glance at us, then trotted away. I checked my watch, something I'd been doing infrequently in recent weeks, and found to my surprise that it was almost midnight. It looked as though St. George might have another sunset, but suddenly I was too tired to wait for it; the months of endless travel were catching up with me, and my bed beckoned.

But I couldn't do it; I hung on my heel for a moment, then instead of walking up the hill toward the hotel, I followed the fox down to the edge of the water, where the waves hissed in over the bouldery beach and murres growled from the low ledges. Seals rolled in the swells, the great seabird flocks poured in across the yellow sky, and I sat quietly, my mind at last empty of worries and hypotheses, washed free even of exhaustion—empty of all but wonder and a sense of privilege as I watched the sun disappear into a wild and wave-tossed ocean.

FIFTEEN

Homeward Bound

When you're a writer, there's always a temptation to tweak history a little bit, to make reality conform to a tidy narrative. As endings go, it would be hard to beat St. George, there at the farthest, wildest shore; I'd like to leave you by the high cliffs and breaking waves, the land bathed in yellow evening light, the birds wheeling, the seals barking.

But that's not what happened. Instead, late on a day when the good weather had finally dissolved into clouds and drizzle, leading me to hope that I might be marooned for a few extra days, a PenAir flight managed to squeeze through a hole in the fog and pick me up. It was nip and tuck to the very last moment, with white mist curling back over the ridge and falling toward us even as we taxied, but as I sat gripping a cold plastic bag (fifteen pounds of frozen reindeer steaks, a gift from Art that I was taking to his St. Paul crew), the plane lifted up into the clouds, which lasted almost until we were on the ground in St. Paul.

I was dogged by spitty rain on the ride into town, toward spidery harbor cranes I could see for miles, past wire crab pots racked up like see-through houses, piles of metal culvert pipe, and workmen pouring concrete for new construction. After my idyll on St. George, it was perhaps inevitable that St. Paul would suffer by comparison. Instead of a tiny village of a hundred people, where when I walked into the store or post office folks looked up, smiled, and said, "Oh, you're the bird guy I heard about," the town of St. Paul was a good deal larger, more commercial, not as openly friendly, with five times the population and constant truck and ATV traffic stirring up enormous clouds of dust from the unpaved roads that the rain had not yet settled.

The King Eider Hotel sat not far from the waterfront, and we tourists—there were maybe a dozen of us—took our meals of institutional food in the commissary of the cavernous Trident Seafood plant, where everything tasted vaguely of haddock, walking there past derelict boats and shabby industrial buildings of corroded metal. The whole town felt disheveled and down at the heel, and I had to remind myself that St. George, as pretty and quaint as it might have seemed in contrast, was on the verge of financial collapse. Quaint doesn't pay the bills.

The island of St. Paul itself was beautiful, though, once I was out of town—lower and gentler than St. George, the vegetation more heavily grazed by reindeer. In terms of volume, there are far fewer seabirds here, and the colonies are some miles from town, unlike the cliffs just a long stone's toss from the hotel in St. George. Maybe one of the Aleut kids from St. George, who had visited his neighboring island on a stewardship camp field trip, put it best: "In St. Paul, nature's just so far away from the village." Not entirely; I had to keep my ground-level window in the King Eider closed unless I was sitting at the desk, lest the Arctic foxes I saw sniffing around just outside jump in, grab my food bag, or foul my gear.

I spent a couple of days in the rain and fog, part of the time with my fellow tourists under the very capable wing of Rick Knight, a professional birding guide from Tennessee hired by the village corporation for the season. What St. Paul lacks in quantity of birdlife it more than makes up for in variety—if you're here at the right time, that is, which I wasn't. In spring, St. Paul is one of the premier birding hot spots in the world, thanks to dozens of species of Asian vagrants that wind up here during migration, priceless additions to a hard-core birder's heavily padded North American life list. But the migrants were gone for the summer, and even Rick admitted he was having a hard time working up much enthusiasm for the small seabird colonies he showed visitors.

The seals, however, were a saving grace. St. Paul has always had the Pribs' biggest fur seal colonies, and the island has done a fine job of making a few of the rookeries accessible to guided visitors. Near Reef Point, we stood in the driving rain behind wooden barricades that shielded us from seals barely arm's length away, especially one young bull that roared again and again at us in a frenzy, his rain-matted hair standing up like punk spikes on his head, white slobber hanging from his lower lip, his dark eyes glassy with testosterone.

My last morning, the rain and fog still thick, I met Art's two seabird biologists, young women named Sadie Wright and Alexis Will, to whom I'd delivered the reindeer meat a couple of days earlier. We hiked a few miles up the western side of the island, to cliffs several hundred feet high, where Sadie unfolded a spotting scope and began checking red-faced cormorant nests while I followed Alexis and helped her census black-legged kittiwakes, huddled on their grassy nests. It took patience; binoculars to our eyes, our fingers growing numb in the cold rain, we'd wait for a sleepy bird to rouse itself and stand, affording us a momentary chance to count the eggs or chicks hidden beneath. It took me an hour to work my way through just twelve nests, and Alexis had hundreds on which she was keeping tabs.

My wool gloves were soaked, and I could no longer feel my fingertips. The air was ripe with the smell of guano, and the kittiwake on nest B15 hadn't moved for twenty minutes, despite my psychic pleas that it stretch, preen, pass gas—do *something*. I'm sure this sounds dreadfully tedious, but in fact it was marvelously engrossing. For several hours I was immersed in the sounds and smells of the kittiwakes' lives, coming to sense the individual differences among them, caught up in the excitement of birth as we watched one newly hatched chick, still wet from the egg, struggling between its parent's legs. When it was time to hike back down the bluffs through the swirling mist, so I could gather my gear and catch the plane home, I didn't want to go; I was leaving before I knew the ending.

It is through such painstaking work, year after year, that biologists like Art Sowls hope to understand what is happening in the Bering Sea—by tracking the productivity of black-legged kittiwakes on St. Paul and the red-leggeds on St. George, tiny puzzle pieces to be tapped in alongside research into stress hormones in sea lions, pup counts on the fur seal rookeries, sample trawls for groundfish in the Golden Triangle, measurements of sea ice, assessments of springtime phytoplankton bloom, thousands of scientific fragments that will (if we are lucky, if we are wise) form a mosaic of information complete enough to guide us to a fuller understanding of this part of the world, and better ways to protect it.

All across North America, field scientists like Sadie and Alexis are doing this kind of critical work. But we're only scratching at what we need to know, in many respects blindly groping toward some sort of balance. More often than not, we still make decisions based on greed or expediency, and even when we do try to let knowledge and the long-term good guide our choices, we often make the wrong one anyway, from simple ignorance of the biosphere's surpassing complexity. But because we lack the luxury of time, we're forced to take our best shot, even in the absence of certainty.

When I set out on my own wild America sojourn, many of the friends and colleagues to whom I explained it gave me a pitying look and said, "Well, that'll be depressing." And there were times when I felt despair welling up, when I saw development run amok gnawing out the heart of once-lovely places or realized that the simple solitude that was once a birthright has become a scarce commodity. The challenges are enormous. There are too many of us to sustain the hugely consumptive lifestyle we lead, which taxes not only this continent but the rest of the world. The reckless folly that passes for land-use planning in America—meaning essentially no planning at all, except to facilitate the cramming together of as much redundant commercial and residential sprawl as possible—may be the biggest existing hurdle to the survival of anything truly wild in most of the country, polarizing the land into developed or protected, with nothing in between. New invasive species arrive almost daily to undermine the functional integrity of our ecosystems. What lies ahead may be even worse. Climate change, driven by our petrochemical addiction, poses the biggest unknown for natural systems in even the most remote parts of North America, and beyond.

That includes the Pribs. What I did not know, as I hiked back along the foggy cliffs, was that warmer-than-usual temperatures in the waters around St. Paul were already driving food too deep for the birds; the kittiwakes, murres, cormorants, and others would, in the weeks ahead, suffer a near-complete breeding failure, and the parakeet auklets would starve to death in large numbers. Such bad seasons are not unusual among colonial seabirds, and with their long life spans they have evolved a boom-or-bust strategy, recouping their losses when conditions improve. But what if conditions don't improve; what if they are a taste of something much larger, something more pervasive and perhaps permanent? Even as I was getting ready to leave the Pribilofs, reports were emerging of a catastrophic breeding collapse at seabird colonies in Great Britain, in the same island colonies where Fisher spent much of his professional life. After decades of persistent food shortages, this was the worst breeding season in living memory, with many species not even laying eggs; great skuas were so hungry they turned to cannibalism. Scientists blamed warming waters, which were causing a regime shift, much as happened in the Bering Sea, and the sand eels on which the birds feed have largely disappeared. I could only hope it was not a glimpse at the future of the Pribilofs.

So yes, at times the journey was a depressing one. And yet I found, to my surprise, a far more optimistic picture emerging from my travels than I ever expected. By giving me a fifty-year perspective, Peterson and Fisher's journey allowed me to see what we conservationists often overlook—that we've made enormous progress in the last five decades. In 1953 you could, with almost complete impunity, foul the air or waters, poison grizzlies, kill a jaguar, shoot hawks and owls, clear-cut millennia-old forests on public land, harpoon great whales for profit, traffic in rare species, spray any pesticide on the market (and market whatever you could concoct), or build dams to flood priceless canyons. There was no national wilderness system, no Wild and Scenic Rivers Act, no federal legislation protecting wetlands, waterways, national forests, or air quality, no Endangered Species or Marine Mammal Protection acts, no public groundswell to bring back wolves or save old growth, no requirement that the government at least try to assess the environmental impact of big projects before pouring millions from the public coffers into them.

And to be sure, some of that is still true today. We're still fighting unnecessary dams and redundant highways, still battling boondoggles that subsidize logging on federal forests, still arguing about how stringently to regulate pollution. But the very conversation has changed, in ways so profound and fundamental that few would have thought it possible in the 1950s. The idea of taking out a multimillion-dollar dam—for any reason, much less for the environment—was simply laughable; Americans built dams, and only fools fought what was clearly progress. Even a few years ago, I suspect, anyone predicting that the U.S. Army Corps of Engineers would spend hundreds of millions of dollars to put the oxbows back *into* the Kissimmee River would have been institutionalized.

I believe that the next fifty years will see a strengthening of that ethic, not its diminution. The difficulty remains getting the average person to focus less on his immediate desires and more on the greater good, but compared with where we were fifty or a hundred years ago, the differences couldn't be more stark. A century ago, we came within a whisker of losing much of our natural heritage—not just the marquee extinctions like passenger pigeons and Carolina parakeets, but species that we take for granted today, like white-tailed deer. The game was gone, the forests felled, the rivers laid waste. But America woke up, and we've learned from our mistakes. The past fifty years have been a time when we've stopped much of the hemorrhaging, but we shouldn't be content with saving only the pieces that remain. Instead, we must work toward rebuilding the vivid, vibrant whole.

We can, if we choose, bind up the tattered edges of the continent's wild mantle and start to make right some of the mistakes of the past. We've shown we're willing to gamble on restoring the Everglades and to embark on a long-term plan to replant the thornscrub forests of the lower Rio Grande. We could do more. In the East we could, if we chose, create reborn wilderness from the fragmented forests; with time, even a bare field can become an ancient forest. In the Midwest, prairie reclamation has moved from backyard plots to the open landscape as conservationists try their hand at tallgrass restoration across tens of thousands of acres. In parts of the Great Plains, where the human population is lower than it was a century ago, there is even talk of taking down the fences, moving out the cattle, and restoring the buffalo ecosystem. It's an idea that raises as many hackles as hopes—but we're talking about it, where once the reaction would have been derision.

We have, I think, a responsibility to stretch beyond what common sense tells us is possible, because if the past half century carries any lesson for conservationists, it's that we can do far more than we think we can. We will never have a pre-Columbian America, complete in all its too-much splendor, but such is the resiliency of wild America that mostly what we need is the courage to dream big and to set goals that are equal to this majestic land. The key is hope, because hope, when paired with the ferocious love Americans feel for their land, becomes action.

The barriers are great and the stakes are high, but the tide is finally running with us. That said, the very fact that we've made such progress also leaves me furious when I see how it is being undercut by an administration and a Congress that seem far more interested in rewarding their cronies than in safeguarding the country's natural heritage. We're seeing a fire-sale mentality at the federal level, a damn-the-environment attitude on a scale we haven't witnessed in generations. As an example, consider the recent proposal to open virtually the entire National Petroleum Reserve in Alaska for drilling, including such sensitive wildlife areas as Teshekpuk Lake, which in the 1980s Reagan's interior secretary James Watt deemed too precious to develop. When Jim Watt starts looking like a model of environmental moderation, you know we're in trouble.

We will need action and vigilance in the years to come, and wild America's defenders will have their work cut out for them. But the despoilers should not gloat, for history is against them. If you doubt that, just look back a few decades.

For one last time, I jammed myself into the little PenAir plane, which took off in the rain from St. Paul and carried me out into the blank Bering Sea, headed for home at last. At the end of his own long American trip, James Fisher said that never had he "seen such wonders, or met landlords so worthy of their land." That is heady praise, and I could only question whether he'd find us so deserving today.

I'd like to think the answer would still be yes.

Notes and Bibliography

PREFACE

CITATIONS

xiv "Wherever one goes": James Fisher in Roger Tory Peterson and James Fisher, *Wild America* (Boston: Houghton Mifflin, 1955), p. 13.

xvii "various never-before-seen objects": Ibid., p. 26.

xvii "There is another book to be written": Roger Tory Peterson, "Evolution of a Field Guide," *Defenders*, Oct. 1980, http://www.defenders.org/rtpeter4.html.

xx "Never have I seen such wonders": Peterson and Fisher, *Wild America*, p. 418.

1. ATLANTIC GATEWAY

CITATIONS

3 "seeking the differences": Roger Tory Peterson and James Fisher, *Wild America* (Boston: Houghton Mifflin, 1955), p. 11.

4 "a voice as familiar to me": Ibid., p. 14.

4 "The most spectacular New World [gannet] colony": Ibid., p. 17.

17 "So much had I seen of wild Europe": Ibid., p. 2.

17 "a more complete cross section": Ibid.

17 "Roger recently spent": Brooks to Fisher, Dec. 15, 1952, James Fisher Collection, Natural History Museum, London.

17 "The trip is only three months away": Peterson to Fisher, Jan. 30, 1953, Fisher Collection.

18 "I was gradually beginning to realize": Peterson and Fisher, *Wild America*, p. 250.

18 "If I had realized": Peterson to Fisher, Aug. 26, 1953, Fisher Collection.

19 "Sometimes in Britain": Peterson and Fisher, *Wild America*, p. 267.

19 "James' account of the Grand Canyon": Brooks to Peterson, Jan. 18, 1955, Fisher Collection.

20 "It occurred to me": Peterson and Fisher, *Wild America*, p. 15.

BIBLIOGRAPHY

Brubaker, Elizabeth. "Cod Don't Vote: How Politics Destroyed Atlantic Canada's Fisheries." *Next City* (Winter 1998–99). http://www.nextcity.com/contents/winter98-99/14fish.html.

Bundy, Alida. "Fishing on Ecosystems: The Interplay of Fishing and Predation in Newfoundland–Labrador." *Canadian Journal of Fisheries and Aquatic Sciences* 58, no. 6 (June 2001), pp. 1153–67.

Canadian Broadcasting Company. "The Cod Wars." *National*, 2003. http://www.tv.cbc.ca/national/pgminfo/fish/.

———. "To the Last Fish: The Codless Sea," 2004. http://stjohns.cbc.ca/features/CodFisheries/.

Cape St. Mary's Seabird Ecological Reserve Management Plan. St. John's, Nfld: Department of Tourism and Culture, 1994.

Devlin, John C., and Grace Naismith. *The World of Roger Tory Peterson*. New York: Times Books, 1977.

Fisheries and Oceans Canada, and Government of Newfoundland and Labrador. *Estimating the Value of the Marine, Coastal, and Ocean Resources of Newfoundland and Labrador*. St. John's, Nfld.: Government of Canada, Government of Newfoundland and Labrador, 2001.

Heckscher, Jurretta Jordan, ed. *Evolution of the Conservation Movement, 1850–1920*. Washington, D.C.: Library of Congress, 1996. http://memory.loc.gov/ammem/amrvhtml/conshome.html.

Kurlansky, Mark. *Cod*. New York: Walker, 1997.

Line, Les, ed. *The National Audubon Society: Speaking for Nature*. New York: National Audubon Society, 1999.

Meades, Susan J. "The Barrens." *Wildflower* 9, no. 1 (Winter 1993), pp. 32–35.

Montevecchi, William A., and Leslie M. Tuck. *Newfoundland Birds*. Cambridge, Mass.: Nuttall Ornithological Club, 1987.

Mowbray, Thomas B. *Northern Gannet* (Morus bassanus). The Birds of North America, ed. A. Poole and F. Gill, no. 693. Philadelphia: Birds of North America Inc., 2002.

Palmer, Ralph S., ed. *Handbook of North American Birds*. Vol. 1. New Haven, Conn.: Yale University Press, 1962.

Titford, Bill, and June Titford. *A Traveller's Guide to Wild Flowers of Newfoundland, Canada*. St. John's, Nfld: Flora Frames, 1995.

2. WOODS IN THE CITY

CITATIONS

28 "The Northeast can hardly be called *wild* America": Roger Tory Peterson and James Fisher, *Wild America* (Boston: Houghton Mifflin, 1955), p. 36

28 "The little pockets": Ibid., p. 47.

29 "one of the nation's most radical patterns": *Back to Prosperity: A Competitive Agenda for Renewing Pennsylvania* (Washington, D.C.: Brookings Institution, 2003), p. 9.

30 "When James and I drove south": Peterson and Fisher, *Wild America*, p. 47.

36 "a weather-beaten, barracky, amphibious structure": G. H. Ballou, "Monomoy," *Harper's*, Dec. 1863, p. 307.

36 "and doubtless even the less favored portions": Ibid., p. 306.

39 "The gulls took to the air": Bruce Babbitt, "ADR Concepts: Reshaping the Way Natural Resource Decisions Are Made," in *Into the 21st Century: Thought Pieces on Lawyering, Problem Solving, and ADR* (New York: CPR Institute for Dispute Resolution, 2001), p. 13.

40 "because it was dull": Roger Tory Peterson, undated notes, James Fisher Collection, Natural History Museum, London.

40–41 "in strong, distilled form": Peterson and Fisher, *Wild America*, p. 42.

47 "Another party": James Fisher, unpublished journal, April 26, 1953, Fisher Collection.

47 "a very satisfying city for a naturalist . . . wild country almost at the city's door": Peterson and Fisher, *Wild America*, p. 57.

48 "a refuge": William O. Douglas, quoted in Linda Rancort, ed., *Chesapeake and Ohio Canal National Historic Park: A Resource Assessment* (Fort Collins, Colo.: National Parks Conservation Association, 2004), p. 6.

50 "this park which has served as almost a place of worship": Peterson and Fisher, *Wild America*, p. 55.

BIBLIOGRAPHY

Achenbach, Joel. "The River View? Frequently Gorges." *Washington Post*, May 9, 2004, p. D1.

Ambler, Allan, et al. "Cover Me!" *Spanning the Gap* 25, no. 1 (Spring 2003). http://www.nps.gov/dewa/InDepth/Spanning/stgAMPHI.html.

Beach, Dana. *Coastal Sprawl: The Effects of Urban Design on Aquatic Ecosystems in the United States*. Arlington, Va.: Pew Oceans Commission, 2002.

Brown, Patricia Leigh. "Defining Sprawl: From A to Z." *New York Times*, June 17, 2004.

Dean, Cornelia. "Policy to Preserve Coastline Runs into Reality on Nation's Beaches." *New York Times*, Oct. 26, 2004.

Elvidge, Christopher D. "U.S. Constructed Area Approaches the Size of Ohio." *Eos* 85, no. 24 (June 15, 2004), pp. 233–40.

Gochfeld, Michael, Joanna Burger, and Ian C.T. Nisbet. *Roseate Tern* (Sterna dougallii). The Birds of North America, ed. A. Poole and F. Gill, no. 370. Philadelphia: Academy of Natural Sciences; Washington, D.C.: American Ornithologists' Union, 1998.

Good, Thomas P. *Great Black-Backed Gull* (Larus marinus). The Birds of North America, ed. A. Poole and F. Gill, no. 330. Philadelphia: Academy of Natural Sciences; Washington, D.C.: American Ornithologists' Union, 1998.

Goodrich, Laurie, Margaret Brittingham, Joseph Bishop, and Patricia Barber. *Wildlife Habitat in Pennsylvania: Past, Present, and Future*. Harrisburg: Pennsylvania Department of Conservation of Natural Resources, 2002.

Gray Seal (Halichoerus grypus): *Western North Atlantic Stock*. Woods Hole, Mass.: Northeast Fisheries Science Center, 2001.

Hayden, Dolores. *Field Guide to Sprawl*. New York: W. W. Norton, 2004.

Katona, Steven K., Valerie Rough, and David T. Richardson. *Field Guide to the Whales, Porpoises, and Seals of the Gulf of Maine and Eastern Canada.* New York: Charles Scribner's Sons, 1983.

Macintosh, Barry. *Rock Creek Park: An Administrative History.* Washington, D.C.: National Park Service, 1985. http://www.nps.gov/rocr/cultural/history/adhi4b.htm.

McAdams, E. J. "Wilderness on 68th Street." *Topic,* no. 3 (Winter 2003), pp. 141–46.

McElhenny, John. "'Mansionization' Tied to Loss of Open Space." *Boston Globe,* Nov. 10, 2003.

McKibben, Bill. "Serious Wind." *Orion* 22 (July/Aug. 2003), pp. 14–15.

National Park Service. "Antiquities Act of 1906." http://www.cr.nps.gov/history/hisnps/NPSHistory/antiq.htm.

Nisbet, Ian C.T. *Common Tern* (Sterna hirundo). The Birds of North America, ed. A. Poole and F. Gill, no. 618. Philadelphia: Academy of Natural Sciences; Washington, D.C.: American Ornithologists' Union, 2002.

Perkins, Simon, et al. *A Survey of Tern Activity Within Nantucket Sound, Massachusetts, During the 2003 Breeding Season.* Lincoln: Massachusetts Audubon Society, 2004.

Pierotti, R. J., and Thomas P. Good. *Herring Gull* (Larus argentatus). The Birds of North America, ed. A. Poole and F. Gill, no. 124. Philadelphia: Academy of Natural Sciences; Washington, D.C.: American Ornithologists' Union, 1994.

Py-Lieberman, Beth. "Wise Old Owl." *Smithsonian,* Aug. 2003. http://www.smithsonianmag.si.edu/smithsonian/issues03/aug03/mall.html#one.

Salvesen, David, and David R. Godschalk. *Development on Coastal Barriers: Does the Coastal Barrier Resources Act Make a Difference?* Washington, D.C.: Coast Alliance, 1998.

Tingley, Morgan. "Wind, Bay, and Birds." *Sanctuary* 43 (Autumn 2003), pp. 6–7.

Trivedi, Bijal P. "Bald Eagle's Manhattan Return Turns Turbulent." *National Geographic Today,* Aug. 28, 2002. http://news.nationalgeographic.com/news/2002/08/0828_020828_TVeagle.htm.

Veit, Richard R., and Wayne R. Petersen. *Birds of Massachusetts.* Lincoln: Massachusetts Audubon Society, 1993.

Wheeler, Linda. "Beaver Chomps into Cherry Blossom Season." *Washington Post,* April 7, 1999, p. B1.

3. FORESTS OF LOSS AND RESILIENCY

CITATIONS

54 "Our journey along its serpentine length": Roger Tory Peterson and James Fisher, *Wild America* (Boston: Houghton Mifflin, 1955), p. 60.

55 "If the East is to have wilderness": Ibid., p. 61.

58 "No greater catastrophe": Ibid., p. 63.

70 "reached heights of field craft": Unpublished excerpt from *Wild America* manuscript, James Fisher Collection, Natural History Museum, London.

71 "Even though the cutover forests": Peterson and Fisher, *Wild America,* p. 61.

BIBLIOGRAPHY

Bair, Mary Willeford. *Eastern Hemlock (*Tsuga canadensis*) Mortality in Shenandoah National Park*. Luray, Va.: Shenandoah National Park, 2002.

Braasch, Gary. "All Things Great and Even Microscopic." *Audubon* 102 (May/June 2000), pp. 54–59.

Cate, Matthew S.L. "Federal Standards Show Smokies as Unhealthy Area." *Chattanooga Times Free Press*, April 15, 2004.

Herring, Hal. "Back Home on the Range." *Orion Afield* (Spring 2002), pp. 20–23.

Hopey, Don. "Virulent Oak Fungus Could Quickly Imperil Pa. Forests." *Pittsburgh Post-Gazette*, April 11, 2004.

Horton, Tom. "The New Old Growth." *Orion*, Nov./Dec. 2003, pp. 18–25.

Houk, Rose. *Great Smoky Mountains National Park*. Boston: Houghton Mifflin, 1993.

Johnson, Kristine D. "IPM—How It Works in the Smokies." In *Exotic Pests of Eastern Forests, Conference Proceedings, April 8–10, 1997*, ed. Kerry O. Britton. Nashville: USDA Forest Service and Tennessee Exotic Pest Plant Council, 1998.

Kloepfer, Deanne. *Shenandoah National Park: A Resource Assessment*. Fort Collins, Colo.: National Parks Conservation Association, 2003.

McClure, Mark S., and Carole S.-J. Cheah. "Establishing *Psuedoscymnus tsugae* for Biological Control of Hemlock Woolly Adelgid, *Adelges tsugae* in the Eastern United States." In *1st International Symposium on Biological Control of Arthropods*. Washington, D.C.: U.S. Forest Service, 2003.

Rancort, Linda, ed. *Great Smoky Mountains National Park: A Resource Assessment*. Fort Collins, Colo.: National Parks Conservation Association, 2004.

"Rangers Launch Attack to Save Hemlock Trees." *Smokies Guide*. Great Smoky Mountains National Park, Spring 2004, p. 1.

Ray, Janisse. "On the Bosom of This Grave and Wasted Land I Will Lay My Head." *Orion* 21 (Summer 2002), pp. 107–15.

Roberson, Mary-Russell. "The New Elk on the Block." *ZooGoer* 32 (May/June 2003). http://natzoo.si.edu/Publications/ZooGoer/2003/3/Elk.cfm.

Silver, Timothy. *Mount Mitchell and the Black Mountains*. Chapel Hill: University of North Carolina Press, 2003.

Simmons, Morgan. "Michigan Girl Scouts Pitch In to Aid Smokies," *Knoxville News Sentinel*, Nov. 24, 2003.

Sudden Oak Death Syndrome: Protecting America's Woodlands from Phytophthora ramorum. FS-794. Arlington, Va.: U.S. Forest Service, Aug. 2004.

Sullivan, Rose. "New Threats to Mountain State Trees Emerge." *WV Wildlife Diversity News* 19 (Winter 2003), pp. 1–2.

Tourtellot, Jonathan B. "Conservation vs. Development: A Tale of Two Parks," *National Geographic Traveler*, Sept. 26, 2003. http://news.nationalgeographic.com/news/2003/09/0926_030926_travelerbanff.html.

Weidensaul, Scott. "The Return of the Elk." *Smithsonian* 30 (Dec. 1999), pp. 82–94.

Wilson, Paul J., et al. "DNA Profiles of the Eastern Canadian Wolf and the Red Wolf Provide Evidence for a Common Evolutionary History Independent of the Gray Wolf." *Canadian Journal of Zoology* 78 (2000), pp. 2156–66.

Young, John, et al. *Modeling Stand Vulnerability and Biological Impacts of the Hemlock Wooly Adelgid*. Kearneysville, W.Va.: U.S. Geological Survey, 2004.

4. REPLUMBING FLORIDA

CITATIONS

81 "Many tourists see only these show places": Roger Tory Peterson and James Fisher, *Wild America* (Boston: Houghton Mifflin, 1955), p. 101.
83 "The southern Florida wilderness scenery": Daniel B. Beard, *Wildlife Reconnaissance: Everglades National Park Project* (Washington, D.C.: National Park Service, 1938), p. 100.
83 "It is true that in many other parts of the world": Peterson and Fisher, *Wild America*, p. 124.
86 "It was as if humanity": Alexander Sprunt, Jr., quoted in Ted Levin, *Liquid Land* (Athens: University of Georgia Press, 2003), p. 171.
87 "I had spent a day slapping mosquitoes": Peterson and Fisher, *Wild America*, p. 128.
87 "like an overloaded Christmas tree": Ibid.
92 "The cattle egret, beautiful and beneficial": Ibid., p. 113.
92–93 "But somehow I would not expect": Ibid., p. 112.
96 "clearly one of the biggest threats": Francois B. LaRoche, ed., *Melaleuca Management Plan*, 3rd ed. (Florida Exotic Pest Plant Council, 1999), p. 1.
98 "approximately 1,800 miles": South Florida Water Management District, "Agency Overview," Nov. 15, 2004. http://www.sfwmd.gov/site/index.php?id=61.

BIBLIOGRAPHY

Derr, Mark. *Some Kind of Paradise*. Gainesville: University of Florida Press, 1998.
Dougherty, Ryan. "Invasive Fern Smothers Plants." *National Parks*, May/June 2003, pp. 12–13.
Douglas, Marjory Stoneman. *The Everglades: River of Grass*. Covington, Ga.: Mockingbird Books, 1947.
Fisher, James. *The Shell Bird Book*. London: Ebury Press, 1966.
"Frequently Asked Questions About the Asian Swamp Eel." U.S. Geological Survey. http://cars.er.usgs.gov/Nonindigenous_Species/Swamp_eel_FAQs/swamp_eel_faqs.html.
Goodnough, Abby. "Forget the Gators: Exotic Pets Run Wild in Florida." *New York Times*, Feb. 29, 2004.
———. "On a Silent Landscape, an Environmental War Endures." *New York Times*, Nov. 4, 2003.
Grunwald, Michael. "Kissimmee River's Strange Twists." *Washington Post*, June 26, 2002.
Kaufman, Kenn. "New Bird on the Block." *Audubon* 101, no. 5 (Sept./Oct. 1999), pp. 124–27.
McAlly, David. *The Everglades: An Environmental History*. Gainesville: University of Florida Press, 1999.

Morgan, Curtis. "Invasion of the Everglades: Giant Snakes Have a New Hangout." *Miami Herald*, Dec. 22, 2002.

Nordeen, Deborah. "South Florida's Watery Wilderness Park Nears 50." Everglades National Park, Jan. 6, 1999. http://www.nps.gov/ever/eco/nordeen.htm.

Semenza, Jan C., Paige E. Tolbert, Carol H. Rubin, Louis J. Guillette, Jr., and Richard J. Jackson. "Reproductive Toxins and Alligator Abnormalities at Lake Apopka, Florida." *Environmental Health Perspectives* 105, no. 10 (Oct. 1997). http://ehp.niehs.nih.gov/members/1997/105-10/semenza-full.html.

5. THE SOUTH'S WILD SOUL

CITATIONS

104 "We drifted, motors cut off": Roger Tory Peterson and James Fisher, *Wild America* (Boston: Houghton Mifflin, 1955), p. 154.

104 "No sound": Ibid.

105 "suited only for the most adventurous of naturalists": Gil Nelson, *Exploring Wild Northwest Florida* (Sarasota, Fla.: Pineapple Press, 1995), p. 69.

108 "one of the least settled parts": Peterson and Fisher, *Wild America*, p. 152.

109 "air as refreshing as a mint julep": Ibid., p. 134.

109 "a charming inland town": Ibid., p. 98.

109 "Was there ever such conspicuous consumption": Ibid., p. 132.

BIBLIOGRAPHY

Bell, C. Ritchie, and Bryan J. Taylor. *Florida Wild Flowers and Roadside Plants*. Chapel Hill, N.C.: Laurel Hill Press, 1982.

Buchheister, Carl W. "The Acquisition and Development of the Corkscrew Swamp Sanctuary, 1952–1967." http://www.audubon.org/local/sanctuary/corkscrew/Information/Buchheister.html

Coulter, M. C., J. A. Rodgers, J. C. Ogden, and F. C. Depkin. *Wood Stork* (Mycteria americana). The Birds of North America, ed. A. Poole and F. Gill, no. 409. Philadelphia: Academy of Natural Sciences; Washington, D.C.: American Ornithologists' Union, 1999.

"Current Specimen List for *Guzmania monostachia*." Missouri Botanical Garden, Feb. 25, 2004. http://mobot.mobot.org/cgi-bin/search_vast.

"Forests." Big Cypress National Preserve, National Park Service. http://www.nps.gov/bicy/pphtml/subnaturalfeatures32.html.

Frank, Howard. "Florida's Native Bromeliads Imperilled by Exotic Evil Weevil." *Palmetto* 19, no. 4 (Winter 1999–2000). http://fcbs.org/articles/weevil-frank.htm.

Jackson, Jerome A. *In Search of the Ivory-Billed Woodpecker*. Washington, D.C.: Smithsonian Institution, 2004.

———. *Ivory-Billed Woodpecker* (Campephilus principalis). The Birds of North America, ed. A. Poole and F. Gill, no. 711. Philadelphia: Academy of Natural Sciences; Washington, D.C.: American Ornithologists' Union, 2002.

Jewell, Susan D. *Exploring Wild South Florida*. Sarasota, Fla.: Pineapple Press, 2002.

Orlean, Susan. *The Orchid Thief*. New York: Random House, 1998.

Sprunt, Alexander, Jr. "Emerald Kingdom." *Audubon*, Jan. 1961, pp. 24–40.

Tanner, James T. *The Ivory-Billed Woodpecker*. New York: National Audubon Society, 1942.

"Waters of the Big Cypress Swamp." Big Cypress National Preserve, National Park Service. http://www.nps.gov/bicy/Brochure.pdf.

White, Peter S., et al. "Regional Trends of Biological Resources: Southeast." In *Status and Trends of the Nation's Biological Resources*. Vol. 1, ed. Michael J. Mac et al. Washington, D.C.: U.S. Geological Survey, 1998.

Williams, John G., and Andrew E. Williams. *Field Guide to Orchids of North America*. New York: Universe Books, 1983.

6. LAS TORTUGAS

CITATIONS

127 "With the name of Wreckers": Maria R. Audubon, ed., *Audubon and His Journals*, vol. 2 (New York: Charles Scribner's Sons, 1897), p. 345.

127 "On landing, I felt for a moment": John James Audubon, *Ornithological Biographies*, http://www.audubon.org/bird/BoA/BOA_index.html.

128 "Don't try it unless you know": Robert R. Budlong, letter, Oct. 30, 1946, quoted in *City on the Sea: A Collection of Dry Tortugas Personal Histories*, CD-ROM (Key West, Fla.: Dry Tortugas National Park, 2003).

128 "the number one ornithological spectacle of the continent": Roger Tory Peterson, *Birds Over America* (New York: Dodd, Mead & Co., 1948), p. 312.

130 "not to run down": Roger Tory Peterson and James Fisher, *Wild America* (Boston: Houghton Mifflin, 1955), p. 135.

137 "The multispecies reef fisheries": *Final General Management Plan Amendment Environmental Impact Statement, Dry Tortugas National Park* (Washington, D.C.: Department of the Interior, Dec. 2000), p. 143.

143 "in June the gulls always came in thousands": Emily Holder, "At the Dry Tortugas During the War: A Lady's Journal," *Californian Illustrated* (1892), repr. in *City on the Sea*.

143 "Oh! how I wished you could have been there": Sgt. Calvin Shedd, letter, April 19, 1862, Calvin Shedd Papers collection, University of Miami, http://www.library.miami.edu/archives/shedd/62apr19.htm.

BIBLIOGRAPHY

Dry Tortugas National Park: Strategic Plan 2001–2005. Homestead, Fla.: Dry Tortugas National Park, 2000.

Final General Management Plan Amendment/Environmental Impact Statement, Dry Tortugas National Park. Washington, D.C.: National Park Service, 2000.

Landrum, Wayne. *Fort Jefferson and the Dry Tortugas National Park*. Big Pine Key, Fla.: Wayne Landrum, 2003.

National Oceanic and Atmospheric Administration. *Florida Keys National Marine Sanctuary: Tortugas 2000 Ecological Reserve*, http://www.fknms.nos.noaa.gov/tortugas/.

Schreiber, E. A., et al. *Sooty Tern* (Sterna fuscata). The Birds of North America, ed. A. Poole and F. Gill, no. 665. Philadelphia: Academy of Natural Sciences; Washington, D.C.: American Ornithologists' Union, 2002.

South Florida Regional Planning Council. *Florida Keys Carrying Capacity Study*, Sept. 14, 2004. http://www.sfrpc.com/gis/fkccs.htm.

2000 Annual Report: Dry Tortugas National Park. Homestead, Fla.: Dry Tortugas National Park, 2001.

Willits, Stacy. "Spring Breakers Tax Cops." *Key West Citizen*, March 15, 2004, p. 1.

7. WILDERNESS LOST (AND FOUND)

CITATIONS

147 "a country of recent, rapid, far-reaching changes": Roger Tory Peterson and James Fisher, *Wild America* (Boston: Houghton Mifflin, 1955), p. 178.

157 "I was so overwhelmed": Ibid., p. 181.

BIBLIOGRAPHY

Chapman, Duane C., Diana M. Papoulias, and Chris P. Onuf. "Environmental Change in South Texas." In *Status and Trends of the Nation's Biological Resources*. Vol. 1, ed. Michael J. Mac et al. Washington, D.C.: U.S. Geological Survey, 1998.

Conover, Adele. "Not a Lot of Ocelots." *Smithsonian*, June 2002, pp. 64–69.

Davis, William B., and David J. Schmidly. *The Mammals of Texas*. Austin: Texas Parks and Wildlife Department, 1994.

Gehlbach, Frederick R. *Mountain Islands and Desert Seas*. College Station: Texas A&M University Press, 1981.

Jahrsdoerfer, Sonja E., and David M. Leslie. *Tamaulipan Brushland of the Lower Rio Grande Valley of South Texas: Description, Human Impacts, and Management Options*. Biological Report 88 (36). Albuquerque, N.M.: U.S. Fish and Wildlife Service, 1988.

Laack, Linda L. "Ecology of the Ocelot (*Felis pardalis*) in South Texas." Master's thesis, Texas A&I University, 1991.

———. "Observing the Elusive Ocelot." *Rio Grande Valley Nature* (Spring 2002), pp. 6–8.

Laguna Atascosa National Wildlife Refuge: Proposed Refuge Expansion Plan. Albuquerque, N.M.: U.S. Fish and Wildlife Service, 1999.

Lockwood, Mark, et al. *A Birder's Guide to the Rio Grande Valley*. Colorado Springs, Colo.: American Birding Association, 1999.

Williams, Wendy. "The Ghost Cat's Ninth Life." *Audubon* 102, no. 4 (July–Aug. 2002), pp. 70–77.

8. THE REAL TREASURE OF THE SIERRA MADRE

CITATIONS

176 "the finest bit of jungle": Roger Tory Peterson and James Fisher, *Wild America* (Boston: Houghton Mifflin, 1955), p. 196.

177 "a primitive village": Ibid., p. 198.

181 "We were becoming sated": Ibid., p. 200.

BIBLIOGRAPHY

BirdLife International. *Threatened Birds of the World*. Barcelona: Lynx Edicions; Cambridge, U.K.: BirdLife International, 2000.

Emmons, Louise H. *Neotropical Rainforest Mammals*. Chicago: University of Chicago Press, 1990.

Howell, Steve N.G. *A Bird-Finding Guide to Mexico*. Ithaca, N.Y.: Comstock Press, 1999.

Howell, Steve N.G., and Sophie Webb. *A Guide to the Birds of Mexico and Northern Central America*. Oxford: Oxford University Press, 1995.

Knopf, Fritz L. *Mountain Plover* (Charadrius montanus). The Birds of North America, ed. A. Poole and F. Gill, no. 211. Philadelphia: Academy of Natural Sciences; Washington, D.C.: American Ornithologists' Union, 1996.

9. THE SKY ISLANDS

CITATIONS

194 "They are as much a true archipelago": Roger Tory Peterson and James Fisher, *Wild America* (Boston: Houghton Mifflin, 1955), p. 221.

196 "is so unlikely": David E. Brown and Carlos A. López González, *Borderland Jaguars* (Salt Lake City: University of Utah Press, 2001), p. 141.

196 "Should this population disappear": Ibid.

201 "an intangible and spiritual resource": Wallace Stegner, *The Sound of Mountain Water* (Garden City, N.Y.: Doubleday, 1969), p. 153.

205 "The big blue Chiricahuas": Peterson and Fisher, *Wild America*, p. 220.

206 "the most beautiful warbler": Ibid., p. 223.

210 "Well, have it your way": Ibid., p. 241.

213 "The world ended": Ibid., p. 258.

214 "a perfect park": Ibid., p. 266.

214 "satisfied, refreshed, emotionally stirred": Ibid.

214 "the great American public": Ibid., p. 267.

BIBLIOGRAPHY

Ames, Norma, et al. *Mexican Wolf Recovery Plan*. Albuquerque, N.M.: U.S. Fish and Wildlife Service, 1982.

Barber, D. A. "To the Bat Cave." *Tucson Weekly*, Oct. 2, 2003.

Bass, Rick. *The New Wolves*. New York: Lyons Press, 1998.

Bennett, Peter S., R. Roy Johnson, and Michael R. Kunzmann. *An Annotated List of Vascular Plants of the Chiricahua Mountains*. Tucson: U.S. Geological Survey and University of Arizona, 1996.

Brown, David E., ed. *Biotic Communities: Southwestern United States and Northwestern Mexico*. Salt Lake City: University of Utah Press, 1994.

Brown, Jerram L. *Mexican Jay* (Aphelocoma ultramarina). The Birds of North America, ed. A. Poole and F. Gill, no. 118. Philadelphia: Academy of Natural Sciences; Washington, D.C.: American Ornithologists' Union, 1994.

Buecher, Robert H. "Microclimate Study of Kartchner Caverns, Arizona." *Journal of Cave and Karst Studies* 61, no. 2 (Aug. 1999), pp. 108–20.

Epple, Anne Orth. *A Field Guide to the Plants of Arizona*. Guilford, Conn.: Globe Pequot Press, 1995.

Grahame, John D., and Thomas D. Sisk, eds. "Canyons, Cultures, and Environmental Change: An Introduction to the Land-Use History of the Colorado Plateau," Sept. 6, 2004. http://www.cpluhna.nau.edu/.

Grand Canyon National Park. "Program Goals for the Overflights and Natural Soundscape Program." http://www.nps.gov/grca/overflights/goals/programgoals.htm.

Hatten, J. R., A. Averill-Murray, and W. E. Van Pelt. *Characterizing and Mapping Potential Jaguar Habitat in Arizona.* Nongame and Endangered Wildlife Program Technical Report 203. Phoenix: Arizona Game and Fish Department, 2002.

Henle, Mark. "Lack of Funding Hurts Grand Canyon Experience." *Arizona Republic,* May 19, 2003.

Hopi Tribe home page, http://www.hopi.nsn.us/default.asp.

Johnson, T. B., and W. E. Van Pelt. *Conservation Assessment and Strategy for the Jaguar in Arizona and New Mexico*. Nongame and Endangered Wildlife Program Technical Report 105. Phoenix: Arizona Game and Fish Department, 1997.

Kammer, Jerry. "Sacred Land, Bitter Battle." *Arizona Republic,* Jan. 30, 2000. http://www.azcentral.com/news/navahopi/navahopi1.shtml.

Kaufman, Kenn. *Kingbird Highway*. Boston: Houghton Mifflin, 1997.

Knickerbocker, Brad. "National Parks Fast Falling into Disrepair." *Christian Science Monitor,* May 25, 2004. http://www.csmonitor.com/2004/0525/p01s02-usgn.html.

Koenig, W. A., P. B. Stacey, M. T. Stanback, and R. L. Mumme. *Acorn Woodpecker* (Melanerpes formicivorus). The Birds of North America, ed. A. Poole and F. Gill, no. 194. Philadelphia: Academy of Natural Sciences; Washington, D.C.: American Ornithologists' Union, 1995.

Muldavin, E. H., et al. "Regional Trends of Biological Resources: Southwest." In *Status and Trends of the Nation's Biological Resources*. Vol. 2, ed. Michael J. Mac et al. Washington, D.C.: U.S. Geological Survey, 1998.

National Parks Conservation Association. *The Bush Administration and America's National Park System: A Two-Year Analysis and Rating*. Washington, D.C., June 2003.

Peterson, Roger Tory. *Birds Over America*. New York: Dodd, Mead & Co., 1948.

Schmidt, Jeremy. *Grand Canyon*. Boston: Houghton Mifflin, 1993.

Tufts, Randy, and Gary Tenen. "Discovery and History of Kartchner Caverns, Arizona." *Journal of Cave and Karst Studies* 61, no. 2 (Aug. 1999), pp. 44–48.

Warshall, Peter. "Southwestern Sky Island Ecosystems." In *Our Living Resources*, ed. E. T. LaRoe, G. S. Farris, C. E. Puckett, P. D. Doran, and M. J. Mac. Washington, D.C.: Department of the Interior, 1995. http://biology.usgs.gov/s+t/noframe/r119.htm.

10. THE GOLDEN COAST

CITATIONS

222 "pumps seesawing and clanking": Roger Tory Peterson and James Fisher, *Wild America* (Boston: Houghton Mifflin, 1955), p. 294.

222 "I had expected ostentation and vulgarity": Ibid., p. 295.

236 "the species was considered doomed": A. W. Anthony, 1924, quoted in B. S. Stewart et al., "History and Present Status of the Northern Elephant Seal Population," in *Elephant Seals*, ed. B. J. LeBoeuf and R. M. Laws (Berkeley: University of California Press, 1994), pp. 31–32.

239 "Had you reported": Howard Granville Sharpe, "The Discovery of the 'Extinct' Sea Otter," http://www.seaotters.org/Otters/index.cfm?DocID=8.

BIBLIOGRAPHY

Anderson, C. G., J. L. Gittleman, K. Koepfli, and R. K. Wayne. "Sea Otter Systematics and Conservation: Which Are Critical Subspecies?" *Endangered Species Update* 13, no. 12 (Dec. 1996), pp. 6–10.

"Condor Population History." Ventura, Calif.: U.S. Fish and Wildlife Service, Jan. 12, 2004. http://hoppermountain.fws.gov/cacondor/recovery.html.

DeMaster, Douglas P., Catherine Marzin, and Ron J. Jameson. "Estimating the Historical Abundance of Sea Otters in California." *Endangered Species Update* 13, no. 12 (Dec. 1996), pp. 79–81.

"Detailed Information on California Condors Released in Arizona." Boise, Idaho: Peregrine Fund. http://www.peregrinefund.org/released_condorsinfo.asp.

Estes, J. A., B. B. Hatfield, K. Ralls, and J. Ames. "Causes of Mortality in California Sea Otters During Periods of Population Growth and Decline." *Marine Mammal Science* 19, no. 1 (2003), pp. 198–216.

Goetz, Peggy. "Parrots Are His Passion." *Irvine World News*, Jan. 9, 2003.

Graham, Frank, Jr. "The Day of the Condor." *Audubon* 102, no. 1 (Jan./Feb. 2000), pp. 46–53.

Hamber, Jan, and Bronwyn Davey. "AC8, AC9, and the Last Days of Wild California Condors." Hopper Mountain National Wildlife Refuge. http://hoppermountain.fws.gov/cacondor/AC8&AC9.html.

Knickerbocker, Brad. "Military Gets Break from Environmental Rules." *Christian Science Monitor*, Nov. 24, 2003.

Koford, Carl B. *The California Condor*. National Audubon Society Research Report, no. 4. New York: National Audubon Society, 1953.

Marine Mammal Protection Act of 1972. Washington, D.C.: National Oceanic and Atmospheric Administration. http://www.nmfs.noaa.gov/prot_res/laws/MMPA/MMPA.html.

Miller, M. A., I. A. Gardner, C. Kreuder, D. M. Paradies, et al. "Coastal Freshwater Runoff Is a Risk Factor for *Toxoplasma gondii* Infection of Southern Sea Otters (*Enhydra lutris nereis*)." *International Journal for Parasitology* 32, no. 8 (2002), pp. 997–1006.

Perkins, Sid. ". . . And the Big Bird That Didn't." *Science News* 166, no. 21 (Nov. 20, 2004), p. 334.

Pyle, Peter. "The White Shark at Southeast Farallon Island." *PRBO Observer* (1992). http://www.prbo.org/cms/index.php?mid=172.

Pyle, Peter, Scot Anderson, and Adam Brown. *White Shark Research at Southeast Farallon Island, 2002*. Stinson Beach, Calif.: Point Reyes Bird Observatory, Feb. 4, 2003.

Ritter, John. "Big Cats Seek Place to Prowl in Urban Areas." *USA Today*, June 14, 2004, p. 13A.

Snyder, Noel F.R., and N. John Schmitt. *California Condor* (Gymnogyps californianus). The Birds of North America, ed. A. Poole and F. Gill, no. 610. Philadelphia: Academy of Natural Sciences; Washington, D.C.: American Ornithologists' Union, 2002.

Snyder, Noel F.R., and Helen Snyder. *Birds of Prey: Natural History and Conservation of North American Raptors*. Stillwater, Minn.: Voyageur Press, 1991.

VanBlaricom, Glenn R. "Saving the Sea Otter Population in California: Contemporary Problems and Future Pitfalls." *Endangered Species Update* 13, no. 12 (Dec. 1996), pp. 85–91.

11. FIRE AND WATER

CITATIONS

248 "A pack of white pelicans": Roger Tory Peterson and James Fisher, *Wild America* (Boston: Houghton Mifflin, 1955), p. 334.

249 "bodies low in the water": Ibid.

252 "unappetizing garbage recyclers": David Klinghoffer, "What Suckers!" *National Review Online*, Sept. 10, 2001, http://www.nationalreview.com/comment/comment-klinghoffer091001.shtml.

254 "The fundamental fact of life": Bill Kier, quoted in John Driscoll, "Feds, States to Work Together on Klamath Conundrum," *Eureka Times-Standard*, Oct. 14, 2004.

255 "528 very good fish": Meriwether Lewis, quoted in *The Journals of Lewis and Clark*, ed. Bernard DeVoto (Boston: Houghton Mifflin, 1953), p. 220.

257 "It's a plot": Peterson and Fisher, *Wild America*, p. 343.

258 "Swinging around the wide curve": Ibid., p. 351.

258 "healthy green growth": Ibid., p. 352.

BIBLIOGRAPHY

Bailey, Eric. "U.S., States Vow to Fix River Use." *Los Angeles Times*, Oct. 13, 2004.

Barker, Rocky. "Drought Threatens Salmon as Lemhi River Reaches 50-Year Low." *Idaho Statesman*, April 20, 2004.

Barnard, Jeff. "Wildlife Refuge Asks Farmers for Water for Ducks." Associated Press, Aug. 27, 2002.

Barringer, Felicity. "Bush Seeks Shift in Logging Rules." *New York Times*, July 13, 2004.

Biscuit Fire Recovery Project Final Environmental Impact Statement. Medford, Oreg.: U.S. Forest Service and U.S. Bureau of Land Management, June 4, 2004. http://www.biscuitfire.com.

"Cheesman Conservation Area (320,000 Acres)—Southern Rocky Mountains Ecoregion." The Nature Conservancy, 2002. http://www.nature.org/initiatives/fire/files/cheeseman.pdf.

Cooperman, Michael S., and Douglas F. Markle. "The Endangered Species Act and the National Research Council's Interim Judgment in Klamath Basin." *Fisheries* 28, no. 3 (March 2003), pp. 10–19.

"C'waam and Qapdo." Klamath Tribes. http://www.klamathtribes.org/suckers.htm.

Daly, Matthew. "States, Federal Government Agree on Klamath Basin Plan." Associated Press, Oct. 14, 2004.

Duncan, Sally. "When the Forest Burns: Making Sense of Fire History West of the Cascades." *Pacific Northwest Research Station Science Findings* 46 (Sept. 2002), pp. 1–5.

Fattig, Paul. "Biscuit Fire Salvage Numbers Drop." *Jackson County (Oreg.) Mail Tribune*, June 2, 2004.

———. "Protesters Interfere with Biscuit Fire Salvage." *Jackson County (Oreg.) Mail Tribune*, Oct. 5, 2004.

Frost, Evan J., and Rob Sweeney. *Fire Regimes, Fire History, and Forest Conditions in the Klamath-Siskiyou Region: An Overview and Synthesis of Knowledge*. Ashland, Oreg.: World Wildlife Fund, Dec. 2000.

Healthy Forests Initiative and Healthy Forests Restoration Act: Interim Field Guide, FS-799. Washington, D.C.: U.S. Forest Service, 2004. http://www.fs.fed.us/projects/hfi/field-guide/web/toc.php.

Kelbie, Paul. "Native Americans v. Scots Industrial Giant: A $1Bn Lawsuit Is Filed in War over Salmon." *Independent*, June 21, 2004.

Kricher, John C. *Western Forests*. Boston: Houghton Mifflin, 1993.

Larson, Ron. "*C'waam* and *Q'pado* and Sacred Ceremonies." *Birdscapes* (Spring/Summer 2003), p. 10.

Lewis, William M., Jr. "Klamath Basin Fisheries: Argument Is No Substitute for Evidence." *Fisheries* 28, no. 3 (March 2003), pp. 20–25.

Lewis, William M., Jr., et al. *Endangered and Threatened Fishes in the Klamath River Basin*. Washington, D.C.: National Academies Press, 2004.

Lichatowich, Jim. *Salmon Without Rivers*. Washington, D.C.: Island Press, 1999.

Lindenmayer, D. B., et al. "Salvage Harvesting Policies After Natural Disturbance." *Science* 303 (Feb. 27, 2004), p. 1303.

Milstein, Michael. "Judge Lifts Biscuit Fire Logging-Stop Order." *Oregonian*, Aug. 31, 2004.

———. "Klamath Basin Deal Signed." *Oregonian*, Oct. 14, 2004.

Nash, J. Madeleine. "Why the West Is Burning." *Time*, Aug. 16, 2004, pp. 46–51.

Preusch, Matthew. "Amid a Forest's Ashes, a Debate over Logging Profits Is Burning." *New York Times*, April 15, 2004.

Robbins, Jim. "Critics Say Forest Service Battles Too Many Fires." *New York Times*, Feb. 8, 2004.

Russell, Betsy. "Lemhi Salmon Catch a Break." *Spokesman-Review*, July 20, 2001.

Service, Robert F. "'Combat Biology' on the Klamath." *Science* 300 (April 4, 2003), pp. 36–39.

Siskiyou County Sesquicentennial Committee. "Siskiyou History 1900–1924." http://www.siskiyouhistory.org/1900.html.

Siskiyou-Klamath National Forest. "Biscuit Fire Chronology: Initial Attack," May 2003. http://www.biscuitfire.com/chronology.htm.

Strittholt, James R., and Heather Rustigan. *Ecological Issues Underlying Proposals to Conduct Salvage Logging in the Areas Burned by the Biscuit Fire*. Corvallis, Oreg.: Conservation Biology Institute, Jan. 2004.

"The Tillamook Burn." *Tillamook County Online*, 2002. http://www.tillamoo.com/burn.html.

Trout Unlimited. *Idaho Crossroads: The Challenge for Idaho's Rivers and Streams in the 21st Century*. Idaho Falls, Idaho, 2004.

Viers, Stephen D., Jr., Paul A. Opler, et al. "Regional Trends of Biological Resources: California." In *Status and Trends of the Nation's Biological Resources*. Vol. 2, ed. Michael J. Mac et al. Washington, D.C.: U.S. Geological Survey, 1998.

Williams, Ted. "Salmon Stakes." *Audubon* 105, no. 1 (March 2003), pp. 42–52.

12. IN THE KINGDOM OF CONIFERS

CITATIONS

276 "When a huge diesel-powered truck": Roger Tory Peterson and James Fisher, *Wild America* (Boston: Houghton Mifflin, 1955), p. 307.

282 "small dumpy birds": Ibid., p. 361.

284 "we had the same feeling": Ibid., p. 354.

285 "How long": Ibid.

286 "It's time to stop fighting": Mike Dombeck and Jack Ward Thomas, "Declare Harvest of Old-Growth Forests Off-Limits and Move On," *Seattle Post-Intelligencer*, Aug. 24, 2003.

288 "it would be criminal": Peterson and Fisher, *Wild America*, p. 355.

BIBLIOGRAPHY

Barnard, Jeff. "Federal Agency Moves to Remove Protection of Murrelet." *Seattle Post-Intelligencer*, Sept. 2, 2004.

Barringer, Felicity. "U.S. Rules Out Dam Removal to Aid Salmon." *New York Times*, Dec. 1, 2004.

Committee on Environmental Issues in Pacific Northwest Forest Management, Board on Biology and National Research Council. *Environmental Issues in Pacific Northwest Forest Management*. Washington, D.C.: National Academy Press, 2000.

Forest Ecosystem Management Assessment Team. *Forest Ecosystem Management: An Ecological, Economic, and Social Assessment*. Portland, Oreg.: U.S. Department of Agriculture and U.S. Department of the Interior, 1993.

Forsman, Eric D. "Northern Spotted Owl." In *Status and Trends of the Nation's Biological Resources*. Vol. 2, ed. Michael J. Mac et al. Washington, D.C.: U.S. Geological Survey, 1998.

Franklin, Jerry F., and Thomas A. Spies. "Composition, Function, and Structure of Old-Growth Douglas-Fir Forests." In *Wildlife and Vegetation of Unmanaged Douglas-Fir Forests*, ed. Leonard F. Ruggiero, Keith B. Aubry, Andrew B. Carey, and Mark H. Huff. General Technical Report PNW-GTR-285. Portland, Oreg.: U.S. Forest Service, May 1991.

Gutiérrez, Ralph J., Alan B. Franklin, and William S. LaHaye. *Spotted Owl* (Strix occidentalis*)*. The Birds of North America, ed. A. Poole and F. Gill, no. 179. Philadelphia: Academy of Natural Sciences; Washington, D.C.: American Ornithologists' Union, 1995.

Houston, Douglas B., Edward S. Schreiner, and Andrea Woodward. "Roosevelt Elk and Forest Structure in Olympic National Park." In *Status and Trends of the Nation's Biological Resources*. Vol. 2.

Johnsgard, Paul A. *North American Owls: Biology and Natural History*. Washington, D.C.: Smithsonian Institution Press, 1988.

Mathews, Daniel. *Cascade-Olympic Natural History*. 2nd ed. Portland, Oreg.: Raven Editions, 1999.

Mazur, Kurt M., and Paul Clive James. *Barred Owl* (Strix varia*)*. The Birds of North America, ed. A. Poole and F. Gill, no. 508. Philadelphia: Academy of Natural Sciences; Washington, D.C.: American Ornithologists' Union, 2000.

McGrath, Susan. "Spawning Hope." *Audubon*, Sept. 2003, pp. 60–66.

McNulty, Tim. *Olympic National Park*. Boston: Houghton Mifflin, 1996.

McShane, C., et al. "Evaluation Report for the 5-Year Status Review of the Marbled Murrelet in Washington, Oregon, and California." Unpublished report. Seattle: EDAW, Inc. Prepared for the U.S. Fish and Wildlife Service, March 2004.

Milstein, Michael. "Bush Officials Order Rewrite of Protected Seabird Report." *Oregonian*, Sept. 2, 2004.

Morgan, Murray. *The Last Wilderness*. New York: Viking Press, 1955.

Morrison, Peter H. *Old Growth in the Pacific Northwest: A Status Report*. Seattle: Wilderness Society, 1988.

Nelson, S. Kim. *Marbled Murrelet* (Brachyramphus mamoratus*)*. The Birds of North America, ed. A. Poole and F. Gill, no. 276. Philadelphia: Academy of Natural Sciences; Washington, D.C.: American Ornithologists' Union, 1997.

Noss, Reed, et al. "Review of Scientific Material Relevant to the Occurrence, Ecosystem Role, and Tested Management Options for Mountain Goats in Olympic National Park." Unpublished report. Corvallis, Oreg.: Conservation Biology Institute, May 30, 2000. Prepared for the U.S. Department of the Interior.

Olympic National Park. "Elwha River Restoration." Port Angeles, Wash. http://www.nps.gov/olym/elwha/home.htm.

"Olympic National Park History." Olympic National Park, Nov. 3, 2004. http://www.olympic.national-park.com/info.htm#his.

Parmesan, Camille, and Gary Yohe. "A Globally Coherent Fingerprint of Climate Change Impacts Across Natural Systems." *Nature* 421 (Jan. 2, 2003), pp. 37–42.

Perkins, Sid. "On Thinning Ice." *Science News* 16 (Oct. 4, 2003), pp. 215–16.

Pojar, Jim, and Andy McKinnon. *Plants of the Pacific Northwest Coast.* Vancouver, B.C.: Lone Pine Press, 1994.

Scharf, Janet. "Elwha River: Running an Obstacle Course from Mountains to Sea," n.d. http://www.nps.gov/olym/issues/isselwha2.htm.

Smith, Jeff P., Michael W. Collopy, et al. "Regional Trends of Biological Resources: Pacific Northwest." In *Status and Trends of the Nation's Biological Resources.* Vol. 2.

U.S. Department of Agriculture, Forest Service; U.S. Department of the Interior, Bureau of Land Management. *Record of Decision for Amendments to Forest Service and Bureau of Land Management Planning Documents Within the Range of the Northern Spotted Owl.* Washington, D.C., 1994.

U.S. Environmental Protection Agency. "Global Warming," Oct. 2, 2002. http://yosemite.epa.gov/OAR/globalwarming.nsf/content/index.html.

U.S. Geological Survey. "Geology of Olympic National Park," Jan. 13, 2004. http://wrgis.wr.usgs.gov/docs/parks/olym/olym1.html.

13. SUPER-TUNDRA

CITATIONS

304 "stiff with northern birds": Roger Tory Peterson and James Fisher, *Wild America* (Boston: Houghton Mifflin, 1955), p. 372.

304–305 "out fourteen hours a day": Ibid.

317 "the potential for large, intense wildfires": Warren Oja, Ron Freeman, Mark Black, and Bill Schuster, *Kenai Peninsula Spruce Bark Beetle Management Strategies and Five-Year Action Plan*, July 8, 1999, http://www.fs.fed.us/r10/chugach/revision/pdfs/beetle_plan_final.pdf.

BIBLIOGRAPHY

Appenzeller, Tim. "The Case of the Missing Carbon." *National Geographic* 205, no. 2 (Feb. 2004), pp. 88–117.

Balogh, Greg. "Lead and the Spectacled Eider." *Endangered Species Bulletin* 24, no. 1 (Jan./Feb. 1999), pp. 6–7.

Barker, James H. *Always Getting Ready/Upterrlainarluta: Yup'ik Eskimo Subsistence in Southwest Alaska.* Seattle: University of Washington Press, 1993.

Hayhoe, Katharine, et al. "Emissions Pathways, Climate Change, and Impacts on California." *Proceedings of the National Academy of Sciences* 101, no. 34 (Aug. 24, 2004). http://www.pnas.org_cgi_doi_10.1073_pnas.0404500101.

Houghton, J. T., Y. Ding, D. J. Griggs, M. Noguer, P. J. van der Linden, X. Dai, K. Maskell, and C. A. Johnson, eds. *Climate Change 2001: The Scientific Basis: Contribution of Working Group I to the Third Assessment Report of the Intergovernmental Panel on Climate Change*. Intergovernmental Panel on Climate Change. Cambridge, U.K., and New York: Cambridge University Press, 2001.

Krauss, Clifford. "Eskimos Fret as Climate Shifts and Wildlife Changes." *New York Times*, Sept. 6, 2004.

Kristof, Nicholas D. "Baked Alaska on the Menu." *New York Times*, Sept. 13, 2003.

McCaffery, Brian, and Robert Gill. *Bar-Tailed Godwit* (Limosa lapponica). The Birds of North America, ed. A. Poole and F. Gill, no. 581. Philadelphia: Academy of Natural Sciences; Washington, D.C.: American Ornithologists' Union, 2001.

Pacific Flyway Council. *Pacific Flyway Management Plan for the Cackling Canada Goose*. Portland, Oreg.: U.S. Fish and Wildlife Service, 1999.

Savory, Eve. "The Shrinking Polar Bears." *CBC News Online*, Oct. 1999. http://www.cbc.ca/news/background/polarbears/.

U.S. Global Change Research Program. *U.S. National Assessment of the Potential Consequences of Climate Variability and Change: Regional Paper: Alaska*, Oct. 12, 2003. http://www.usgcrp.gov/usgcrp/nacc/education/alaska/ak-edu-3.htm.

Watson, R. T., et al., eds. *Climate Change 2001: Synthesis Report: A Contribution of Working Groups I, II, and III to the Third Assessment Report of the Intergovernmental Panel on Climate Change*. Intergovernmental Panel on Climate Change. Cambridge, U.K., and New York: Cambridge University Press, 2001.

Wohlforth, Charles. "On Thin Ice." *Orion*, March/April 2004, pp. 46–53.

———. *The Whale and the Supercomputer*. New York: North Point Press, 2004.

14. THE WILDEST SHORE

CITATIONS

320 "they looked at him": Roger Tory Peterson and James Fisher, *Wild America* (Boston: Houghton Mifflin, 1955), p. 397.

340 "not liking scenes of carnage": Ibid., p. 395.

340 "Undersize and oversize animals": Ibid.

BIBLIOGRAPHY

Arctic Research Consortium of the United States. *Bering Ecosystem Study: Draft Science Plan*, 2003. http://www.arcus.org/bering/science_plan.html.

Banks, David, et al. *Ecoregion-Based Conservation in the Bering Sea*. Washington, D.C.: World Wildlife Fund; Anchorage: Nature Conservancy of Alaska, 1999.

Barth, Tom F. *Geology and Petrology of the Pribilof Islands, Alaska*. Washington, D.C.: Department of the Interior, 1956.

Bourne, Joel K., Jr. "Alaska's Wild Archipelago." *National Geographic* 204, no. 2 (Aug. 2003), pp. 72–95.

Bureau of Indian Affairs. *Pribilof Island Survey Reports*, Oct. 28, 1949.

Byrd, G. Vernon, and Jeffrey C. Williams. *Red-Legged Kittiwake* (Rissa brevirostris). The Birds of North America, ed. A. Poole and F. Gill, no. 60. Philadelphia: Academy of Natural Sciences; Washington, D.C.: American Ornithologists' Union, 1993.

Dragoo, Donald E., G. Vernon Byrd, and David B. Irons. *Breeding Status, Population Trends, and Diets of Seabirds in Alaska, 2001*. U.S. Fish and Wildlife Service Report AMNWR 03/05. Homer: Alaska Maritime National Wildlife Refuge, 2003.

Gaston, Anthony J., and J. Mark Hipfner. *Thick-Billed Murre* (Uria lomvia). The Birds of North America, ed. A. Poole and F. Gill, no. 497. Philadelphia: Academy of Natural Sciences; Washington, D.C.: American Ornithologists' Union, 2000.

Hatch, Scott A., and John F. Piatt. "Seabirds in Alaska." In *Our Living Resources*, ed. E. T. LaRoe, G. S. Farris, C. E. Puckett, P. D. Doran, and M. J. Mac. Washington, D.C.: Department of the Interior, 1995. http://biology.usgs.gov/s+t/noframe/b023.htm.

Jones, Ian L. *Least Auklet* (Aethia pusilla). The Birds of North America, ed. A. Poole and F. Gill, no. 69. Philadelphia: Academy of Natural Sciences; Washington, D.C.: American Ornithologists' Union, 1993.

Jones, Ian L., and Catherine M. Gray, Johanne Dusureault, and Arthur L. Sowls. "Auklet Demography and Norway Rat Distribution and Abundance at Sirius Point, Aleutian Islands, Alaska, in 2001," Sept. 28, 2001. http://www.mun.ca/acwern/finalkiskaREP.pdf.

Kitaysky, Alexander S., John F. Piatt, and John C. Wingfield. *Are Seabirds Breeding in the Southeastern Bering Sea Food-Limited?* Fairbanks: North Pacific Marine Research Program, University of Alaska–Fairbanks, Oct. 2002.

Major, Heather L., and Ian L. Jones. "Impacts of the Norway Rat on the Auklet Breeding Colony at Sirius Point, Kiska Island, Alaska, in 2003." http://www.mun.ca/acwern/kiskaRATREP2003.pdf.

National Marine Mammal Laboratory, "Northern Fur Seal Research." http://nmml.afsc.noaa.gov/AlaskaEcosystems/nfshome/nfs.htm.

Nowak, Ronald M. *Walker's Mammals of the World*. 5th ed. Vol. 2. Baltimore: Johns Hopkins University Press, 1991.

Pendleton, Catherine. "Aleut Independence Movement," July 21, 2002. http://www.beringsea.com/communities/Saint_Paul/museum/aleutindmov/ALIMR.php.

Piatt, John F., and Alexander S. Kitaysky. *Horned Puffin* (Fratercula corniculata). The Birds of North America, ed. A. Poole and F. Gill, no. 603. Philadelphia: Academy of Natural Sciences; Washington, D.C.: American Ornithologists' Union, 2002.

———. *Tufted Puffin* (Fratercula cirrhata). The Birds of North America, ed. A. Poole and F. Gill, no. 708. Philadelphia: Academy of Natural Sciences; Washington, D.C.: American Ornithologists' Union, 2002.

Pyle, Peter. "Northern Fur Seal Born on the Farallones: Recent Returns." *Observer*, no. 108 (1997).

Robson, Bruce W., et al. "Separation of Foraging Habitat Among Breeding Sites of a Colonial Marine Predator, the Northern Fur Seal (*Callorhinus ursinus*)." *Canadian Journal of Zoology* 82 (2004), pp. 20–29.

Rusch, Donald H., Richard E. Malecki, and Robert Trost. "Canada Geese in North America." In *Our Living Resources*. http://biology.usgs.gov/s+t/noframe/b011.htm.

Sanders, Eli. "More Oil Is Thought Spilled from Freighter off Alaska." *New York Times*, Dec. 26, 2004.

Springer, Alan M., et al. "Sequential Megafaunal Collapse in the North Pacific Ocean: An Ongoing Legacy of Industrial Whaling?" *Proceedings of the National Academy of Sciences* 100, no. 21 (Oct. 2, 2003), pp. 12223–28.

Stolzenberg, William. "Danger in Numbers." *Nature Conservancy* 54, no. 2 (Summer 2004), pp. 42–50.

U.S. Fish and Wildlife Service. "Final Rule to Remove the Aleutian Canada Goose from the Federal List of Endangered and Threatened Wildlife." *Federal Register* 66, no. 54 (March 20, 2001).

West, George C. *A Birder's Guide to Alaska*. Colorado Springs, Colo.: American Birding Association, 2002.

Woods, Bruce. "'Sea Otter' and the Geese." *Endangered Species Bulletin* 28, no. 4 (July/Dec. 2003), pp. 26–27.

15. HOMEWARD BOUND

BIBLIOGRAPHY

Royal Society for the Protection of Birds. "Disastrous Year for Scotland's Birds," July 29, 2004. http://www.rspb.org.uk/scotland/action/disaster/index.asp.

Acknowledgments

If I lacked quite the reach of a Roger Tory Peterson in planning my own continental circumnavigation, I still depended on a large number of kind and generous people who made the months of travel both productive and efficient.

At the start, I had invaluable support and advice from my friend Sally Conyne, from Jim Berry of the Roger Tory Peterson Institute, and from Kenn Kaufman, who extended much enthusiasm for a project that, in an ideal world, would have been his to do. I am particularly grateful to Dr. Clemency Thorne Fisher, James Fisher's daughter. Clem was wonderful in helping to arrange archive research in England, and her friends Gwen Takoradi and Iain Bishop opened their home to me there; thanks also to Anne-Louise Fisher and Patrick Janson-Smith, and to Michael and Antje Learoyd, who welcomed Clem and me on a visit to the former Fisher home at the Old Rectory in Northamptonshire, a short distance from where James Fisher is buried.

At the Natural History Museum in London, particular thanks to natural history librarian Carol Gokce and assistant librarian Paul Cooper for access to the James Fisher Collection.

In Newfoundland, thanks to William A. Montevecchi at Memorial University, and to Tony Power and Chris Mooney at Cape St. Mary's Ecological Reserve. In the Northeast and the mid-Atlantic, particular thanks to Norman Smith of Massachusetts Audubon; Dean Steeger of Friends of Monomoy; Wayne R. Petersen of Swarovski Birding Community; the staff of the Manomet Center for Conservation Sciences in Manomet, Massachusetts; E. J. McAdams of the New York City Audubon Society; and education specialist Maggie Zadorozny of Rock Creek Park in Washington, D.C. In the southern Blue Ridge, thanks again to Charlie and Sue Staines, and to Jeanie Hilten of Discover Life in America.

In Florida, I had help from Dr. David Maehr of the University of Kentucky; Tim Downey at Everglades National Park; Greg Toppin and Mike Owen of Fakahatchee Strand Preserve; Ed Carlson at Corkscrew Swamp Sanctuary; Phil Manor at the Florida Fish and Wildlife Conservation Commission; and Oron "Sonny" Bass and Elsa Alvear of the National Park Service. My particular thanks to Mike Ryan of Dry Tortugas National Park for his wealth of information on the park and its history.

In Texas, my "winter Texan" friends Guy and Barbara Ubaghs provided hospitality; thanks also to Karen and Phil Hunke; to Ron and Sharron Smith; to Jeff Rupert and Mike Carlo at Lower Rio Grande Valley/ Santa Ana NWR; and, at Laguna Atascosa NWR, to its manager, John Wallace, and to biologist Linda Laack.

In Mexico, in addition to my friend Bob McCready of the Nature Conservancy, sincere thanks to TNC's northeastern Mexico program director, Jeff Weigel, for logistical support, and to Dave Mehlman, also of TNC. My thanks to Roberto Pedraza Ruiz of Grupo Ecológico Sierra Gorda, and to Miguel Ángel Cruz Nieto, Armando Jiménez, and Mario A. Morales Loa of Pronatura Noreste for showing me La Soledad and allowing Bob and me to stay at the group's house in San Rafael.

In the Southwest and the Pacific coast, special thanks to Sheri Williamson and Tom Wood at the Southeastern Arizona Bird Observatory for not only bumming around the Huachucas with me for several days but also giving this eastern hummingbird bander a chance to work with some sexy western species. My thanks also to Graham Chisholm at the Nature Conservancy in California; Denise Stockton of Hopper Mountain NWR; Ellie Cohen of Point Reyes Bird Observatory; Ken Peterson, Andrew Johnson, and Teri Nicholson of the Monterey Bay Aquarium; and Joelle Buffa of Farallon NWR.

I am once again indebted to my old friend Stan Senner of Audubon Alaska. Thanks also to Dave Cline for advice; to manager Mike Rearden and supervisory wildlife biologist Fred Broerman at Yukon Delta NWR for support, and to Brian McCaffery, Alice Nunes, Grace Leacock, and James McCallum at Old Chevak for friendship and the best week of birding ever; to manager Greg Siekaniec and biologist Art Sowls at Alaska Maritime NWR, along with Sadie Wright and Alexis Will; to Greg McGlashan of the St. George Traditional Council and Aquilina D.

Lestenkof of the St. Paul Island Tribal Government Ecosystem Conservation Office; and to Karin Holser of the Pribilof Islands Stewardship Program.

In its formative phases, this book was shaped by my longtime editor Ethan Nosowsky and by Rebecca Saletan and, most substantively, by Eric Chinski, who took on the task midway to completion and handled it with grace and sensitivity. As always, thanks to my gifted agent, Peter Matson at Sterling Lord Literistic, and his more-than-able associates Saskia Cornes and Jim Rutman.

Finally, love to my then fiancée, now wife, Amy, who didn't see much of me for long periods of time, with gratitude for her patience and her enthusiasm for this project, and her advice and guidance along the way.

Index

xiv- xvi History of US environmental Law / movement
xviii - Bush admin
xix - Prescience